Advances in Celiac Disease

Jorge Amil-Dias · Isabel Polanco
Editors

Advances in Celiac Disease

Improving Paediatric and Adult Care

 Springer

Editors
Jorge Amil-Dias
Department of Pediatrics
São João University Hospital
Porto, Portugal

Isabel Polanco
Department of Pediatrics
Autonoma University
Madrid, Spain

ISBN 978-3-030-82403-7   ISBN 978-3-030-82401-3   (eBook)
https://doi.org/10.1007/978-3-030-82401-3

This Springer imprint is published by the registered company Springer Nature Switzerland AG
The registered company address is: Gewerbestrasse 11, 6330 Cham, Switzerland

Preface

Celiac Disease (CD) is now recognized as one of the most common diseases in the world. Clinical recognition started in Paediatric patients only less than a century ago but epidemiology and research knowledge has progressed immensely.

The many faces of CD in all age groups are now well known and this condition is a fascinating example of a food-related immune disease clearly different from all other known forms of food allergy.

It is also amazing that a staple food so essential in the evolution of mankind can induce so severe lesions and complications in susceptible patients. At the same time, even severe disease can be managed just with dietary measures.

In this book we gathered contributions from world-known experts and researchers in CD, pooling updated information on the many aspects of CD that may be of interest for the practitioners, regardless of their area of clinical intervention.

Non-celiac gluten intolerance is still a mysterious condition with evolving knowledge, that is also addressed in this textbook.

We are grateful to all authors and hope to meet the interest of readers that may need to be aware of CD in any of its multiple clinical presentations.

Porto, Portugal   Jorge Amil-Dias
Madrid, Spain   Isabel Polanco

Contents

Celiac Disease: Background

Isabel Polanco and Jorge Amil-Dias

Celiac disease (CD) is known since ancient times. It was given its name by Aretaeus the Cappadocian in the first or second century A.D (Fig. 1) [1]. It was not until 1887, Samuel Gee described the disease in children [2] (Fig. 2). Later, in 1950, Willem Karel Dicke presented his thesis (Celiac Disease: investigation of harmful effects of certain types of cereal on patients with celiac disease) and the role of gluten was discovered [3].

The development of devices to obtain jejunal biopsies in children (Fig. 3) and interpreting the histology [4], the identification of serological markers [5, 6] and the association between CD and the HLA complex [7, 8] are only some of the milestones that have allowed the progress in the knowledge of pathogenesis, diagnosis and treatment of CD. Later, endoscopy replaced the use of biopsy capsules, providing the additional value of observing the macroscopic aspect of the mucosa and allowing for multiple biopsies from different locations.

CD is an autoimmune disorder characterized by enteropathy in response to intestinal exposure to gluten in genetically predisposed individuals. Just a few decades ago, it was thought to be an uncommon disease of childhood affecting predominantly European populations. It has since been shown to be present universally and can develop at any age in individuals consuming gluten-containing foods [9–11]. Furthermore, it is also now clear that CD may have variable presentation patterns ranging from no symptoms to a wide range of gastrointestinal or extra-digestive signs and symptoms. CD affects ∼ 1% of the global population but, despite its rising prevalence, the majority of patients remain undiagnosed [11]. Currently, most patients with suspected CD are screened serologically for antibody

I. Polanco
Faculty of Medicine, Department of Pediatrics, Autonoma University, Madrid, Spain
e-mail: ipolanco@telefonica.net

J. Amil-Dias (✉)
Centro Hospitalar Universitário S. João, Porto, Portugal
e-mail: jorge.amil@outlook.pt

© The Author(s), under exclusive license to Springer Nature Switzerland AG 2022
J. Amil-Dias and I. Polanco (eds.), *Advances in Celiac Disease*,
https://doi.org/10.1007/978-3-030-82401-3_1

1

Fig. 1 Aretaeus of Cappadocia

Fig. 2 Special edition of envelope and stamps on the occasion of the Samuel Gee symposium (London, 1988, organized by John Walker-Smith)

positivity, in particular for IgA antibodies to tissue transglutaminase 2 (tTG) and antiendomysial (EMA) IgA antibodies [9]. As these serological markers are not 100% specific for detecting intestinal lesions compatible with CD, positive celiac serology is confirmed by duodenal biopsies demonstrating the hallmark pathological changes of mucosal remodelling, such as villous atrophy, crypt hyperplasia and intraepitelial lymphocytosis [12].

Fig. 3 Paediatric Crosby capsule with tube passed through a pacifier for infants

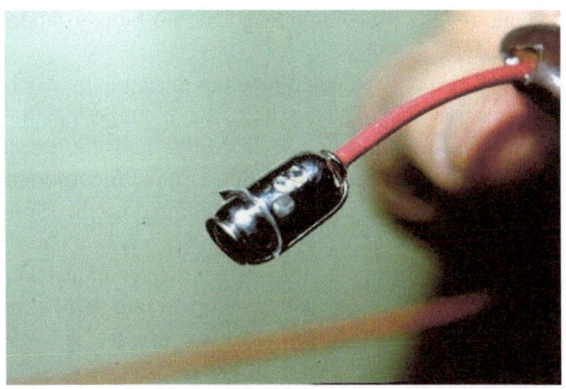

In 1969 in Interlaken, during the annual meeting of the then called ESPGAN (European Society for Paediatric Gastroenterology and Nutrition), strict diagnostic criteria for the diagnosis of celiac disease in the paediatric population were established for the first time. These were called the "Interlaken criteria (1969)", also known as the "three-biopsy rule" [13]. In fact, these criteria established that it was mandatory to demonstrate a typical intestinal lesion (hyperplasic jejunal villous atrophy) while the patient was on a gluten-containing diet, followed by complete histological recovery after removing gluten from the diet and subsequent histologic damage upon gluten re-exposure (challenge test). The demand for the three biopsies aimed at differentiating CD from other frequent causes of enteropathy and to demonstrate the permanent nature of the intolerance to gluten.

The wide application of these criteria by paediatric gastroenterologists in the subsequent years allowed to accumulate large experience. The additional result from relevant multicentre studies and the identification of a biological parameter—antigliadin antibodies (AGA)—as a marker for active CD led to a revision of these criteria 20 years later [14]. According to these, a single biopsy might be enough for a solid diagnosis of CD. Additional biopsies would only be needed in cases that were classified as CD without previous biopsy or the clinical response to gluten elimination was unclear.

The identification of more specific antibodies—anti-transglutaminase IgA (tTG IgA)—allowed further refinement in the diagnostic criteria. In 2012, the European Society for Paediatric Gastroenterology, Hepatology and Nutrition (ESPGHAN) published new guidelines for the diagnosis of CD challenging the wide necessity for duodenal biopsies in paediatric patients [15]. It was then suggested that anti-tTG IgA antibody titre greater than 10 times the upper limit of normal (ULN), in combination with a positive EMA antibody test and compatible human leucocyte antigen (HLA) genotype, is sufficient to support the diagnosis of CD in symptomatic individuals. This eliminated the need for endoscopy and its associated costs/risks in selected paediatric patients.

The recent 2020 guidelines removed the requirement for the presence of symptoms and HLA testing in the diagnostic pathway [16]. This underscores the specificity of a serology-based or 'no-biopsy' approach for the diagnosis of pae-diatric CD.

Studies have evaluated whether this strategy can be applied in symptomatic adult patients [17–27]. These studies have suggested that tTG levels of $\geq 10 \times$ ULN could be predictive of CD in adults, and the recently published Finnish national guidelines for the diagnosis of CD have incorporated this diagnostic pathway into their practice (Working group appointed by the Finnish Medical Society Duodecim and the Finnish Gastroenterology Society, Celiac disease. Current care guidelines, 2018. Available at https://www.kaypahoito.fi/en/ccs00086). However, this approach has not been widely adopted into adult clinical practice or guidelines [28].

The clinical spectrum of celiac disease is wide, including cases with either classical intestinal (e.g. chronic diarrhoea, weight loss) or extraintestinal (e.g. anaemia, osteoporosis, neurological disturbances) features, as well as silent forms that are occasionally discovered because of serological screening [29, 30].

Clinical features of CD differ considerably depending on the age at presentation. Intestinal symptoms and failure to thrive are common in children diagnosed within the first years of life. Presentation of the disease later in childhood is characterized by the prevalence of extraintestinal signs, among which are short stature, delayed puberty, anaemia, enamel hypoplasia, osteopenia or bilateral occipital calcifica-tions, related to the presence of gluten in the diet. In adults, all the above signs and symptoms may occur as well as osteoporosis and infertility [31, 32]. The broad spectrum of symptoms contributes to the large proportion of undiagnosed cases found in screening-studies.

Celiac disease is the only known treatable autoimmune disease, provided that a correct diagnosis is achieved and a strict, lifelong gluten-free diet is implemented, as this has become the cornerstone of the management of CD patients and must be recommended for life in both, symptomatic and asymptomatic individuals [33].

Implementation of a GFD should be monitored by a dietician, in order to ensure nutritional adequacy and prevent potential risks including micronutrient deficien-cies, high fat, sugar and salt intake [34–37].

Family members of CD patients or those suffering from another immune-mediated disease are at higher risk of developing CD. Unrecognized and therefore untreated CD patients have a greater risk of developing associated com-plications or other immune-mediated diseases (e.g. type 1 diabetes, autoimmune hepatitis or thyroid disease) [38]. The antibodies used as markers for CD have a relatively high sensitivity and specificity, but those with mild lesions, partial villous atrophy, or children younger than 2 years may be missed by these tests [39, 40], so careful clinical judgement by a paediatric gastroenterologist is needed to evaluate all possible patients.

There is ongoing investigation in all fields of aetiology, pathogenesis, genetics and additional therapeutic options in this fascinating unique example of controllable auto-immune disease. This book aims at providing the reader with an updated overview of current status and future prospects in Celiac Disease.

Conflicts of Interest: The authors declare no conflict of interest.

References

1. Adams F. The extant works of Aretaeus, the Cappadocian. London: London: Printed for the Sydenham Society; 1856.
2. Gee SJ. On the coeliac affection. St Bartholomew's Hospital Reports 1988;24:4.
3. Dicke WK. Coeliakie. Een onderzoek naar de nadelige invloed van sommige graansoorten op de lijder aan coeliakie (Coeliac disease. Investigation of the harmful effects of certain types of cereal on patients with coeliac disease). Utrecht: Utrecht; 1950.
4. Shiner M. Duodenal biopsy. Lancet. 1956;270(6906):17–9.
5. Berger E. [Allergic pathogenesis of celiac disease with studies of the splitting up of pathogenic antigens by enzymes]. Bibl Paediatr 195867):1–55.
6. Burgin-Wolff A, Gaze H, Hadziselimovic F, et al. Antigliadin and antiendomysium antibody determination for coeliac disease. Arch Dis Child. 1991;66(8):941–7.
7. Falchuk ZM, Rogentine GN. Strober W Predominance of histocompatibility antigen HL-A8 in patients with gluten-sensitive enteropathy. J Clin Invest. 1972;51(6):1602–5.
8. Stokes PL, Asquith P, Holmes GK, et al. Histocompatibility antigens associated with adult coeliac disease. Lancet. 1972;2(7769):162–4.
9. Lebwohl B, Sanders DS. Green PHR coeliac disease. Lancet. 2018;391(10115):70–81.
10. Leonard MM, Sapone A, Catassi C, et al. Celiac disease and nonceliac gluten sensitivity: a review. JAMA. 2017;318(7):647–56.
11. Singh P, Arora A, Strand TA, et al. Global prevalence of celiac disease: systematic review and meta-analysis. Clin Gastroenterol Hepatol 2018;16(6):823–36 e2.
12. Marsh MN. Gluten, major histocompatibility complex, and the small intestine. A molecular and immunobiologic approach to the spectrum of gluten sensitivity ('celiac sprue'). Gastroenterology 1992;102(1):330–54.
13. Meeuwisse GW. Diagnostic criteria in coeliac disease. Acta Paediatr Scand 1970;59:461–63.
14. Walker-Smith JA, Guandalini S, Schmitz J et al. Revised criteria for diagnosis of coeliac disease. Report of Working Group of European Society of Paediatric Gastroenterology and Nutrition. Arch Dis Child 1990;65(8):909–11.
15. Husby S, Koletzko S, Korponay-Szabo IR, et al. European society for Pediatric gastroenterology, hepatology, and nutrition guidelines for the diagnosis of coeliac disease. J Pediatr Gastroenterol Nutr. 2012;54(1):136–60.
16. Husby S, Koletzko S, Korponay-Szabo I, et al. European society paediatric gastroenterology, hepatology and nutrition guidelines for diagnosing coeliac disease 2020. J Pediatr Gastroenterol Nutr. 2020;70(1):141–56.
17. Alessio MG, Tonutti E, Brusca I, et al. Correlation between IgA tissue transglutaminase antibody ratio and histological finding in celiac disease. J Pediatr Gastroenterol Nutr. 2012;55 (1):44–9.
18. Beltran L, Koenig M, Egner W, et al. High-titre circulating tissue transglutaminase-2 antibodies predict small bowel villous atrophy, but decision cut-off limits must be locally validated. Clin Exp Immunol. 2014;176(2):190–8.
19. Efthymakis K, Serio M, Milano A, et al. Application of the biopsy-sparing ESPGHAN guidelines for celiac disease diagnosis in adults: a real-life study. Dig Dis Sci. 2017;62 (9):2433–9.
20. Hill PG. Holmes GK Coeliac disease: a biopsy is not always necessary for diagnosis. Aliment Pharmacol Ther. 2008;27(7):572–7.
21. Holmes GKT, Forsyth JM, Knowles S, et al. Coeliac disease: further evidence that biopsy is not always necessary for diagnosis. Eur J Gastroenterol Hepatol. 2017;29(6):640–5.
22. Oyaert M, Vermeersch P, De Hertogh G, et al. Combining antibody tests and taking into account antibody levels improves serologic diagnosis of celiac disease. Clin Chem Lab Med. 2015;53(10):1537–46.

23. Previtali G, Licini L, D'Antiga L, et al. Celiac disease diagnosis without biopsy: Is a 10x ULN Antitransglutaminase result suitable for a chemiluminescence method? J Pediatr Gastroenterol Nutr. 2018;66(4):645–50.
24. Sugai E, Moreno ML, Hwang HJ, et al. Celiac disease serology in patients with different pretest probabilities: is biopsy avoidable? World J Gastroenterol. 2010;16(25):3144–52.
25. Tortora R, Imperatore N, Capone P, et al. The presence of anti-endomysial antibodies and the level of anti-tissue transglutaminases can be used to diagnose adult coeliac disease without duodenal biopsy. Aliment Pharmacol Ther. 2014;40(10):1223–9.
26. Wakim-Fleming J, Pagadala MR, Lemyre MS, et al. Diagnosis of celiac disease in adults based on serology test results, without small-bowel biopsy. Clin Gastroenterol Hepatol. 2013;11(5):511–6.
27. Zanini B, Magni A, Caselani F, et al. High tissue-transglutaminase antibody level predicts small intestinal villous atrophy in adult patients at high risk of celiac disease. Dig Liver Dis. 2012;44(4):280–5.
28. Al-Toma A, Volta U, Auricchio R, et al. European Society for the Study of Coeliac Disease (ESsCD) guideline for coeliac disease and other gluten-related disorders. United European Gastroenterol J. 2019;7(5):583–613.
29. Maki M, Mustalahti K, Kokkonen J, et al. Prevalence of Celiac disease among children in Finland. N Engl J Med. 2003;348(25):2517–24.
30. Myleus A, Ivarsson A, Webb C, et al. Celiac disease revealed in 3% of Swedish 12-year-olds born during an epidemic. J Pediatr Gastroenterol Nutr. 2009;49(2):170–6.
31. Lasa JS, Zubiaurre I. Soifer LO Risk of infertility in patients with celiac disease: a meta-analysis of observational studies. Arq Gastroenterol. 2014;51(2):144–50.
32. Leffler DA, Green PH. Fasano A Extraintestinal manifestations of celiac disease. Nat Rev Gastroenterol Hepatol. 2015;12(10):561–71.
33. Tack GJ, Verbeek WH, Schreurs MW, et al. The spectrum of celiac disease: epidemiology, clinical aspects and treatment. Nat Rev Gastroenterol Hepatol. 2010;7(4):204–13.
34. Potter MDE, Brienesse SC, Walker MM, et al. Effect of the gluten-free diet on cardiovascular risk factors in patients with coeliac disease: a systematic review. J Gastroenterol Hepatol. 2018;33(4):781–91.
35. Skodje GI, Minelle IH, Rolfsen KL, et al. Dietary and symptom assessment in adults with self-reported non-coeliac gluten sensitivity. Clin Nutr ESPEN 2019;31:88–94.
36. Vici G, Belli L, Biondi M, et al. Gluten free diet and nutrient deficiencies: a review. Clin Nutr. 2016;35(6):1236–41.
37. Wild D, Robins GG, Burley VJ, et al. Evidence of high sugar intake, and low fibre and mineral intake, in the gluten-free diet. Aliment Pharmacol Ther. 2010;32(4):573–81.
38. Rubio-Tapia A, Kyle RA, Kaplan EL, et al. Increased prevalence and mortality in undiagnosed celiac disease. Gastroenterology. 2009;137(1):88–93.
39. Sanders DS, Hurlstone DP, McAlindon ME, et al. Antibody negative coeliac disease presenting in elderly people–an easily missed diagnosis. BMJ. 2005;330(7494):775–6.
40. Sweis R, Pee L, Smith-Laing G. Discrepancies between histology and serology for the diagnosis of coeliac disease in a district general hospital: is this an unrecognised problem in other hospitals? Clin Med (Lond) 2009;9(4):346–8.

Epidemiology of Celiac Disease

Mahendra Singh Rajput, Ashish Chauhan, and Govind K. Makharia

1 Introduction

The journey of celiac disease (CD) from its first description by Samuel Gee to a great breakthrough discovery of wheat being the cause of CD, based on diligent clinical observation and clinical enquiry of five young patients, by Willem Karel Dicke has been very inspiring [1, 2]. CD is a unique in the sense that the treatment of the disease has been discovered decades before understanding or unravelling of its pathophysiology. While the introduction of gastrointestinal endoscopic techniques in 1970s for taking biopsies from the intestinal mucosa and identification of two human leukocyte antigen (HLA) molecules (HLA-DQ2 and HLA-DQ8) in late 1980s led to the understanding of the pathology and pathophysiology of CD, the discovery of serologic tests such as anti-endomysial antibody (EMA), anti-tissue transglutaminase antibody (IgA tTG Ab), or anti-deamidated gliadin peptide antibody (anti-DGP Ab) has not only allowed screening of high-risk group for CD, but also made it possible to estimate the true prevalence of CD in the general population [3–8].

While the abovementioned discoveries eased the making of a diagnosis, certain other factors in our understanding of the distribution of the disease and its clinical characteristics have led to an increase in the rate of the diagnosis of CD globally. Firstly, while CD has been thought traditionally to be a disease of children and seen

M. S. Rajput
Department of Gastroenterology and Human Nutrition, All India Institute of Medical Sciences, New Delhi, India

A. Chauhan
Department of Gastroenterology, Indira Gandhi Medical College and Hospital, Shimla, India

G. K. Makharia (✉)
Department of Gastroenterology and Human Nutrition, All India Institute of Medical Sciences, New Delhi 110029, India
e-mail: govindmakharia@aiims.edu

© The Author(s), under exclusive license to Springer Nature Switzerland AG 2022
J. Amil-Dias and I. Polanco (eds.), *Advances in Celiac Disease*,
https://doi.org/10.1007/978-3-030-82401-3_2

only by paediatricians, the realization of the fact that CD is a disease of life-long, patients with CD started getting diagnosed in all age groups including adults and elderly [9–11]. Secondly, once believed to be affecting people of European origin predominantly, studies from other continents later confirmed that the CD also affects non-Caucasian population including Africans and Asians [12, 13]. Thirdly, once thought that the gluten hypersensitivity in CD is limited to the small intestinal mucosa and thus only those having gastrointestinal manifestations were screened for CD. It later became apparent that CD affects many other organs and it has a wide spectrum of clinical manifestations. Hence patients having non-gastrointestinal manifestations even in the absence of gastrointestinal symptoms started getting diagnosed as CD [14–17]. The last factor which led to an increase in the rate of diagnosis of CD is the ease of making a diagnosis by simplification of the diagnostic criteria. Prior to the revised diagnostic criteria for CD in 1990, now of a historical importance, making a diagnosis of CD required three sequential intestinal mucosal biopsies [18]. Now with further simplification of diagnostic criteria, a diagnosis of CD can be made solely on the basis of presence of high titre of anti-tTG Ab alone [19]. With advancement created by fundamental and clinical research, CD has now become the commonest autoimmune diseases of humans.

2 Origin of Epidemiology of CD

Initial epidemiological studies conducted in 1950, when the diagnosis of CD was based entirely on the presence of typical gastrointestinal symptoms, showed a cumulative prevalence of 1 in 8000 in England and 1 in 4000 in Scotland [20]. With the invention of more specific tests for malabsorption, advent of intestinal biopsy, and increase in awareness about CD, the prevalence of CD increased in 1970s to 1 in 450 in Ireland, Scotland, and Switzerland [21, 22].

3 Modern Epidemiology of CD

A multicentre study from Italy involving school children aged 6–15 years, using the three-layered strategy of clinical screening, serological tests, and intestinal biopsies gave birth to the modern epidemiology of CD. Among 17,201 healthy Italian students, the overall prevalence of CD was found to be 1 in 184. More interestingly, only 1 in 7 was previously diagnosed as CD, highlighting a big iceberg phenomenon, where clinically detectable patients were just a few and a larger number of subjects remained clinically undiagnosed [23]. This landmark serology-based study catalyzed the exploration of epidemiology of CD in different parts of the world.

3.1 The Global Perspective

A real-time assessment of the prevalence of CD is denoted via seroprevalence of CD (proportion of people having a positive anti-tTG Ab and /or anti-endomysial Ab) and prevalence of biopsy-confirmed CD (proportion of individuals with villous abnormalities of modified Marsh grade 2 or more along with a positive serological test).

3.2 Global Seroprevalence of CD

The pooled global seroprevalence of CD in the general population is 1.4% (95% CI 1·1%, 1·7%), as shown by a systematic review and meta-analysis of population-based studies, including 275,818 [13]. The seroprevalence of CD varies from continent to continent, and the highest seroprevalence has been reported in Europe and Asia (Table 1). Furthermore, the seroprevalence of CD also varies from country to country, the highest being in Algeria, Czech Republic, India, Israel, Mexico, Saudi Arabia, Sweden, Portugal, and Turkey and lowest in Estonia, Germany, Iceland, Libya, Poland, Republic of San Marino, and Spain [13] (Fig. 1).

3.3 Global Prevalence of Biopsy-Confirmed CD

The global pooled prevalence of biopsy-confirmed CD has been shown to be 0·7% (95% CI 0·5%, 0·9%) in a systematic review and meta-analysis of population-based studies [13]. On stratification of countries into quintiles based on the prevalence of biopsy-confirmed CD, countries with the highest prevalence (76–100th quintile) include Argentina, Egypt, Hungary, Finland, India, New Zealand, and Sweden and the countries with the lowest prevalence (0–25th quintile) include Brazil, Germany, Republic of San Marino, Russia and Tunisia (Fig. 2).

Table 1 Continent wise seroprevalence and prevalence of biopsy-confirmed CD disease

Continent	Seroprevalence of CD (CI)	Prevalence of Biopsy confirmed CD (CI)
Europe	1.3 (1.1–1.5)	0.8 (0.6–1.1)
North America	1.4 (0.7–2.2)	0.5
South America	1.3 (0.5–2.5)	0.4 (0.1–0.6)
Africa	1.1 (0.4–2.2)	0.5 (0.2–0.9)
Asia	1.8 (1–2.9)	0.6 (0.4–0.8)
Oceania	1.4 (1.4–1.8)	0.8 (0.2–1.7)

CD: Celiac disease; CI: Confidence interval

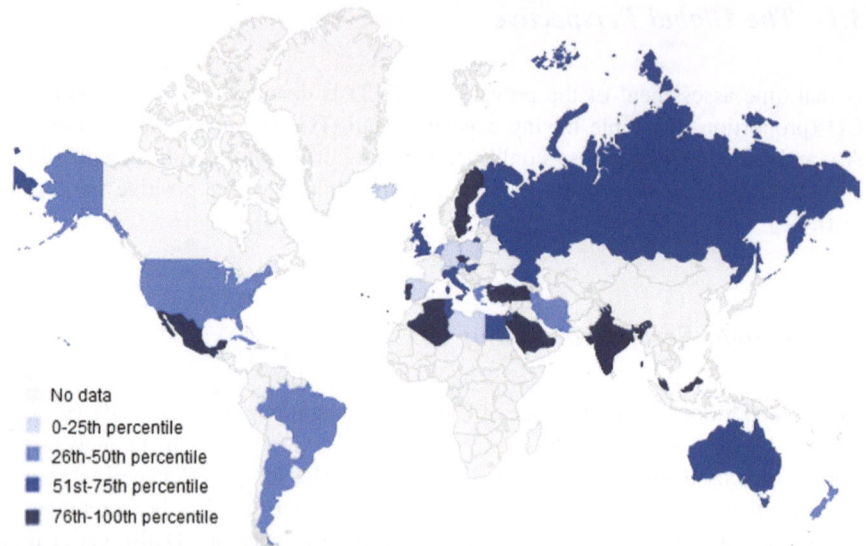

Fig. 1 Worldwide celiac disease seroprevalence rates for the countries reporting data. Prevalence values were stratified into 4 groups of percentiles representing the 0–25th percentile (light grey) to the 76–100th percentile (dark black). The lowest and highest percentiles include countries with pooled national prevalence ranging from 0.2% to 0.8% and 2.1% to 8.5%, respectively (Reprinted from the Clinical Gastroenterology and Hepatology volume 16,issue 6, June 01,2018,Singh et al., Global prevalence of celiac disease—systemic review and meta-analysis, P823-836,2021,with permission from Elsevier)

Most population-based epidemiological studies to assess the prevalence of CD are based on a positive celiac serological test, and the diagnosis of CD in all seropositive patients has not been confirmed by intestinal mucosal biopsies, which likely is the explanation of the differences in the population-based seroprevalence and prevalence of biopsy-confirmed CD [13]. Furthermore, the population-based prevalence data is still not available from many countries and thus the presently observed prevalence data may not reflect the real global prevalence of CD.

3.4   Continent-Wise Prevalence of CD

Prevalence of CD in Europe

Most of the initial studies on the prevalence of CD has risen from European countries such as Italy, UK, Finland. In the first multinational European study including Finland, Germany, Italy, and the UK, 29,212 subjects were screened for CD using anti-tTG antibody, and all those who had either a positive or a borderline titre of anti-tTG Ab were further tested for EMA in their serum. The overall

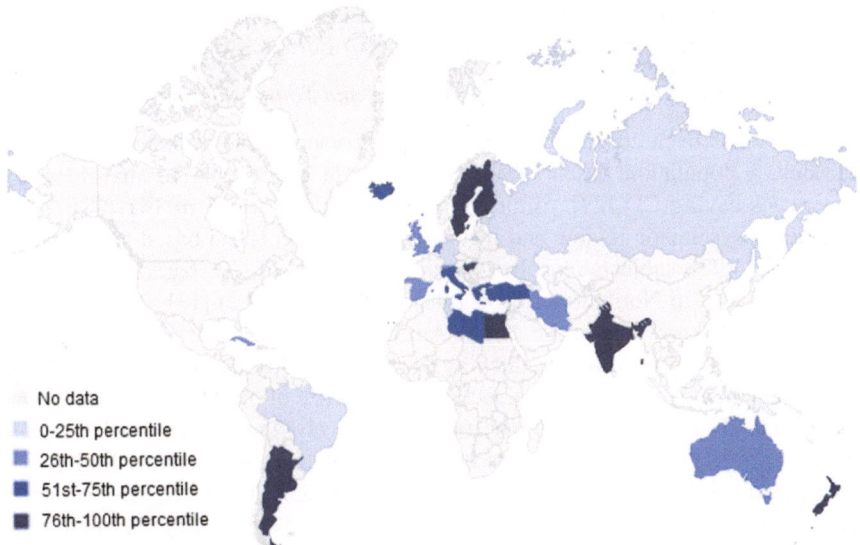

Fig. 2 Worldwide celiac disease prevalence rates (based on biopsy) for the countries reporting data. Prevalence values were stratified into 4 groups of percentiles representing the 0–25th percentile (light grey) to the 76–100th percentile (dark black). The lowest and highest percentiles include countries with a pooled national prevalence ranging from 0.2% to 0.4% and 0.9% to 2.4%, respectively. (Reprinted from the Clinical Gastroenterology and Hepatology volume 16, issue 6, June 01,2018, Singh et al., Global prevalence of celiac disease - systemic review and meta-analysis, P823-836,2021,with permission from Elsevier)

prevalence CD in this multinational European study has been reported to be 1.0% (95% CI 0.9–1.1) [24]. Interestingly, the prevalence of CD was not uniform in the four participating European countries, despite sharing of a similar distribution of causal factors (level of gluten intake and frequency of HLA-DQ2 and DQ8 genotype). The prevalence of CD was 2.0% (95% CI 2.0–2.8) in Finland, 1.5 (95% CI 1.1–1.9) in UK, 0.7% (95% CI 0.4–1.0) in Italy, and the lowest prevalence of 0.3% (95% CI 0.1–0.5) in Germany [24]. Interestingly another study later from Germany including 12,741 participants, aged 1–17 years, has shown the prevalence of CD to be 0.9%, which is much higher than that reported earlier in the multicentre European study [25].

Taken together, the seroprevalence and prevalence of biopsy-confirmed CD in Europe has been reported to be 1.3% (95% 1.1–1.5) and 0.8% (95% 0.6–1.1), respectively in a systemic review and meta-analysis of 33 studies conducted in 2018 [13]. While the abovementioned systematic review has included studies including both adults and children, the systematic review and meta-analysis in 2021 has focused on the prevalence of biopsy-confirmed CD in children in Europe. The prevalence of biopsy-confirmed CD in children in Europe in this study has been reported to be 0.7%, but varying widely between 0.10 and 3.03%. Furthermore, a regional variation was noted in the prevalence of CD in children and the prevalence is reported to be significantly higher in northern Europe (1.6%) than that in eastern (0.98%), southern (0.69%), and western (0.60%) Europe [26].

Prevalence of CD in America (North America and South America)

While CD has been considered to be an uncommon disease in America in earlier decades, a population based prevalence study in 2003 reported that 1 in 133 Americans having CD [27]. Similarly in other study it was 1 in 141 [28]. Taken together, a systematic review and meta-analysis of population-based seven studies including 17,778 subjects revealed that the prevalence of CD, based on a positive serological test, in North America is 1.4% (95% CI 0.7–2.2) [13].

CD is well-known in those South American countries that are populated by individuals of European origin, such as Brazil. In a study including 4405 subjects from Brazil, the overall seroprevalence and prevalence of biopsy-confirmed CD has been reported to be 3.6 and 3.4 per 1000, respectively. Prevalence of CD in adults and children has been reported to be 2.1 and 5.4 per 1000, respectively [29]. As per a systematic review of the studies from South America, the pooled seroprevalence and prevalence of biopsy-confirmed CD was 1.3% (95% 0.5–2.5) (11 studies and 20,245 subjects screened) and 0.4% (0.1–0.6) (5 studies and 16,550 subjects), respectively [13].

Prevalence of CD in Oceania

As in the European countries, a population-based study from Australia including 3011 subjects showed the seroprevalence and prevalence of biopsy-confirmed CD to be 1 in 251 and 1 in 430, respectively [30]. A similar population-based study from New Zealand including 1064 subjects has shown the prevalence of CD to be 1.1% [31].

Prevalence of CD in Africa

The pooled seroprevalence (7 studies and 15,775 subjects) and prevalence of biopsy-confirmed CD (4 studies and 7902 subjects) in African continent has been shown to be 1.1% (95% CI 0.4–2.2) and 0.5% (95% CI 0.2–0.9), respectively [13]. The Saharawi population of Arab-Berber origin, originally living in western Sahara, has the highest prevalence of CD in the world. A study of 990 Saharawi children showed that the prevalence of CD in this population is 5.6% [32]. Specifically in other regions of Africa, the prevalence of CD has been reported to be 0.5% in Egypt [33], 0.8% in Libya [34] and 0.6% in Tunisia [35].

Prevalence of CD in Asia

Until recent times, CD was considered to be a rare disease in Asia and patients presenting with diarrhoea and malabsorption were diagnosed usually as having tropical sprue [36]. After the widespread availability of serological tests, multiple screening studies have been performed in many Asian countries such as Turkey, Iran, Israel, Jordan, and India and almost all of them summarily show that CD is not an uncommon and it most often remains underdiagnosed in Asia [37]. Due to the heterogeneity of the population, genetics, economic conditions, and the dietary habits, the epidemiology of CD is different in different parts of Asia.

In India, CD has been recognized mainly in the northern part of India, where wheat is the predominant cereal consumed and a population-based study including 2879 subjects showed a prevalence of CD to be 1.04% (1 in 96) [38]. Later, a pan-India study including 23,331 healthy adults from three different regions of India, showed a regional variation in the prevalence of CD. While the age-adjusted seroprevalence of CD in Northern, North-Eastern regions were 1.23%, 0.87%, respectively, it was only 0.10% in the Southern region, showing Northern and Southern region gradients in the prevalence of CD [39].

The epidemiology of CD in China, the largest country, was largely unknown until recent years, except for a small case series. In a cross-sectional study including 19,778 Chinese adolescents and young adults (age 16–25 years) from 27 geographic regions in China showed that more than 2% (2.19%) of them had at least one of the serological test positive including 1.8% for IgG anti-DGP Ab and 0.36% for IgA anti-tTG Ab [40]. The prevalence of people with a positive antibody varied remarkably among different regions of China and it was 12 times higher in the Northern provinces, such as Shandong, Shaanxi, and Henan, where wheat was the staple diet [40]. In another recent study, including 2277 inpatients with gastrointestinal symptoms in four major ethnic groups of Xinjiang Uyghur Autonomous Region of China, the seroprevalence and prevalence of biopsy-confirmed CD was observed to be 1.27% (95% CI, 0.81–1.73%), and 0.35% (95% Cl, 0.11–0.59%), respectively [41]. Interestingly, among 246 patients with diarrhoea-predominant irritable bowel syndrome in China, 2.85% were reported to have CD [42]. These preliminary studies have established the foundation for the exploration of the exact prevalence of CD and regional geographical differences in the prevalence of CD in China.

In a pilot study, including 562 young healthy volunteers from Malaysia, the seroprevalence of CD has been reported to be 1.25% (95% CI 0.78–1.72%) [43]. In a study from Japan including 2008 subjects, anti-tTG Ab was found to be positive in a high proportion (8%), however, none of them was EMA positive and only one showed celiac-type alterations at the small intestinal biopsy [44]. Similarly, in a study including 1961 Vietnamese children, the seroprevalence, based on anti-tTG Ab, was observed to be 1%, but none of them was positive for EMA [45].

Summarizing the prevalence studies addressing low-risk groups from Asian Pacific region, a recent systemic review and meta-analysis has shown that the pooled sero-prevalence of CD among low-risk groups is 1.2% and that of biopsy-confirmed CD is 0.61% [46]. Furthermore, the authors also segregated and reported the prevalence of CD in the middle east (Iran, Turkey, Saudi Arabia, Israel, Jordan), south-east Asia (India, Malaysia, and Egypt) and Eastern Asia. The pooled seroprevalence and prevalence of biopsy-confirmed CD in the Middle East region and South-East region of Asia are 1.6% (95% CI 1.2–2.1) and 0.6% (95% CI 0.4–0.8); and 2.6% (95% CI 0.3–7.2) and 0.8% (0.4–1.4), respectively, which are quite similar to that reported from many European countries. Interestingly, the seroprevalence of CD is found to be lowest (0.06%; 95% CI 0.03–0.09%) in the East-Asian countries [46].

3.5 Prevalence of CD Over Time

Looking at the time trends of the prevalence of CD, a systematic review and meta-analysis stratified studies reporting prevalence of CD into two time periods: January 1991 to December 2000 and January 2001 onward. The result of the systematic review shows that the prevalence of CD has increased over time from 0.6% in 1991 to 2000 to 0.8% between 2001 and 2016 [13].

4 Variations in the Prevalence of CD as Per Age, Gender, Geographical Distribution

4.1 Children Versus Adults

While CD was described originally in paediatric patients and believed to be a disease of children only, over time but it has been realized that CD can be diagnosed at any age group including elderly [13]. A systematic review including 43 studies has reported the prevalence of biopsy-confirmed CD in the paediatric and adult patients. The pooled prevalence of biopsy- confirmed CD is higher in children in comparison to that in adults (0.9% vs. 0.5%). While the prevalence of CD is higher in children, the absolute number of patients with CD globally and in each country, is likely to be higher in the adult age-group because of much higher proportion of adults in any country compared to children in that country [13].

4.2 Men Versus Women

As with many other autoimmune diseases, CD is more common in women as compared to men. A systematic review and meta-analysis has also confirmed that the pooled prevalence of biopsy-confirmed CD higher in women (0.6%; 95% CI, 0.5%–0.8%) in comparison to that in men (0.4%, 95% CI, 0.3%–0.5%) [13].

4.3 Geographical Location

A higher prevalence of many autoimmune diseases such as multiple sclerosis, rheumatoid arthritis and inflammatory bowel disease has been reported at higher geographical latitudes [47–49]. The associations between the autoimmune diseases and the latitude has been linked to less solar exposure and resultant vitamin D deficiency in them. In a systematic review involving 128 studies, with 155 prevalence estimates representing 40 countries, the prevalence of CD has been reported to be higher at higher latitudes of 51 to 60° (relative risk of 1.62) and 61 to 70° (relative risk 2.30), in comparison to prevalence at latitudes of 41 to 50° as reference level [50]. In this study, when latitudes were categorized into intervals of 10° latitudinal increments, the prevalence of CD increased incrementally at latitude higher than 40°.

5 Incidence of CD

The incidence of CD is expressed as a rate, i.e. the number of new clinically diagnosed patients with CD per 100,000 subjects over one year. Due to the diffusion of CD serological tests in clinical practice and the improved awareness about the clinical polymorphism of CD, CD incidence has greatly increased in many western countries during the last decades [51, 52]. For instance, during 2010–2014, twenty times more patients were diagnosed in UK than that during 1975–1979 [53]. In the US (Olmsted County, Minnesota) the overall age and sex-adjusted incidence of CD increased from 11.1 per 100,000 persons/year in 2000–2001 to 17.3 in 2008–2010 [54].

   While there is paucity of population-based study for incidence from many parts of the world except industrialized and developed countries, in a recent systematic review and meta-analysis, King et al. reported the differences in incidence of CD before the year 2000 and that after year 2000. The pooled average annual incidence of CD has been estimated to be rising by 7.5% (95% CI: 5.8, 9.3) per year over the past several decades [55]. The systematic review showed that the pooled incidence of CD in women and men is 17.4 (95% CI: 13.7, 21.1) and 7.8 (95% CI: 6.3, 9.2) per 100,000 person-years, respectively. Children specific incidence of CD is higher (21.3 per 100,000 person-years) in comparison to that of the adults (12.9 per 100,000 person-years) [55].

   Another systematic review and meta-analysis of incidence of CD in children in Europe showed a large increase in the incidence of diagnosed CD across Europe and it has reached 50 per 100 000 person-years in Scandinavia, Finland, and Spain [26]. The median age at diagnosis of CD has increased from 1.9 years before 1990 to 7.6 years since 2000.

   As discussed above, while the incidence rates for CD are increasing in many countries such as UK [53], USA [56], and New Zealand [57], the incidence rate in Finland and Sweden has reached peaked and it is stabilizing [58, 59]. This increase in incidence of CD is not likely only due to improvement in the rate of diagnosis because of ease of diagnosis and increase in the awareness of the disease amongst physicians but also due to changes in our environment and eating practices [60, 61].

6 Risk Factors for CD

CD occurs because of interaction between both environmental (gluten) and genetic factors (HLA and non-HLA genes), and the distribution of these two components can guide to identify the areas of the world at risk for CD [62].

7 Wheat, Barley, and Rye

During the very early part of the evolution, men led a nomadic life and obtained food by hunting, fishing and collecting fruits and vegetables. Therefore, we can infer that CD did not exist during the Palaeolithic age, as the diet of hunter-gatherers consisted of only meat, vegetables, seeds and fruits and was gluten-free by its origin. About 10,000 years ago in a small region of South-Western Asia, called the "Fertile Crescent" including Southern Turkey, Lebanon, Syria, Palestine and Iraq, the local community started cultivating wild grains due to the special environmental conditions created by the flooding. In the Fertile Crescent, some tribes changed their lifestyle from nomadic to a stable settlement because land cultivation permitted them to store food [63]. The first wheat varieties, which were successfully domesticated, were Einkorn (diploid wheat) and Emmer wheat (tetraploid wheat) [64]. The progressive spread of agriculture to Europe took place through the migration of farmers and their mixing with and partially replacing the indigenous European population. The agricultural spread was stimulated by population growth (as a result of the increasing availability of food) and local migratory activity [33, 65, 66].

In the evolutionary process, the genome of wheat has changed from diploid (14 chromosomes) to hexaploid genome (42 chromosomes) [67]. The genome of the most ancient wheat is diploid and is called AA, BB, DD. These grass-like wheat species had a very low seed yield and their seed dropped easily. Natural hybridization between two of these diploid species led to birth of the tetraploid *Triticum* species having AABB genome. Finally, around 4000 BC, natural hybridization *T. turgidum* (dicoccum) carrying the AABB genome and a wild diploid species *Aegilops tauschii* carrying the D genome led to origin of Bread wheat (*Triticum aestivum*). The introduction of the D genome in the wheat improved the bread-making properties of the wheat [68, 69].

The protein content of wheat grain varies between 8 and 17% of its total mass. Gluten comprises of 78–85% of the total wheat endosperm protein. Gluten proteins can be divided into two main fractions according to their solubility in aqueous alcohols: the soluble gliadins and the insoluble glutenins. Gliadins are mainly monomeric proteins with molecular weights (MWs) around 28,000–55,000 and can be classified according to their different primary structures into the *a*-, *b*-, *g*- and *w*-type. Glutenin consists of glutenin subunits of high (MW 67,000–88,000) or low MW (MW 32,000–35,000) that are connected by intermolecular SS bonds.

Noncovalent bonds such as hydrogen bonds, ionic bonds and hydrophobic bonds are important for the aggregation of gliadins and glutenins and implicate structure and physical properties of dough. Glutenins confer elasticity, while gliadins mainly confer viscous flow and extensibility to the gluten complex. Thus, gluten is responsible for most of the viscoelastic properties of wheat flour dough, and it is the main factor dictating the use of a wheat variety in bread and pasta making.

Gliadins and glutenins have a unique amino acid composition with a high content of proline (15%) and glutamine (35%). Moreover, they contain domains with numerous repetitive sequences rich in these amino acids. The incomplete digestion of gliadin by digestive tract enzymes leads to the generation of peptides, many of which are immunogenic for patients with CD [64, 70]

Over the past five decades, several changes in the pattern of wheat consumption have been observed including an increase in per capita consumption of wheat, an increase in the use of gluten in food processing and an increase in the consumption of processed foods. Furthermore, an increase in CD-related T-cell stimulatory epitopes has also been observed in wheat. It is conceivable that these changes in the wheat consumption pattern and increase in T-cells stimulatory epitopes in wheat may be the reasons for an increase in the incidence of CD world over [71].

7.1 Genetic Risk Factors

CD is considered to be a polygenic disease with a complex non-Mendelian pattern of inheritance, involving both MHC and non-MHC genes. The strong genetic predisposition is demonstrated by concordance rate of 80% in monozygotic twins and 20% in dizygotic twins [72, 73]. Furthermore, the prevalence of CD in the first-degree relatives of patients with CD has been reported to vary from 1.6 to 38% [74–76]. A systematic review and meta-analysis have shown that 7.5% of first-degree relatives and 2.3% of second-degree relatives have CD. The risk of CD is 1 in 7 in sisters,1 in 8 in daughters, 1 in 13 in sons, 1 in 16 in brothers, 1 in 32 in mothers, and 1 in 33 in fathers [77].

CD is a multigenic disorder, in which the most dominant genetic risk factors are the genotypes encoding the HLA class II molecules HLA-DQ2 (encoded by HLA-DQA1*0501 and HLA-DQB1*02) and HLA-DQ8 (encoded by HLA-DQA1*0301 and HLA-DQB1*0302) [78, 79]. About 90–95% of individuals with CD carry the HLA-DQ2 heterodimer encoded either in cis or in trans, and HLA-DQ8 [80, 81]. Approximately 20–30% of the general population of Europe, America, Australasia certain part of Asia also carry HLA-DQ2 or DQ8 haplotype [82]. Interestingly, this most of these people do not develop CeD even if they consume gluten.

While ingestion of gluten and HLA-DQ2 or HLA-DQ8 are essential factors, there however are many other factors which likely play a role in the development of CD. Currently, 57 susceptibility loci, not related to HLA, have also been identified by genome-wide association studies, each of which is estimated to be associated with small risk of developing CD [83].

Table 2 Risk factor for celiac disease

Essential factors	Risk factor modifier
Gluten	Amount of gluten ingestion Timing of gluten introduction during weaning Gluten processing
Genetic MHC gene: HLA-DQ2, HLA-DQ8	Non MHC genes Epigenetic factor
	Breastfeeding Childhood infection, Use of antibiotics in childhood Socioeconomic status

The relevance of HLA and other relevant environmental factors such as age of the introduction and amount of gluten, infant feeding, infection in childhood, antibiotics use in childhood and socioeconomic factors that play a relevant role in the epidemiology of CD and are addressed in detail in other part of this book (Table 2).

8   Conclusions

While CD is now a global disease, and approximately 40–60 million people around the world are estimated to have CD. Of them only a proportion of patients are diagnosed, and a majority still remains undetected. There is a need to increase the awareness of CD amongst the general population and the physicians.

References

1. Losowsky MS. A history of coeliac disease. Dig Dis Basel Switz. 2008;26(2):112–20.
2. Yan D, Holt PR. Willem Dicke. Brilliant clinical observer and translational investigator. Discoverer of the toxic cause of celiac disease. Clin Transl Sci. 2009;2:446–8.
3. Crosby WH, Kugler HW. Intraluminal biopsy of the small intestine; the intestinal biopsy capsule. Am J Dig Dis. 1957;2:236–41.
4. Sollid LM, Markussen G, Ek J, Gjerde H, Vartdal F, Thorsby E. Evidence for a primary association of celiac disease to a particular HLA-DQ alpha/beta heterodimer. J Exp Med. 1989;169:345–50.
5. Kivel RM, Kearns DH, Liebowitz D. Significance of antibodies to dietary proteins in the serums of patients with nontropical sprue. N Engl J Med. 1964 O;271:769–72.
6. Chorzelski TP, Sulej J, Tchorzewska H, Jablonska S, Beutner EH, Kumar V. IgA class endomysium antibodies in dermatitis herpetiformis and coeliac disease. Ann N Y Acad Sci. 1983;420:325–34.
7. Dieterich W, Ehnis T, Bauer M, Donner P, Volta U, Riecken EO, et al. Identification of tissue transglutaminase as the autoantigen of celiac disease. Nat Med. 1997;3:797–801.

8. Korponay-Szabó IR, Vecsei Z, Király R, Dahlbom I, Chirdo F, Nemes E, et al. Deamidated gliadin peptides form epitopes that transglutaminase antibodies recognize. J Pediatr Gastroenterol Nutr. 2008;46:253–61.

9. Swinson CM, Levi AJ. Is coeliac disease underdiagnosed? Br Med J. 1980;281:1258–60.

10. Beaumont DM, Mian MS. Coeliac disease in old age: "a catch in the rye." Age Ageing. 1998;27:535–8.

11. Murray JA, Van Dyke C, Plevak MF, Dierkhising RA, Zinsmeister AR, Melton LJ. Trends in the identification and clinical features of celiac disease in a North American community, 1950–2001. Clin Gastroenterol Hepatol Off Clin Pract J Am Gastroenterol Assoc. 2003;1:19–27.

12. Singh P, Arora S, Singh A, Strand TA, Makharia GK. Prevalence of celiac disease in Asia: a systematic review and meta-analysis. J Gastroenterol Hepatol. 2016;31:1095–101.

13. Singh P, Arora A, Strand TA, Leffler DA, Catassi C, Green PH, et al. Global prevalence of celiac disease: systematic review and meta-analysis. Clin Gastroenterol Hepatol Off Clin Pract J Am Gastroenterol Assoc. 2018;16:823-836.e2.

14. Hadjivassiliou M, Sanders DS, Grünewald RA, Woodroofe N, Boscolo S, Aeschlimann D. Gluten sensitivity: from gut to brain. Lancet Neurol. 2010;9:318–30.

15. Antiga E, Caproni M, Pierini I, Bonciani D, Fabbri P. Gluten-free diet in patients with dermatitis herpetiformis: not only a matter of skin. Arch Dermatol. 2011;147:988–9.

16. Marsh MN. Gluten, major histocompatibility complex, and the small intestine. A molecular and immunobiologic approach to the spectrum of gluten sensitivity ('celiac sprue'). Gastroenterology. 1992;102:330–54.

17. Makharia G. Where are Indian adult celiacs? Trop Gastroenterol DDS. 2006;27:1–3.

18. Revised criteria for diagnosis of coeliac disease. Report of Working Group of European Society of Paediatric Gastroenterology and Nutrition. Arch Dis Child. 1990;65:909–11.

19. Husby S, Koletzko S, Korponay-Szabó I, Kurppa K, Mearin ML, Ribes-Koninckx C, et al. European society paediatric gastroenterology, hepatology and nutrition guidelines for diagnosing coeliac disease 2020. J Pediatr Gastroenterol Nutr. 2020;70:141–56.

20. Davidson LSP, Fountain JR. Incidence of the sprue syndrome; with some observations on the natural history. Br Med J. 1950;1:1157–61.

21. van Stirum J, Baerlocher K, Fanconi A, Gugler E, Tönz O, Shmerling DH. The incidence of coeliac disease in children in Switzerland. Helv Paediatr Acta. 1982;37:421–30.

22. Mylotte M, Egan-Mitchell B, McCarthy CF, McNicholl B. Incidence of coeliac disease in the West of Ireland. Br Med J. 1973;1:703–5.

23. Catassi C, Fabiani E, Rätsch IM, Coppa GV, Giorgi PL, Pierdomenico R, et al. The coeliac iceberg in Italy. A multicentre antigliadin antibodies screening for coeliac disease in school-age subjects. Acta Paediatr Oslo. 1996;412:29–35.

24. Mustalahti K, Catassi C, Reunanen A, Fabiani E, Heier M, McMillan S, et al. The prevalence of celiac disease in Europe: results of a centralized, international mass screening project. Ann Med. 2010;42:587–95.

25. Laass MW, Schmitz R, Uhlig HH, Zimmer K-P, Thamm M, Koletzko S. The prevalence of celiac disease in children and adolescents in Germany. Dtsch Arzteblatt Int. 2015;112:553–60.

26. Roberts SE, Morrison-Rees S, Thapar N, Benninga MA, Borrelli O, Broekaert I, et al. Systematic review and meta-analysis: the incidence and prevalence of paediatric coeliac disease across Europe. Aliment Pharmacol Ther. 2021.

27. Fasano A, Berti I, Gerarduzzi T, Not T, Colletti RB, Drago S, et al. Prevalence of celiac disease in at-risk and not-at-risk groups in the United States: a large multicenter study. Arch Intern Med. 2003;163:286–92.

28. Rubio-Tapia A, Ludvigsson JF, Brantner TL, Murray JA, Everhart JE. The prevalence of celiac disease in the United States. Am J Gastroenterol. 2012;107:1538–44.

29. Pratesi R, Gandolfi L, Garcia SG, Modelli IC, Lopes de Almeida P, Bocca AL, et al. Prevalence of coeliac disease: unexplained age-related variation in the same population. Scand J Gastroenterol. 2003;38:747–50.

30. Hovell CJ, Collett JA, Vautier G, Cheng AJ, Sutanto E, Mallon DF, et al. High prevalence of coeliac disease in a population-based study from Western Australia: a case for screening? Med J Aust. 2001;175:247–50.

31. Cook HB, Burt MJ, Collett JA, Whitehead MR, Frampton CM, Chapman BA. Adult coeliac disease: prevalence and clinical significance. J Gastroenterol Hepatol. 2000;15:1032–6.

32. Catassi C, Rätsch IM, Gandolfi L, Pratesi R, Fabiani E, El Asmar R, et al. Why is coeliac disease endemic in the people of the Sahara? Lancet Lond Engl. 1999;354:647–8.

33. Abu-Zekry M, Kryszak D, Diab M, Catassi C, Fasano A. Prevalence of celiac disease in Egyptian children disputes the east-west agriculture-dependent spread of the disease. J Pediatr Gastroenterol Nutr. 2008;47:136–40.

34. Alarida K, Harown J, Ahmaida A, Marinelli L, Venturini C, Kodermaz G, et al. Coeliac disease in Libyan children: a screening study based on the rapid determination of anti-transglutaminase antibodies. J Ital Soc Gastroenterol. 2011;43:688–91.

35. Bdioui F, Sakly N, Hassine M, Saffar H. Prevalence of celiac disease in Tunisian blood donors. Gastroenterol Clin Biol. 2006;30:33–6.

36. Baker SJ, Mathan VI. Tropical enteropathy and tropical sprue. Am J Clin Nutr. 1972;25:1047–55.

37. Makharia GK, Catassi C. Celiac Disease in Asia. Gastroenterol Clin North Am. 2019;48:101–13.

38. Makharia GK, Verma AK, Amarchand R, Bhatnagar S, Das P, Goswami A, et al. Prevalence of celiac disease in the northern part of India: a community based study. J Gastroenterol Hepatol. 2011;26:894–900.

39. Ramakrishna BS, Makharia GK, Chetri K, Dutta S, Mathur P, Ahuja V, et al. Prevalence of adult celiac disease in India: regional variations and associations. Am J Gastroenterol. 2016;111:115–23.

40. Yuan J, Zhou C, Gao J, Li J, Yu F, Lu J, et al. Prevalence of celiac disease autoimmunity among adolescents and young adults in China. Clin Gastroenterol Hepatol Off Clin Pract J Am Gastroenterol Assoc. 2017;15:1572–9.

41. Zhou C, Gao F, Gao J, Yuan J, Lu J, Sun Z, et al. Prevalence of coeliac disease in Northwest China: heterogeneity across Northern Silk road ethnic populations. Aliment Pharmacol Ther. 2020;51:1116–29.

42. Kou GJ, Guo J, Zuo XL, Li CQ, Liu C, Ji R, Liu H, Wang X, Li YQ. Prevalence of celiac disease in adult Chinese patients with diarrhea-predominant irritable bowel syndrome: a prospective, controlled, cohort study. Journal DDS. 2018;19:136–43.

43. Yap TW, Chan WK, Leow AH, Azmi AN, Loke MF, Vadivelu J, Goh KL. Prevalence of serum celiac antibodies in a multiracial Asian population-a first study in the young Asian adult population of Malaysia. PloS one. 2015;10:e0121908.

44. Fukunaga M, Ishimura N, Fukuyama C, Izumi D, Ishikawa N, Araki A, et al. Celiac disease in non-clinical populations of Japan. J Gastroenterol. 2018;53:208–14.

45. Zanella S, De Leo L, Nguyen-Ngoc-Quynh L, Nguyen-Duy B, Not T, Tran-Thi-Chi M, Phung-Duc S, Le-Thanh H, Malaventura C, Vatta S, Ziberna F. Cross-sectional study of coeliac autoimmunity in a population of Vietnamese children. BMJ Open. 2016;6:e011173.

46. Ashtari S, Najafimehr H, Pourhoseingholi MA, Rostami K, Asadzadeh-Aghdaei H, Rostami-Nejad M, Tavirani MR, Olfatifar M, Makharia GK, Zali MR. Prevalence of celiac disease in low and high risk population in Asia-Pacific region: a systematic review and meta-analysis. Sci Rep. 2021;11:1–3.

47. Simpson S, Blizzard L, Otahal P, Van der Mei I, Taylor B. Latitude is significantly associated with the prevalence of multiple sclerosis: a meta-analysis. J Neurol Neurosurg Psychiatry. 2011;82:1132–41.

48. GEO-RA Group. Latitude gradient influences the age of onset of rheumatoid arthritis: a worldwide survey. Clin Rheumatol. 2017;36:485–97.

49. Khalili H, Huang ES, Ananthakrishnan AN, Higuchi L, Richter JM, Fuchs CS, et al. Geographical variation and incidence of inflammatory bowel disease among US women. Gut. 2012;61:1686–92.

50. Celdir MG, Jansson-Knodell CL, Hujoel IA, Prokop LJ, Wang Z, Murad MH, et al. Latitude and celiac disease prevalence: a meta-analysis and meta-regression. Clin Gastroenterol Hepatol Off Clin Pract J Am Gastroenterol Assoc. 2020;30:31382–3.
51. Kang JY, Kang AHY, Green A, Gwee KA, Ho KY. Systematic review: worldwide variation in the frequency of coeliac disease and changes over time. Aliment Pharmacol Ther. 2013;38:226–45.
52. Altobelli E, Paduano R, Petrocelli R, Di Orio F. Burden of celiac disease in Europe: a review of its childhood and adulthood prevalence and incidence as of September 2014. Ann Ig Med Prev E Comunita. 2014;26:485–98.
53. West J, Fleming KM, Tata LJ, Card TR, Crooks CJ. Incidence and prevalence of celiac disease and dermatitis herpetiformis in the UK over two decades: population-based study. Am J Gastroenterol. 2014;109:757–68.
54. Almallouhi E, King KS, Patel B, Wi C, Juhn YJ, Murray JA, et al. Increasing incidence and altered presentation in a population-based study of pediatric celiac disease in North America. J Pediatr Gastroenterol Nutr. 2017;65:432–7.
55. King JA, Jeong J, Underwood FE, Quan J, Panaccione N, Windsor JW, et al. Incidence of celiac disease is increasing over time: a systematic review and meta-analysis. Am J Gastroenterol. 2020;115:507–25.
56. Ludvigsson JF, Rubio-Tapia A, van Dyke CT, Melton LJ, Zinsmeister AR, Lahr BD, et al. Increasing incidence of celiac disease in a North American population. Am J Gastroenterol. 2013;108:818–24.
57. Cook B, Oxner R, Chapman B, Whitehead M, Burt M. A thirty-year (1970–1999) study of coeliac disease in the Canterbury region of New Zealand. N Z Med J. 2004;117:U772.
58. Kivelä L, Kaukinen K, Lähdeaho M-L, Huhtala H, Ashorn M, Ruuska T, et al. Presentation of celiac disease in Finnish children is no longer changing: a 50-year perspective. J Pediatr. 2015;167:1109-1115.e1.
59. Virta LJ, Saarinen MM, Kolho K-L. Declining trend in the incidence of biopsy-verified coeliac disease in the adult population of Finland, 2005–2014. Aliment Pharmacol Ther. 2017;46:1085–93.
60. Lohi S, Mustalahti K, Kaukinen K, Laurila K, Collin P, Rissanen H, et al. Increasing prevalence of coeliac disease over time. Aliment Pharmacol Ther. 2007;26:1217–25.
61. Catassi C, Kryszak D, Bhatti B, Sturgeon C, Helzlsouer K, Clipp SL, et al. Natural history of celiac disease autoimmunity in a USA cohort followed since 1974. Ann Med. 2010;42:530–8.
62. Jabri B, Sollid LM. Tissue-mediated control of immunopathology in coeliac disease. Nat Rev Immunol. 2009;9:858–70.
63. Harlan JR, Zohary D. Distribution of wild wheats and barley. Science. 1966;153:1074–80.
64. Spaenij-Dekking L, Kooy-Winkelaar Y, van Veelen P, Drijfhout JW, Jonker H, van Soest L, et al. Natural variation in toxicity of wheat: potential for selection of nontoxic varieties for celiac disease patients. Gastroenterology. 2005;129:797–806.
65. Cataldo F, Montalto G. Celiac disease in the developing countries: a new and challenging public health problem. World J Gastroenterol. 2007;13:2153–9.
66. Rostami K, Malekzadeh R, Shahbazkhani B, Akbari MR, Catassi C. Coeliac disease in Middle Eastern countries: a challenge for the evolutionary history of this complex disorder? Dig Liver Dis. 2004;36:694–7.
67. Gupta PK, Mir RR, Mohan A, Kumar J. Wheat genomics: present status and future prospects. Int J Plant Genomics. 2008;2008.
68. Matsuoka Y. Evolution of polyploid triticum wheats under cultivation: the role of domestication, natural hybridization and allopolyploid speciation in their diversification. Plant Cell Physiol. 2011;52:750–64.
69. Feldman M, Levy AA. Genome evolution in allopolyploid wheat–a revolutionary reprogramming followed by gradual changes. J Genet Genomics Yi Chuan Xue Bao. 2009;36:511–8.
70. Shan L, Molberg Ø, Parrot I, Hausch F, Filiz F, Gray GM, et al. Structural basis for gluten intolerance in celiac sprue. Science. 2002;297:2275–9.

71. van den Broeck H, Hongbing C, Lacaze X, Dusautoir J-C, Gilissen L, Smulders M, et al. In search of tetraploid wheat accessions reduCD in celiac disease-related gluten epitopes. Mol Biosyst. 2010;6:2206–13.

72. Greco L, Romino R, Coto I, Di Cosmo N, Percopo S, Maglio M, Paparo F, Gasperi V, Limongelli MG, Cotichini R, D'agate C. The first large population based twin study of coeliac disease. Gut. 2002;50:624–8.

73. Nisticò L, Fagnani C, Coto I, Percopo S, Cotichini R, Limongelli MG, et al. Concordance, disease progression, and heritability of coeliac disease in Italian twins. Gut. 2006;55:803–8.

74. Tursi A, Brandimarte G, Giorgetti GM, Inchingolo CD. Effectiveness of the sorbitol H2 breath test in detecting histological damage among relatives of coeliacs. Scand J Gastroenterol. 2003;38:727–31.

75. Rubio-Tapia A, Van Dyke CT, Lahr BD, Zinsmeister AR, El-Youssef M, Moore SB, et al. Predictors of family risk for celiac disease: a population-based study. Clin Gastroenterol Hepatol Off Clin Pract J Am Gastroenterol Assoc. 2008;6:983–7.

76. Högberg L, Fälth-Magnusson K, Grodzinsky E, Stenhammar L. Familial prevalence of coeliac disease: a twenty-year follow-up study. Scand J Gastroenterol. 2003;38:61–5.

77. Singh P, Arora S, Lal S, Strand TA, Makharia GK. Risk of celiac disease in the first- and second-degree relatives of patients with celiac disease: a systematic review and meta-analysis. Am J Gastroenterol. 2015;110:1539–48.

78. Sollid LM, Thorsby E. HLA susceptibility genes in celiac disease: genetic mapping and role in pathogenesis. Gastroenterology. 1993;105:910–22.

79. Kaukinen K, Partanen J, Mäki M, Collin P. HLA-DQ typing in the diagnosis of celiac disease. Am J Gastroenterol. 2002;97:695–9.

80. Lundin KE, Sollid LM, Qvigstad E, Markussen G, Gjertsen HA, Ek J, et al. T lymphocyte recognition of a celiac disease-associated cis- or trans-encoded HLA-DQ alpha/beta-heterodimer. J Immunol Baltim Md. 1950;1990(145):136–9.

81. Tollefsen S, Arentz-Hansen H, Fleckenstein B, Molberg O, Ráki M, Kwok WW, et al. HLA-DQ2 and -DQ8 signatures of gluten T cell epitopes in celiac disease. J Clin Invest. 2006;116:2226–36.

82. Lionetti E, Catassi C. Co-localization of gluten consumption and HLA-DQ2 and-DQ8 genotypes, a clue to the history of celiac disease. Dig Liver Dis. 2014;46:1057–63.

83. Dieli-Crimi R, Cénit MC, Núñez C. The genetics of celiac disease: a comprehensive review of clinical implications. J Autoimmun. 2015;64:26–41.

Classical and Non-classical Forms of CD in Paediatrics

Gemma Castillejo de Villasante

Until a few years ago, celiac disease (CD) was considered a childhood disease, because it was usually diagnosed in young children presenting with malabsorption (chronical diarrhoea, abdominal distension and failure to thrive). In the last decades, there has been a change in the presentation, which started in some countries in the 70s and 80s, although it has been observed more recently in others. This change has been observed at two levels: (1) CD has started to be diagnosed at an older age (in Finland, the average age went from 2 to 8 years of age [1]) and (2) The profile of the diagnosed patients has changed, as they present with more subtle symptoms or are even asymptomatic (between the years 2005–2011 in the UK more than 50% of children showed few or no symptoms at diagnosis [2]). The emergence of new diagnostic techniques has undoubtedly contributed to these changes, especially in the serology field, as well as their universalization, that were initially only available in specialized care, often in hospitals, and that progressively have been implemented in primary care centres.

Samuel Gee did the initial description of the CD in 1888 [3]: a patient "… *preferably between 1 and 5 years of age, cachectic, with distended abdomen, sad, irritable, anorexic, and with soft but not liquid, abundant, pale and foul-smelling stools …*".

Nowadays, the situation has changed from one in which diagnosis was limited to this classical clinical presentation, that was diagnosed using an aggressive method such as peroral intestinal biopsy, to the current situation, in which screening and active case finding can be carried out in a very simple way, with a simple test to determine the presence of specific antibodies.

More cases are also diagnosed due to the routine inclusion of screening in two groups of individuals: (1) among first-degree relatives of celiac (screening is usually requested at the time of diagnosis of the index case), and (2) Among those with a chromosomal abnormality such as Down's or Turner's syndrome, as well as

G. C. de Villasante (✉)
Hospital Universitari Sant Joan de Reus, Unitat Trastorns Relacionats amb el Gluten del Camp de Tarragona, Universitat Rovira i Virgili, Tarragona, Spain

patients with another autoimmune disease, such as diabetes mellitus, thyroiditis or autoimmune hepatitis, among others, since CD patients have been reported to have a 3- to 10-fold higher risk of developing another autoimmune disease than the general population.

Currently, most diagnoses in paediatrics are made at school age, usually with gastrointestinal symptoms (often oligosymptomatic). The main symptom at the time of CD diagnosis is abdominal pain, especially in pre-school and schoolchildren, as shown by recent studies in various parts of the world [4–7]. However, in children under 3 years of age, the typical presentation with malabsorption (chronic diarrhoea, failure to thrive and abdominal distension) is still the most frequent. Positive findings among relatives and patients with another autoimmune or genetic disease have increased the number of asymptomatic children at diagnosis compared to the situation a few years ago.

However, in the same way that it is clear that the highly affected clinical picture has been replaced by a milder presentation with less growth impairment, it seems that in recent years there has been a stabilization and no new changes in the clinical presentation at diagnosis [8].

CD is a systemic disease that is not limited to the gastrointestinal tract. Extraintestinal manifestations can affect almost all organs, with more or less significant manifestations, including the nervous system, liver, reproductive system, skin, cardiovascular system and musculoskeletal system [9]. Some of these manifestations can be seen in childhood and others do not appear until adulthood or even old age.

We shall classify the forms of presentation as asymptomatic or symptomatic. Among the latter, we find the classical (include diarrhoea, steatorrhoea, weight loss or failure to thrive), non-classical/oligosymptomatic (where individuals present without signs and symptoms of malabsorption) and extraintestinal forms [8].

1 Asymptomatic Forms

Usually patients who belong to a risk group (first-degree relatives, affected by another autoimmune disease or chromossomopathies) (Table 1), for whom screening for CD is recommended at diagnosis [10]. If the genetic study is compatible (HLA DQ2/8 positive), screening is performed periodically, especially if they develop symptoms de novo.

Table 1 Risk groups in paediatric patients

First degree relatives with CD
Autoimmune conditions: autoimmune thyroid disease, Type 1 diabetes mellitus, autoimmune hepatitis, Sjögren Sd, primary billiary cholangitis
IgA deficiency
Down syndrome, Turner syndrome, Williams-Beurer syndrome

The remaining asymptomatic patients are usually "casual" findings in children who undergo testing or screening studies [11]. Although it is not a generally recommended practice, as universal screening remains controversial (because there is insufficient scientific evidence), more and more paediatricians are including CD markers when ordering testing for any other reason.

2 Symptomatic Forms

Classical: the classical form of the disease affects children between 1 and 5 years of age (often before the age of 2), who present with loose/shaggy, bulky, foul-smelling, pale stools (often described by parents as vanilla/mustard coloured), abdominal distention and failure to thrive. In addition to the above triad, other symptoms or signs may include anorexia, irritability or behavioural change, anaemia, growth retardation or weight loss, decreased bone mineral density and micronutrient deficiencies.

Occasionally they may present as a celiac crisis, an urgent and potentially fatal complication (often triggered by infectious gastroenteritis in an undiagnosed patient). In these cases, there is a rapid deterioration of the general condition and clinically it is recognized by the presence of profuse watery diarrhoea, severe dehydration, nutritional and electrolyte disturbances, metabolic acidosis, hypotension, renal dysfunction, hypoproteinaemia and oedema, as well as a marked decrease in weight, with cachexia and sarcopenia.

The response to the gluten-free diet (GFD) in children with classical symptoms is usually fast and very apparent.

Non-Classical/Oligosymptomatic: increasingly common form of presentation today (Table 2). It has recently been reported that up to 43% of children present with non-classical symptoms at diagnosis [12]. It is difficult to attribute very common symptoms in the general population (such as headache, asthenia or abdominal pain) to CD, especially if the response to GFD is unsatisfactory, bearing in mind that the symptom could be due to CD alone or be part of an added functional disorder.

Most frequently reported symptoms are:

– Chronic or intermittent diarrhoea: this is common, with almost 20% of CD patients presenting with it at diagnosis. Many other patients present with bowel irregularity, alternating soft/liquid stools with normal or even hard stools.
– Constipation: there are no prospective studies with sufficient evidence to establish with certainty the correlation between constipation and CD. In the few studies, the prevalence of CD among constipated children is similar to that of the general population. However, in retrospective series it appears to be as common a symptom as diarrhoea in children with CD.
– Abdominal pain: as mentioned above, abdominal pain is the most frequently reported symptom in children diagnosed with CD. However, the causal

Table 2 Signs and symptoms suggestives of CD

Gastrointestinal	Abdominal pain, chronic or intermittent diarrhoea, constipation, irregular pattern of defecation, abdominal distention/bloating/excess flatus, recurrent nausea and/or vomiting
Extraintestinal	Weight loss/failure-to-thrive/stunted growth/short estature, chronic iron deficiency anaemia, dermatitis herpetiformis-like rash/alopecia areata, arthritis/arthralgia/osteomuscular recurrent pain, recurrent or chronic headaches/neuropathy/cerebellar ataxia, decreased bone mineralization (osteopenia/osteoporosis)/fractures with inadequate traumes, dental enamel defects, persistent or recurrent aphthous stomatitis, irritability, chronic fatigue, abnormal liver biochemistry, (hypertransaminasemia)

relationship has not been demonstrated in prospective studies, as few studies exist and those that do exist show very low prevalence's of CD, similar to those in the general population.

- Patients with irritable bowel syndrome: prevalence is two to four times higher than in the general population in the few studies conducted.
- Celiac patients often describe other symptoms such as bloating, vomiting and nausea at diagnosis, occasionally as the only complaint.

Extraintestinal: these manifestations are defined as clinical presentations in which GFD causes improvement, especially if initiated early. These may be accompanied by digestive symptoms or occur in isolation, and it is important to be aware of them in order to suspect the existence of CD and to make an early diagnosis and treatment, which will prevent long-term complications. The prevalence of these manifestations is similar in children (60%) and adults (62%).

The pathogenesis of extraintestinal manifestations is unclear, since they might be due to the severity of the disease (in patients with a more serologically and histologically affected presentation) or simply be different phenotypes of presentation. They could also correspond to a complication of untreated CD and as such, their response to diet would not be good, especially if there is a long delay in diagnosis, as would be the case for example in osteopenia/osteoporosis or short stature.

In general, there seems to be a correlation between the delay in the diagnosis of CD and the presence or absence of gastrointestinal symptoms (more delay in those who do not). Probably for this reason, it has been reported that children who debut mainly with extraintestinal symptoms have a higher degree of intestinal atrophy than those with oligosymptomatic or asymptomatic digestive symptoms. Furthermore, recovery at 24 months would also be somewhat lower in extraintestinal patients, due to the higher degree of injury and the more complex pathophysiological mechanisms in these cases [13], which would be due to both atrophy-induced malabsorption and sustained autoimmune response. In this regard, deposits of IgA colocalized with anti-transglutaminase 2 have been found in the liver, lymph nodes, muscle, thyroid, bone and brain, indicating that autoantibodies, originating in the gut, could access transglutaminase 2 in various parts of the body and cause pathological effects locally [14, 15].

Although the response to GFD for all these manifestations is usually good, especially in the paediatric age group, in some cases it may not be complete [9] and symptoms such as fatigue, some neurological manifestations or those compatible with functional digestive disorders may persist in a subgroup of patients for a prolonged period of time. As mentioned above, the best response will be obtained if the diagnosis and treatment are established early, especially before the problem becomes irreversible (for example, CD with short stature, which is not diagnosed before the end of growth; an alteration of tooth enamel before it appears; or a very advanced gluten ataxia, which has caused irreversible cerebellar atrophy). Sometimes, the gluten-free diet is not enough and must be complemented with treatments such as iron supplements in anaemia, dapsone in dermatitis herpetiformis or calcium and vitamin D in severe osteopenia. In all cases, the lack of improvement in symptoms despite the diet should lead us to check that the diet is being followed correctly, using all the resources at our disposal, including the recently discovered gluten immunogenic peptides, which can be determined using ELISA techniques in urine or stools [16].

Stunting and short stature: this is the most common extra-intestinal manifestation in paediatrics (10–40% at diagnosis) [16]. It can be seen at any age, usually in patients with late diagnosis or with more general involvement (with weight loss/ stagnation) but is also seen in children with normal nutritional status and weight. A recent meta-analysis [17] suggests that 1 in 14 patients with all-cause short stature and 1 in 9 patients with idiopathic short stature have celiac disease. Given these findings, the authors would recommend ruling out CD in all patients with short stature.

The pathogenic mechanisms described, in addition to malnutrition, would include dysfunction of the growth hormone (GH)-insulin-like growth factor (IGF-1) and ghrelin axis.

Catch-up growth is usually rapid (<6 months) in the case of weight and although catch-up growth may take longer, it is usually resolved by 24 months from the onset of GFD. If catch-up does not occur within this time, other causes of short stature should be ruled out, such as inflammatory disease, growth hormone deficiency or Turner syndrome (in which CD is also more prevalent). Catch-up may be more difficult in children diagnosed at the end of puberty, who may have missed their window of opportunity, as bone maturation accelerates before mucosal recovery is complete, so early diagnosis and initiation of the diet may minimize the risk of compromising final height.

It is very important to emphasize that a correct nutritional status does not exclude the possibility that the patient may be suffering from CD. In fact, most celiac patients at diagnosis are currently eutrophic, with increasing percentages of overweight and/or obese children, a fact to be considered before ruling out serology based on the patient's appearance. As we have said, nowadays, the most frequent phenotype of celiac disease is that of a "normal" child.

Anaemia: this is a very common extraintestinal manifestation, especially in adult patients (up to 50%). Although less commonly seen in children (between 3–10%

depending on studies), in both cases it is usually associated with a more severe presentation of CD (usually due to late diagnosis), as a complication of CD, or it may also be the only manifestation present. Anaemia is less common among patients with asymptomatic forms, while iron deficiency is a more common finding in all phenotypes.

Iron, and folic acid are mainly absorbed in the duodenum, so malabsorption at this level (due to intestinal atrophy) would be responsible for the anaemia, although it has also been described in individuals with only autoimmunity (without the presence of enteropathy), in which case chronic inflammation would play a role. Despite Vitamin B12 is absorbed in terminal ileum, deficit it's not uncommon in celiac patients, because when celiac disease progresses, the lower part of the small intestine can be damaged, causing vitamin B12 deficiency. Supplementation (together with GFD) usually corrects the deficiency of these micronutrients, in parallel with mucosal healing.

Dermatological Manifestations

Dermatitis herpetiformis: this is a cutaneous form of CD, which is induced and dependent on gluten intake. Although more common in adults, it also occurs in paediatric patients (5%). Its overall incidence is decreasing and unlike CD, it tends to be more common in males.

It should be suspected in the presence of a very pruritic vesicular-papular rash appearing on elbows, knees and buttocks (less frequently on the upper and lower back, scalp and face). It is common to find crusted lesions due to scratching, which ruptures the vesicles and forms lesions that often leave white marks on the skin when healed. The condition may run intermittently.

Gastrointestinal symptoms are rare, although up to 70% of patients present with enteropathy. The remaining 30%, with normal intestinal biopsy, show positivity for anti-transglutaminase 2 and anti-endomysial antibodies in 40% of cases. Autoantibodies against transglutaminase 3 have been reported in the majority (94%) of these patients.

Cases have been described with classic digestive symptoms at diagnosis, in which exposure to gluten causes the disease to manifest itself in the form of DH, and cases of DH in which intensifying gluten intake eventually causes intestinal damage [18].

For diagnosis, a skin biopsy sample of healthy skin near the lesion must be obtained. Direct immunofluorescence will show granular IgA deposits in the papillary dermis.

Although the response to GFD in paediatric patients is close to 100%, in some cases, the response is not complete, thus requiring the administration of dapsone for proper symptom control, and treatment can be discontinued during the following months in most cases.

Alopecia: a rare manifestation in paediatrics (1%). Hair loss can occur in areas of hair (*alopecia areata*), all hair and facial hair (*alopecia totalis*) or all over the body (*alopecia universalis*). The pathophysiology is unclear and may involve cell-mediated autoimmunity, autoantibodies directed against the hair follicle or the presence of DQB1*03. Many patients have another associated autoimmune disorder, especially vitiligo and autoimmune thyroiditis, which is why TSH determination is recommended when *alopecia areata* is diagnosed [19].

The response to topical treatments is poor. Half of the patients respond to GFD, most of them after 2–3 months, although in some patients it takes somewhat longer, up to 12–24 months (in 26% the hair never grows back and in 22% the growth is partial despite normalization of serology and intestinal biopsy [18]). The response is better in younger children and in those for whom the time between the onset of alopecia and the onset of GFD was short.

Psoriasis: no direct association with CD has been established. However, celiac patients have a slightly higher risk of developing psoriasis, so if the condition is difficult to manage, the presence of CD should be ruled out.

Joint/musculoskeletal problems: occasionally (5–10%), CD patients present with joint/muscle pain at diagnosis. In the paediatric age group, these are difficult to differentiate from growth-related pain. Although synovitis has been reported, it is rarely found. The most frequently affected joints are knees, followed by hips and ankles.

If the symptoms persist after starting the diet, other rheumatological autoimmune diseases with joint/muscle pain such as juvenile idiopathic arthritis, lupus, rheumatoid arthritis or Sjögren's disease should be considered.

Although response to diet is usually good, in some patients the pain persists despite diet, with unknown pathogenic mechanisms.

Neuropsychiatric manifestations: the most frequent manifestation in paediatrics is *headache*, present in up to 30% of children at diagnosis. The cause is unclear, being related to vitamin deficiencies, macro elements such as magnesium, low serotonin levels or the presence of pro-inflammatory cytokines that cause changes in vascular tone. In most patients, the headache disappears after the onset of GFD.

In adults, the most common neurological manifestations are peripheral neuropathy and gluten ataxia (in paediatrics 0.1–7.4% and 0.068–1.79% respectively). Pathogenesis of these manifestations is unclear. Many of these patients are only positive for anti-gliadin antibodies, often with low values, and although anti-transglutaminase-6 antibodies have been found in many cases, the presence of these autoantibodies has also been described in celiac and non-celiac children without apparent neurological manifestations [20]. In celiac children, positivity was related to the time of exposure to gluten, disappearing after the start of the diet. It is not known whether transglutaminase 6 (TG6) positivity would be related to the appearance of neurological pathology in the future. TG6 is expressed in neurons, plays an important role in neurogenesis and represents the autoantigen in gluten ataxia.

Many patients with *peripheral neuropathy* have anti-ganglioside antibodies. These patients have a response to GFD that ranges from significant improvement to little or no improvement, in this case, with little response to other treatments, to which non-gluten-sensitive peripheral neuropathies do respond.

Gluten ataxia is clinically indistinguishable from other forms of cerebellar ataxia. It is now considered an autoimmune disorder related to gluten. Digestive symptoms are rare in these patients (<10%) and although many of them do not have enteropathy (but TG2 deposits in the gut), up to 50% have antibodies to TG2 in the blood, usually at low titres and sometimes only IgG or anti-gliadin. Combined TG2 and TG6 positivity is present in up to 85% of patients.

The response to GFD, although variable, is usually good, much better if diagnosis and treatment takes place early (because cerebellar atrophy is not reversible). The occurrence of atrophy in children is exceptional; probably because its severity is related to the time of gluten exposure, (children are more likely to develop unilateral or bilateral focal hyper intense white matter lesions [21]).

Most studies have failed to demonstrate an increased prevalence of epilepsy in CD patients. Epilepsy with brain calcifications in the occipital area has been described and linked to celiac disease, although GFD alone is not sufficient to reverse epilepsy in most patients.

Other gluten-related problems, such as learning disorders, neurodevelopmental delay, attention deficit with/without hyperactivity, autism spectrum disorder in children, as well as depression, anxiety and even schizophrenia, have been described, but there is currently little available evidence.

Dental enamel alterations: this was mainly described in older series with small patients who were clinically very affected, because CD occurred during the critical period of enamel formation. The exact prevalence is unknown, with studies reporting 40–50% of patients at diagnosis and others only <15%.

It is believed that calcium malabsorption, together with genetic and immunological factors would produce an alteration of enamel amelogenesis, with permanent enamel impairment if the damage occurs in the permanent dentition. It has also been related to the time of exposure to gluten (the average age at diagnosis of CD was higher in patients with enamel alterations [22]). In the few series available, it is described as a much more prevalent sign in CD patients than in the general population, highlighting that in celiac patients the defects occur symmetrically and in all quadrants of the dentition, in the form of pitting, complete loss of enamel, discolouration and structural changes.

Liver involvement: a sign described more frequently in older series, probably due to the greater clinical involvement of patients in the past, (currently some series report a prevalence of around 9–14%). Although severe cases with liver failure have been described, hepatitis is generally mild-moderate and reversible soon after starting GFD. Its severity is related to the presence of malabsorption, elevated celiac serology values and increased intestinal damage [23].

It has been related to the translocation of food antigens and bacteria due to increased intestinal permeability secondary to intestinal injury, which would reach the liver via the portal circulation and, once there, trigger a local immune response leading to inflammation. It has also been related to direct autoimmune damage, as the presence of anti-transglutaminase 2 antibody deposits in liver biopsies of affected patients has been demonstrated.

If the inflammation does not resolve within 12–24 months after starting the gluten-free diet, autoimmune hepatitis, primary biliary cholangitis and primary sclerosing cholangitis should be ruled out.

Bone disease: rickets used to be a relatively common finding in young patients diagnosed with the classical form of CD. Nowadays, in developed countries it is quite rare, with osteopenia/osteoporosis being more frequent.

Nutrition plays an important role in bone homeostasis. In CD, due to proximal intestinal atrophy, the absorption of vitamin D, calcium and minerals is compromised (absorption occurs mostly at this level). Decreased serum levels of these minerals lead to parathyroid stimulation, resulting in increased bone turnover and secondarily osteopenia/osteoporosis. Increased circulating inflammatory cytokines may also play a role.

In the paediatric age group, up to 75% of children at diagnosis have osteopenia, but only 30% have osteoporosis [24], which is more common in adults, especially in the elderly and postmenopausal women. Although osteoporosis is considered a complication of CD, it may also be the only finding in asymptomatic patients (detected by screening), even before the development of enteropathy.

The treatment of choice is strict GFD. By 12 months, most children have complete remineralization of the bones, without requiring supplementation, especially if the diagnosis was done early. Children diagnosed during infancy are not at increased risk of developing fractures in the future.

Endocrine problems: delayed puberty has been observed in up to 10% of celiac patients at diagnosis. It is related to malnutrition and possibly the presence of autoimmunity against hormones, their receptors or endocrine organs, so GFD usually resolves the problem in 6–8 months. If it persists after 12–24 months, an endocrinological assessment should be carried out to rule out other processes.

Reproductive system problems: in women of childbearing age, amenorrhoea, infertility, repeated miscarriages, prematurity and intrauterine growth retardation have been reported. In men, changes in the number and motility of spermatozoa have been reported, which could lead to infertility. This may be related to micronutrient deficiencies and may be reversible after initiation of GFD.

Oral thrush: there is an increased prevalence of CD among patients with recurrent oral thrush, so if it is present repeatedly, it should be ruled out (as well as inflammatory bowel disease). The mechanism of production is unknown and malabsorption, changes in the oral microbiota, local leukocyte levels or salivary flow have been proposed. Establishment of GFD usually solves the problem.

Asthenia, tiredness or fatigue: this is currently one of the most frequent forms of presentation of CD, although it is probably because it has been underestimated in the past. It is suggested to ask for CD serology in patients with this symptom, because, in addition, GFD is often effective in resolving it.

Although some of the symptoms and signs described in this chapter do not have sufficient evidence to recommend screening for CD on all occasions, the proposed approach is to maintain a high index of suspicion, which would allow early diagnosis, before the onset of complications. Therefore, if a suggestive sign or symptom is found, the recommendation would be to screen the markers to rule out the presence of CD.

References

1. Mäki M, Kallonen K, Lähdeaho M-L, Visakorpi JK. Changing pattern of childhood coeliac disease in Finland. Acta Paediatr Scand. 1988;77:408–12.
2. Whyte LA, Jenkins HR. The epidemiology of coeliac disease in South Wales: a 28-year perspective. Arch Dis Child. 2013;98(6):405–7.
3. Gee SJ. The coeliac affection. St Bart Hosp Rep. 1888;24:17–20.
4. Riznik P, De Leo L, Dolinsek J, Gyimesi J, Klemenak M, Koletzko B, et al. Clinical presentation in children with coeliac disease in Central Europe. J Pediatr Gastroenterol Nutr. 2021;72(4):546–51.
5. Van Kalleveen MW, de Meij T, Plötz FB. Clinical spectrum of paediatric coeliac disease: a 10-year single-centre experience. Eur J Pediatr. 2018;177(4):593–602.
6. Jansen M, van Zelm M, Groeneweg M, Jaddoe V, Dik W, Schreurs M, et al. The identification of celiac disease in asymptomatic children: the generation R study. J Gastroenterol. 2018;53(3):377–86.
7. Krauthammer A, Guz-Mark A, Zevit N, Marderfeld L, Waisbourd-Zinman O, Silbermintz A, et al. Two decades of pediatric celiac disease in a tertiary referral center: what has changed? Dig Liver Dis. 2020;52(4):457–61.
8. Ludvigsson JF, Leffler DA, Bai JC, et al. The Oslo definitions for coeliac disease and related terms. Gut. 2013;62(1):43–52.
9. Laurikka P, Nurminen S, Kivelä L, Kurppa K. Extraintestinal manifestations of celiac disease: early detection for better long-term outcomes. Nutrients. 2018;10(8):1015.
10. Walker-Smith JA, Guandalini S, Schmitz J, Shmerling DH, Visakorpi JK. Revised criteria for dianosis of coeliac disease. Report of Working Grup of European society of paediatric gastroenterology and nutrition. Arch Dis Child. 1990;65(8):909–11.
11. Rutz R, Ritzler E, Fierz W, Herzog D. Prevalence of asymptomatic celiac disease in adolescents of eastern Switzerland. Swiss Med Wkly. 2002;132(3–4):43–7.
12. Almallouhi E, King KS, Patel B, Wi C, Juhn YJ, Murray JA, et al. Increasing incidence and altered presentation in a population-based study of pediatric celiac disease in North America. J Pediatr Gastroenterol Nutr. 2017;65(4):432–7.
13. Jericho H, Guandalini S. Extra-intestinal manifestation of celiac disease in children. Nutrients. 2018;10(6):755.
14. Yu XB, Uhde M, Green PH, Alaedini A. Autoantibodies in the extraintestinal manifestations of celiac disease. Nutrients. 2018;10(8):1123.
15. Nardecchia S, Auricchio R, Discepolo V, Troncone R. Extra-intestinal manifestations of coeliac disease in children: clinical features and mechanisms. Front Pediatr. 2019;7:56.
16. Morón B, Cebolla A, Manyani H, Alvarez-Maqueda M, Megías M, Thomas Mdel C, et al. Sensitive detection of cereal fractions that are toxic to celiac disease patients by using monoclonal antibodies to a main immunogenic wheat peptide. Am J Clin Nutr. 2008;87 (2):405–14.

17. Singh AD, Singh P, Farooqui N, Strand T, Ahuja V, Makharia GK. Prevalence of celiac disease in patients with short stature: a systematic review and meta-analysis. J Gastroenterol Hepatol. 2021;36(1):44–54.
18. Popp A, Mäki M. Gluten-induced extra-intestinal manifestations in potential celiac disease-celiac trait. Nutrients. 2019;11(2):320.
19. Fessatou S, Kostaki M, Karpathios T. Coeliac disease and alopecia areata in childhood. J Paediatr Child Health. 2003;39(2):152–4.
20. De Leo L, Aeschlimann D, Hadjivassiliou M, Aeschlimann P, Salce N, Vatta S, et al. Not T. Anti-transglutaminase 6 antibody development in children with celiac disease correlates with duration of gluten exposure. J Pediatr Gastroenterol Nutr. 2018;66(1):64–8.
21. Kieslich M, Errázuriz G, Posselt HG, Moeller-Hartmann W, Zanella F, Boehles H. Brain white-matter lesions in celiac disease: a prospective study of 75 diet-treated patients. Pediatrics. 2001;108(2):e21.
22. Majorana A, Bardellini E, Ravelli A, Plebani A, Polimeni A, Campus G. Implications of gluten exposure period, CD clinical forms, and HLA typing in the association between celiac disease and dental enamel defects in children. A case-control study. Int J Paediatr Dent. 2010;20(2):119–24.
23. Äärelä L, Nurminen S, Kivelä L, Huhtala H, Mäki M, Viitasalo A, et al. Prevalence and associated factors of abnormal liver values in children with celiac disease. Dig Liver Dis. 2016;48(9):1023–9.
24. Pantaleoni S, Luchino M, Adriani A, Pellicano R, Stradella D, Ribaldone DG, et al. Bone mineral density at diagnosis of celiac disease and after 1 year of gluten-free diet. Sci World J. 2014:173082.

Immunopathogenesis of Celiac Disease

Eduardo Arranz and José A. Garrote

1 Introduction

The mucosal associated lymphoid tissue is the largest component of the immune system of the body, due to its central role at the interface with the external environment and shows the ability to discriminate between infectious agents and commensal bacteria and food antigens. Under normal conditions, homeostasis at mucosal surfaces depends on mechanisms mediated by secretory IgA antibodies and tolerogenic T and B cells, limiting the entrance of pathogens by immune exclusion, and inducing responses of oral tolerance. After the absorption of food proteins through the intestinal epithelium, lamina propria (LP) antigen presenting cells (APC), particularly dendritic cells (DCs), transport these antigens to draining mesenteric lymph nodes, where they promote gut-homing T-cell responses. Regulatory T cells are responsible for the inhibition of inflammatory responses to food antigens during oral tolerance [1, 2]. In celiac disease (CD), however, dietary gluten drives a T-cell mediated immune response leading to the destruction of the epithelium in the small intestine.

Dietary Gluten and Proteolysis of Gluten Peptides

The wheat protein fraction can be classified, according to their structural properties and solubility, in α, γ, and ω-gliadins (monomeric, alcohol soluble), as well as low- and high-molecular weight glutenins (polymeric, soluble under stronger conditions), though the term gluten is currently used to identify proline- and

E. Arranz (✉) · J. A. Garrote
Mucosal Immunology Lab, Unidad de Excelencia Instituto de Biología y Genética Molecular (IBGM), University of Valladolid-CSIC, Valladolid, Spain
e-mail: earranz@uva.es

J. A. Garrote
Servicio de Análisis Clínicos, Hospital Universitario Rio-Hortega de Valladolid, Gerencia Regional de Salud (SACYL), Valladolid, Spain

glutamine-rich proteins contained in the storage fraction of wheat, barley, and rye grains, collectively referred as prolamines [3, 4]. Under normal conditions, proteins are mostly hydrolyzed by gastric, pancreatic and intestinal brush border proteases, resulting in the formation of smaller peptides or isolated amino acids, which may cross the epithelium more easily. However, the high proline content of gluten proteins makes them resistant to proteolysis, and long fragments containing from 15 to 50 residues are generated in the intestinal lumen [5, 6].

Partially digested gluten peptides, which include several copies of immunogenic epitopes, may cross the intestinal epithelium and translocate into the LP by different pathways, either paracellular, through the tight junctions between enterocytes [7, 8]; or transcellular, by a mechanism involving enterocyte endocytosis and lysosome degradation, which is altered in CD and may allow intact peptides to cross the epithelium [9–12]; but also by the mechanism of retrotranscytosis, which depends on peptide binding to secretory immunoglobulin A1 (SIgA1), a transferrin receptor (CD71) ligand, over-expressed at the apical epithelial membrane in CD mucosa [12, 13]; or following the interaction with C-X-C chemokine receptor type 3 (CXCR3) expressed by enterocytes, leading to the release of zonulin, a potent modulator of the intestinal barrier function [14]. Finally, gluten peptides can reach the LP by direct access through extensions of monocyte-derived DCs, which are sandwiched between epithelial cells [4].

Other Environmental Factors

Only a small percentage of individuals carrying CD-predisposing HLA.DQ alleles actually develops T cell mediated immunity and tissue damage after the ingestion of gluten, and this suggests that other non-MHC genetic variants (most of them shared with other autoimmune diseases) and/or other environmental factors may be also involved in the pathogenesis of CD [15]. Among these environmental factors, microbiota and/or intestinal infections are most cited. The microbiota has an important role in the maintenance of intestinal homeostasis, by promoting epithelial integrity and the generation of tolerogenic cells, and this is mainly mediated by bacterial metabolism of dietary susbstrates with immunomodulatory effects, such as small-chain fatty acids (SCFAs) [16–18]. In this context, the large gluten peptides generated by partial digestion in the small intestine of CD patients are a good substrate for bacterial metabolism.

The changes in the duodenal and faecal microbiota reported in patients with CD have been characterized by the decrease in beneficial species (*Lactobacillus* and *Bifidobacterium*), the expansion of *Proteobacteria*, and the proliferation of opportunistic pathogens, such as *Neisseria*, *E. coli*, or *Pseudomonas* [19–21]. These changes in the composition of gut microbiota could play a role in increasing the permeability of the epithelial barrier [22], but commensal bacteria may also affect the immunogenicity of peptides by modulating the metabolism of gluten and its proteolitic activity. It has been reported that *Lactobacillus*, which is depleted in CD patients, can degrade and detoxify partially-digested immunogenic peptides, whereas the opportunistic pathogen *Pseudomonas aeruginosa* metabolizes these large gluten fragments generating shorter highly-immunogenic peptides that cross the epithelium more easily [23]. Moreover, *Pseudomonas* also triggers a pro-inflammatory response in intestinal epithelial cells, mediated by the upregulation of protein-activated-receptor-2 (PPAR-2) [24]. Finally, the immunogenicity of gluten peptides can be also reduced by the effect of transglutaminase from *Streptomyces mobaraensis* [25].

Enteral infections by viruses such as Adenovirus, Enterovirus, Hepatitis C virus and Rotavirus, have been associated with CD [26]. The effect of Rotavirus infection may be mediated by changes in the permeability of the small intestine, leading to an increased passage of dietary antigens through the epithelium [26, 27]. It has been also suggested that viral infections may determine the upregulation and release of transglutaminase [28]. Moreover, the probably shared effect of some viruses in promoting the loss of tolerance to gluten, and the development of the disease, can be explained by the induction of type 1 interferons. The presence of elevated IFN-α levels in the mucosa of patients with CD may be a critical factor for the proinflammatory differentiation of DCs [29], as suggested by the onset of the disease in patients with hepatitis C treated with interferon (IFN)-α [30, 31].

Reovirus infections are common during early childhood, normally without clinical manifestations [32], but, when the infection occurs at the same time of first dietary intake of the protein, the virus may impair the induction of tolerance to food antigens by suppressing the generation of tolerogenic regulatory T cells, and driving a Th1 response to dietary antigens. In two studies published by the same group, using a HLA-DQ8-transgenic mice model, it has been shown that reovirus, and similar enteric viruses, may promote the disruption of immune homeostasis at inductive sites of oral tolerance (ie. mesenteric lymph nodes) by imprinting a proinflammatory signature in DCs, which involves the upregulation of the transcription factor interferon regulatory factor (IRF)-1 [33, 34].

2 Gluten-Specific Immunity

The predisposition to CD is associated to certain HLA-DQ allotypes, and T-cell mediated immunity is restricted by HLA class II molecules, due to its role in the preferential presentation of gluten epitopes to gluten-reactive CD4 + T cells [28, 35]. This response is determined in the organized gut-associated lymphoid tissue, such as mesenteric lymph nodes or Peyer´s patches, which are the induction sites of the immune response to gluten. Several gliadin and glutenin-derived T-cell epitopes, with different immunogenicity, have been identified and listed according to the definition of an specific T cell clone, the corresponding HLA-restriction element, and the nine-amino acid core of the epitope [36, 37]. Most gluten proteins are proteolyzed by intestinal enzymes, but a few large fragments remain undigested, containing several gluten epitopes which can elicit T cell responses. The differences in the immunogenicity of gluten peptides, and the selection of gluten T-cell epitopes, as occur in HLA-DQ2.5 positive CD patients, is determined by the resistance to proteolytic degradation, the substrate affinity to tissue transglutaminase (TG2), and the binding specificity to HLA-DQ molecules [38].

2.1 Deamidation of Gluten Peptides and Binding Affinity to HLA-DQ Molecules

The resistance of gluten peptides to proteolytic degradation by intestinal digestive enzymes depends on the high proline content of gluten proteins, and the resulting large undigested fragments are also rich in glutamine residues, which make these peptides a very good substrate for the enzyme TG2 [39–41]. TG2 plays a fundamental pathogenic role in CD by catalyzing the deamidation (posttranscriptional modification) of gluten peptides, and the enzyme is also the main antigen of anti-TG2 antibodies [42]. The enzyme recognizes specific glutamine residues in protein sequences of the type glutamine-X-proline (G-X-P), converting them into negative-charged glutamate residues [41]. Deamidation highly increases the binding affinity of gluten peptides to HLA-DQ molecules and, therefore, facilitates the subsequent recognition of these T-cell epitopes by gluten-specific CD4 + T cells [43–45].

It has recently been reported that the source of pathogenic TG2 is luminal TG2 derived from the renewal and shedding of enterocytes [46]. Under normal conditions, TG2 is a cytosolic enzyme, highly expressed in lymphoid tissue [47], though studies in mice have suggested that TG2 is not constitutively active in vivo [48], and other inflammatory or viral stimuli are required for the transient activation of the enzyme [49, 50]. TG2 may also have a role in the mechanism of retrotranscytosis of gluten peptides across the epithelium, following its interaction with the transferrin receptor (CD71) and secretory IgA at the apical surface of enterocytes [51]. Moreover, in this same study it was confirmed that the use of TG2 inhibitors have also the effect of blocking the transport of gliadin peptides through this pathway.

The HLA-DQ2.5 and DQ8 molecules associated to CD show distinct peptide binding groove preferences, according to the different T cell repertoire found in each CD patient. These molecules have a high binding affinity for deamidated peptides with negatively charged anchor residues in specific positions, as well as a high ability of generating stable peptide-HLA complexes. For example, the core structure of the HLA-DQ2.5 peptide pocket shows a preference for bingeing negatively charged glutamate residues at positions P4, P6 and P7 [43, 52], whereas the HLA-DQ8 molecule shows preferences for more external residues, at positions P1 and P9 [53]. The positions of glutamate residues are related to those of proline in the peptide binding groove, particularly for the HLA-DQ2.5 molecule [54, 55]. In HLA-DQ2.5 individuals, the immunodominant epitopes eliciting specific T cell responses are mainly found in α and ω-gliadins, whereas responses to γ-gliadins are less relevant [56].

2.2 Gluten-Reactive CD4 + T Cells and Proinflammatory Cytokines

The recognition of gluten epitopes by gluten-specific CD4 + T cells with HLA-DQ restriction leads to the synthesis of a pro-inflammatory cytokine profile characterized by interferon (IFN)-γ and interleukin [IL]-21, as well as low levels of the immunoregulatory interleukin [IL]-10 and transforming growth factor (TGF)-β [57–59]. The small intestine of untreated CD patients also contains other cytokines, such as interleukin [IL]-15, interleukin [IL]-18 and interferon (IFN)-α [29, 60, 61]. These cytokines contribute to the development of the enteropathy, by promoting the activation of cytotoxic intraepithelial lymphocytes (IEL) and the destruction of epithelial cells, and by providing help to B cells to produce antibodies to gluten and TG2 [62–64]. By using a (DQ-D-villin-IL-15tg transgenic) mouse model overexpressing IL-15 in both the epithelium and the LP, it has been found that IFN-γ is required for the expansion of intraepithelial lymphocytes (IELs) and the development of villous atrophy. Moreover, in this model, the administration of TG2 inhibitors with dietary gluten prevented the lesion [65].

Recent technical developments have allowed the characterization of the T and B cell immune responses elicited by gluten. Gluten-reactive CD4 + T cells are found in the small intestine and peripheral blood from both treated and untreated CD patients [35, 66]. Gluten-specific T cells induced by oral gluten challenge, or expanded from the intestine of untreated CD patients, share the specificity for deamidated, immunodominant T cell epitopes, in both children and adults [67, 68]. Moreover, by using a HLA-DQ-gluten peptide tetramers, it was found that these cells were specific to four immunodominant epitopes of α- and ω-gliadins [69]. After activation, gluten-specific CD4 + T cells were clonally expanded in both the intestine and the peripheral blood of untreated CD patients, and these clonotypes persist in low levels for decades as memory T cells, with the same specificity, even on a gluten-free diet. These gluten-specific T cells undergo a rapid expansion and dominate the subsequent recall response after gluten challenge [69].

The TCR repertoire of gluten-specific CD4 + T cells is polyclonal with a biased use of TCR-Vα chain segments, probably reflecting their preferential interaction with HLA-DQ molecules [70, 71]. By using tetramers constructed with five gluten peptides complexed to HLA-DQ2.5, a small cluster of small intestinal CD4 + T cells was defined, and characterized by a distinct phenotype, similar to that found in peripheral blood from untreated HLA-DQ2.5 + CD patients [72]. These cells, mostly effector memory T cells (CD45RA-, CD62L-), expressed a number of activation markers, such as C-X-C chemokine receptor type 3 (CXCR3), CD38, CD161 and HLA-DR, but also the stimulatory checkpoint molecules OX40 and CD28, CD39, and the programmed cell death protein 1 (PD-1). Moreover, by RNA sequencing analysis, these cells transcribed also markers of follicular B helper T cells, such as CD200, CD84, C-X-C chemokine ligand type 13 (CXCL13) and IL-21, which may indicate a possible role in the differentiation of plasma cells in the inflamed tissue [72].

2.3 Role of B Cells and Production of Autoantibodies

Untreated CD patients produce specific antibodies, and serum antibody levels disappear after gluten withdrawal from the diet. The number of plasma cells In the intestinal LP of patients with active CD is highly increased, and a great proportion of these cells are involved in the production of IgA antibodies specific for gluten or TG2 or both [73, 74]. In active CD, there is a two to threefold increase of these antibodies in the intestinal lesion, and subepithelial TG2-specific IgA deposits have been found in all disease stages, even before the onset of symptoms, or before the intestinal lesion is confirmed [75, 76]. B cells are not only antibody-secreting cells, receiving help from T cells when both share the same antigen specificity, but they are also very efficient APC. Plasma cells have been confirmed as the dominant APC of immunodominant gluten epitopes in the intestinal LP of CD patients [74].

Under normal conditions, the intracellular location of TG2 may be responsible of preventing the induction of B cell tolerance to TG2, and the generation of autor-reactive B cells, which produce auto-antibodies after receiving appropriate T cell help [38, 77]. The initial proliferation of these cells take place outside the CD lesion, and the interaction between B and T cells occur once they enter the intestine. This interaction between gluten-specific T cells and TG2-specific B cells has been confirmed, both in vitro and in vivo [77]. The production of anti-TG2 antibodies depends on the presence of dietary gluten, and these antibodies are only found in individuals expressing HLA-DQ2.5/DQ8, which suggests the involvement of glu-ten-specific CD4 + T cells in this process.

Both the uptake of gluten peptides, and deamidation by active TG2, may depend on the formation of covalent TG2-gluten complexes (acting as hapten-carrier complexes) following B-cell receptor (BCR)-mediated endocytosis by TG2-specific B cells. The internalization of these complexes is linked to the presentation of deamidated gluten epitopes, bound to membrane-linked HLA-DQ molecules, to gluten-specific T CD4 + cells, because the efficiency of B cells as APCs is based on the epitope recognized by the BCR and the activity of BCR-bound TG2 [46]. Activated gluten-specific CD4 + T cells provide cognate help to gluten-specific B cells to differentiate into antibody-producing plasma cells [77, 78], but these CD4 + T cells also control the activity of cytotoxic intraepithelial CD8 + T cells [65, 72, 74]. Therefore, the mutual activation of T and B cells is manifested by both the production of autoantibodies and the release of a pro-inflammatory cytokine profile, leading to the amplification of the immune response to gluten.

3 Epithelial Stress and Tissue Destruction

In CD, mucosal tissue injury is the result of both innate [not necessarily driven by gluten) and adaptive immunity (gluten specific), and the intestinal chronic inflammation permanently reconfigures the tissue-resident TCRγ + IEL compartment. Several gluten peptides, whose paradigm is gliadin peptide p31-42 (but also p31-49 and p31-55), have been identified with pathogenic effects related to the induction of innate immunity and epithelial cell stress, irrespective of CD4 + T-cells and the restriction by HLA-DQ2/DQ8 molecules (revised by Chirdo et al. [79]). These peptides seem to have a direct effect on epithelial cells, though they are not recognized by gliadin-specific T CD4 + cells.

In CD, intestinal IELs lose the expression of inhibitory CD94/NKG2A receptors, while increasing the expression of the activating receptors NKG2D and CD94/NKG2C. At the same time, epithelial cells increase the expression of their ligands MIC and HLA-E [80, 81]. Epithelial damage leads to an increased gut permeability, which may allow the passage of larger, partly-digested gliadin peptides, to the LP mucosa, where they are the target of TG2 and may be presented to gluten-specific T CD4 + cells in the context of HLA-DQ molecules, thereby triggering a positive feedback loop that maintains inflammation and the development of the lesion [28]. The expression of NKG2D is driven by the upregulation of IL-15 by epithelial cells, and cytotoxic IELs stimulated by IL-15 can mediate TCR-independent cytotoxicity [80, 82]. However, in CD in contrast to the adaptive immune response to gluten, innate immunity does not require the presence of gluten peptides, and it may be activated by other intercurrent factors as intestinal infections, dysbiosis or other aggressive events taking place at the intestinal mucosa [83].

3.1   Inflammatory Mediators and Epithelial Stress

The epithelial cell stress induced by gluten peptides is consequence of the activation of the NFκB pathway, and the extracellular TG2, following the inhibition of the chloride channel CFTR [84], as well as the alteration of vesicular trafficking and the activation of the NLRP3 inflammasome, as shown in murine models [85]. The result is the upregulation of IL-15, type I IFNs and other inflammatory mediators, and the release of reactive oxygen species (ROS) and intracellular Ca2 + , due to mitochondrial malfunction, as well as the induction of crypt cell proliferative, probably by acting on the epidermal growth factor pathway [79].

The mechanism by which gluten peptides interact with epithelial cells is still elusive. Several gluten peptides may use the chemokine receptor CXCR3, activating the My88D and NFkB pathways on epithelial cells, and increasing permeability by the release of zonulin, a protein that rearranges the cell cytoskeleton and modifies tight junctions [14]. Furthermore, the interaction of peptide p31-43 with the chloride channel CFTR has recently been described [84]. This membrane channel is involved in the adaptation of enterocytes to oxidative stress, and the interaction with gluten peptides may affect the function of the endocytic pathway in the epithelial cells [86]. This interference results in the activation of extracellular TG2 and the inflammasome NRLP3, as well as in the upregulation of IL-15 by epithelial cells.

The induction of innate immunity and epithelial cell stress responses leads to the upregulation of IL-15, cyclooxygenase (COX)-2, and the expression of CD25 and CD83 activation markers by LP mononuclear cells [87]. IL-15 has become the cornerstone in eliciting the intestinal mucosal injury in CD. The production of this cytokine is not confined to the epithelium, but it is also secreted by DCs and other APCs from the intestinal LP mucosa [88]. However, whether the upregulation of IL-15 induced by gluten peptides in the intestine is a specific phenomenon of CD is still a controversial matter [89].

On the other hand, the source of the stimuli for IL-15 production in CD does not exclusively derive from gluten peptides, as DCs produce also IL-15 in response to type I IFNs. These IFNs (particularly, IFNα) are produced in response to enteroviral infections, and they may participate in the pathogenesis of several immunologically-based diseases, such as Systemic Lupus, Rheumatoid Arthritis or Diabetes Mellitus type 1 [90]. Type I IFNs may have a central role in DCs reprogramming and the loss of tolerance to harmless (dietary) antigens. Particularly, in the intestine, type I IFNs may stimulate the production of IL-15 and IFNγ by DCs, activating antigen presentation and the corresponding adaptive immune response to gluten in the LP, but also the cytotoxic function of CD8 + TCRα and TCRγ cells and innate lymphoid cells (ILC) [65].

IFNα may be involved in Th1 cell differentiation by enhancing IFNγ production. It has been observed that IFN-α administration in susceptible individuals can induce a Th1 response leading to hyperplastic lesions [29]. Although not yet confirmed, IFNα may be secreted by activated fibroblasts and macrophages, and even DCs, in the LP mucosa after an episode of intestinal infection [31]. Moreover, it could contribute to intestinal inflammation by rescuing activated T-cells from apoptosis, maintaining memory T-cells once the stimulus has disappeared, and increasing expression of co-stimulatory molecules in local APCs [29]. IL-18 is a cytokine produced by macrophages, DCs and epithelial cells, which enhances the expression of IL-12- or IFNα-dependent IFNγ on memory and effector cells. Under normal conditions, the intestine expresses IL-18, but this increases in CD at the expense of its mature form, which requires the involvement of the IL-1β converting enzyme (ICE) or local proteinases [61].

Using two in vitro culture models in gluten-sensitive macaques, it has been observed that the IFNγ secreted by activated T-cells in the LP increases gut permeability and promotes immunoreactive α-gliadin (p57-89) peptide 33-mer passage across the epithelium [9, 91, 92]. According to the degree of intestinal inflammation, the paracellular pathway may also affects peptide transport across the epithelium, after binding to the chemokine receptor CXCR3, the activation of the MyD88 adapter, and the release of zonulin [14]. An increased mRNA expression of CXCL10 and CXCL11 has been found in biopsies from patients with active CD, as well as high serum levels of CXCL10 [93]. The study confirmed that CXCL10 is produced by plasma cells and epithelial cells, and its expression increases when IL-15 is present. The expression of CXCR3 is also increased in cells infiltrating the epithelium and LP mucosa, T cells and plasma cells [79, 93].

3.2 Intraepithelial CD8 + T Lymphocytes and Activating NK Receptors

IELs are increased in active CD [94], but these cells are not gluten-specific [95]. Moreover, after NK cell reprogramming, the TCR repertoire of IEL is extremely restricted. Cytotoxic IELs express NK receptors that recognize stress-induced ligands on epithelial cells, leading to the destruction of the small intestinal mucosa, independently from their TCR specificity [96]. In CD, there is an expansion of IELs expressing the activating natural killer receptor NKG2D and the heterodimer NKG2C-CD94, in the absence of inhibitory CD94/NKG2A receptors [80, 81]. These cells recognize non-classical MHC molecules MICA/B and HLA-E expressed by intestinal epithelial cells under stress, which are the main ligands for NKG2D and NKG2C, respectively [80–82]. It has been observed that cytokines from CD4 + T cells control the activity of cytotoxic IELs, and tissue destruction [65].

4 Integrative Model of CD Pathogenesis

Celiac disease develops as the result of a complex interaction between several innate and adaptive immune pathways that culminates in tissue destruction. Several elements, including gluten, TG2, HLA-DQ2, CD4 + T cells, IL-15 and cytotoxic IELs, are cooperatively involved in promoting the destruction of the intestinal epithelium by up-regulating the IFNγ response and the expansion of IELs with a fully activated cytolytic phenotype [65]. The immunopathology of CD is the result of the activation of gluten-reactive CD4 + T cells in the LP, with B cells probably acting as antigen-presenting cells, and stress-induced changes in epithelial cells, which are associated with the upregulation of IL-15 and the expression of non-classical MHC-class I molecules [95]. Cytokines produced by gluten-reactive CD4 + T cells, such as IFNγ and IL-21, upregulate HLA-E expression and increase cytotoxicity on NK and other cytotoxic cells, but also provide help to antibody-producing B cells.

References

1. Mowat AM. To respond or not to respond - a personal perspective of intestinal tolerance. Nat Rev Immunol junio de. 2018;18(6):405–15.
2. Pabst O, Mowat AM. Oral tolerance to food protein. Mucosal Immunol mayo de. 2012;5 (3):232–9.
3. Shewry PR, Halford NG. Cereal seed storage proteins: structures, properties and role in grain utilization. J Exp Bot abril de. 2002;53(370):947–58.

4. Visser J, Rozing J, Sapone A, Lammers K, Fasano A. Tight junctions, intestinal permeability, and autoimmunity: celiac disease and type 1 diabetes paradigms. Ann N Y Acad Sci. mayo de 2009;1165:195–205.

5. Hausch F, Shan L, Santiago NA, Gray GM, Khosla C. Intestinal digestive resistance of immunodominant gliadin peptides. Am J Physiol Gastrointest Liver Physiol. octubre de 2002;283(4):G996–1003.

6. Shan L, Molberg Ø, Parrot I, Hausch F, Filiz F, Gray GM, et al. Structural basis for gluten intolerance in celiac sprue. Science. 27 de septiembre de 2002;297(5590):2275–9.

7. Clemente MG, De Virgiliis S, Kang JS, Macatagney R, Musu MP, Di Pierro MR, et al. Early effects of gliadin on enterocyte intracellular signalling involved in intestinal barrier function. Gut febrero de. 2003;52(2):218–23.

8. Schumann M, Siegmund B, Schulzke JD, Fromm M. Celiac disease: role of the epithelial barrier. Cell Mol Gastroenterol Hepatol marzo de. 2017;3(2):150–62.

9. Schumann M, Richter JF, Wedell I, Moos V, Zimmermann-Kordmann M, Schneider T, et al. Mechanisms of epithelial translocation of the alpha(2)-gliadin-33mer in coeliac sprue. Gut junio de. 2008;57(6):747–54.

10. Zimmer K-P, Fischer I, Mothes T, Weissen-Plenz G, Schmitz M, Wieser H, et al. Endocytotic segregation of gliadin peptide 31–49 in enterocytes. Gut marzo de. 2010;59(3):300–10.

11. Luciani A, Villella VR, Vasaturo A, Giardino I, Pettoello-Mantovani M, Guido S, et al. Lysosomal accumulation of gliadin p31–43 peptide induces oxidative stress and tissue transglutaminase-mediated PPARgamma downregulation in intestinal epithelial cells and coeliac mucosa. Gut marzo de. 2010;59(3):311–9.

12. Ménard S, Lebreton C, Schumann M, Matysiak-Budnik T, Dugave C, Bouhnik Y, et al. Paracellular versus transcellular intestinal permeability to gliadin peptides in active celiac disease. Am J Pathol febrero de. 2012;180(2):608–15.

13. Matysiak-Budnik T, Moura IC, Arcos-Fajardo M, Lebreton C, Ménard S, Candalh C, et al. Secretory IgA mediates retrotranscytosis of intact gliadin peptides via the transferrin receptor in celiac disease. J Exp Med. 21 de enero de 2008;205(1):143–54.

14. Lammers KM, Lu R, Brownley J, Lu B, Gerard C, Thomas K, et al. Gliadin induces an increase in intestinal permeability and zonulin release by binding to the chemokine receptor CXCR3. Gastroenterology julio de. 2008;135(1):194-204.e3.

15. Abadie V, Sollid LM, Barreiro LB, Jabri B. Integration of genetic and immunological insights into a model of celiac disease pathogenesis. Annu Rev Immunol. 2011;29:493–525.

16. Hooper LV, Littman DR, Macpherson AJ. Interactions between the microbiota and the immune system. Science. 8 de junio de 2012;336(6086):1268–73.

17. Brestoff JR, Artis D. Commensal bacteria at the interface of host metabolism and the immune system. Nat Immunol julio de. 2013;14(7):676–84.

18. Tan J, McKenzie C, Vuillermin PJ, Goverse G, Vinuesa CG, Mebius RE, et al. Dietary fiber and bacterial SCFA enhance oral tolerance and protect against food allergy through diverse cellular pathways. Cell Rep. 21 de junio de 2016;15(12):2809–24.

19. Collado MC, Donat E, Ribes-Koninckx C, Calabuig M, Sanz Y. Specific duodenal and faecal bacterial groups associated with paediatric coeliac disease. J Clin Pathol marzo de. 2009;62 (3):264–9.

20. De Palma G, Nadal I, Medina M, Donat E, Ribes-Koninckx C, Calabuig M, et al. Intestinal dysbiosis and reduced immunoglobulin-coated bacteria associated with coeliac disease in children. BMC Microbiol. 24 de febrero de 2010;10:63.

21. D'Argenio V, Casaburi G, Precone V, Pagliuca C, Colicchio R, Sarnataro D, et al. Metagenomics reveals dysbiosis and a potentially pathogenic N. flavescens Strain in duodenum of adult celiac patients. Am J Gastroenterol. junio de 2016;111(6):879–90.

22. Heyman M, Abed J, Lebreton C, Cerf-Bensussan N. Intestinal permeability in coeliac disease: insight into mechanisms and relevance to pathogenesis. Gut septiembre de. 2012;61(9):1355–64.

23. Caminero A, Galipeau HJ, McCarville JL, Johnston CW, Bernier SP, Russell AK, et al. Duodenal bacteria from patients with celiac disease and healthy subjects distinctly affect gluten breakdown and immunogenicity. Gastroenterology octubre de. 2016;151(4):670–83.

24. Caminero A, McCarville JL, Galipeau HJ, Deraison C, Bernier SP, Constante M, et al. Duodenal bacterial proteolytic activity determines sensitivity to dietary antigen through protease-activated receptor-2. Nat Commun. 13 de marzo de 2019;10(1):1198.

25. Zhou L, Kooy-Winkelaar YMC, Cordfunke RA, Dragan I, Thompson A, Drijfhout JW, et al. Abrogation of Immunogenic Properties of Gliadin Peptides through Transamidation by Microbial Transglutaminase Is Acyl-Acceptor Dependent. J Agric Food Chem. 30 de agosto de 2017;65(34):7542–52.

26. Stene LC, Honeyman MC, Hoffenberg EJ, Haas JE, Sokol RJ, Emery L, et al. Rotavirus infection frequency and risk of celiac disease autoimmunity in early childhood: a longitudinal study. Am J Gastroenterol octubre de. 2006;101(10):2333–40.

27. Troncone R, Auricchio S. Rotavirus and celiac disease: clues to the pathogenesis and perspectives on prevention. J Pediatr Gastroenterol Nutr mayo de. 2007;44(5):527–8.

28. Sollid LM, Jabri B. Triggers and drivers of autoimmunity: lessons from coeliac disease. Nat Rev Immunol abril de. 2013;13(4):294–302.

29. Monteleone G, Pender SL, Alstead E, Hauer AC, Lionetti P, McKenzie C, et al. Role of interferon alpha in promoting T helper cell type 1 responses in the small intestine in coeliac disease. Gut marzo de. 2001;48(3):425–9.

30. Cammarota G, Cuoco L, Cianci R, Pandolfi F, Gasbarrini G. Onset of coeliac disease during treatment with interferon for chronic hepatitis C. Lancet. 28 de octubre de 2000;356 (9240):1494–5.

31. Di Sabatino A, Pickard KM, Gordon JN, Salvati V, Mazzarella G, Beattie RM, et al. Evidence for the role of interferon-alfa production by dendritic cells in the Th1 response in celiac disease. Gastroenterology octubre de. 2007;133(4):1175–87.

32. Tai JH, Williams JV, Edwards KM, Wright PF, Crowe JE, Dermody TS. Prevalence of reovirus-specific antibodies in young children in Nashville, Tennessee. J Infect Dis. 15 de abril de 2005;191(8):1221–4.

33. Bouziat R, Biering SB, Kouame E, Sangani KA, Kang S, Ernest JD, et al. Murine Norovirus Infection Induces TH1 Inflammatory Responses to Dietary Antigens. Cell Host Microbe. 14 de noviembre de 2018;24(5):677–688.e5.

34. Bouziat R, Hinterleitner R, Brown JJ, Stencel-Baerenwald JE, Ikizler M, Mayassi T, et al. Reovirus infection triggers inflammatory responses to dietary antigens and development of celiac disease. Science. 7 de abril de 2017;356(6333):44–50.

35. Lundin KE, Scott H, Hansen T, Paulsen G, Halstensen TS, Fausa O, et al. Gliadin-specific, HLA-DQ(alpha 1*0501,beta 1*0201) restricted T cells isolated from the small intestinal mucosa of celiac disease patients. J Exp Med. 1 de julio de 1993;178(1):187–96.

36. Sollid LM, Qiao S-W, Anderson RP, Gianfrani C, Koning F. Nomenclature and listing of celiac disease relevant gluten T-cell epitopes restricted by HLA-DQ molecules. Immunogenetics junio de. 2012;64(6):455–60.

37. Sollid LM, Tye-Din JA, Qiao S-W, Anderson RP, Gianfrani C, Koning F. Update 2020: nomenclature and listing of celiac disease-relevant gluten epitopes recognized by CD4+ T cells. Immunogenetics febrero de. 2020;72(1–2):85–8.

38. du Pré MF, Sollid LM. T-cell and B-cell immunity in celiac disease. Best Pract Res Clin Gastroenterol. junio de 2015;29(3):413–23.

39. Dørum S, Arntzen MØ, Qiao S-W, Holm A, Koehler CJ, Thiede B, et al. The preferred substrates for transglutaminase 2 in a complex wheat gluten digest are Peptide fragments harboring celiac disease T-cell epitopes. PLoS One. 19 de noviembre de 2010;5(11):e14056.

40. Dørum S, Qiao S-W, Sollid LM, Fleckenstein B. A quantitative analysis of transglutaminase 2-mediated deamidation of gluten peptides: implications for the T-cell response in celiac disease. J Proteome Res abril de. 2009;8(4):1748–55.

41. Vader LW, de Ru A, van der Wal Y, Kooy YMC, Benckhuijsen W, Mearin ML, et al. Specificity of tissue transglutaminase explains cereal toxicity in celiac disease. J Exp Med. 4 de marzo de 2002;195(5):643–9.
42. Dieterich W, Ehnis T, Bauer M, Donner P, Volta U, Riecken EO, et al. Identification of tissue transglutaminase as the autoantigen of celiac disease. Nat Med julio de. 1997;3(7):797–801.
43. van de Wal Y, Kooy Y, van Veelen P, Peña S, Mearin L, Papadopoulos G, et al. Selective deamidation by tissue transglutaminase strongly enhances gliadin-specific T cell reactivity. J Immunol. 15 de agosto de 1998;161(4):1585–8.
44. Molberg O, Mcadam SN, Körner R, Quarsten H, Kristiansen C, Madsen L, et al. Tissue transglutaminase selectively modifies gliadin peptides that are recognized by gut-derived T cells in celiac disease. Nat Med junio de. 1998;4(6):713–7.
45. Arentz-Hansen H, Körner R, Molberg O, Quarsten H, Vader W, Kooy YM, et al. The intestinal T cell response to alpha-gliadin in adult celiac disease is focused on a single deamidated glutamine targeted by tissue transglutaminase. J Exp Med. 21 de febrero de 2000;191(4):603–12.
46. Iversen R, Roy B, Stamnaes J, Høydahl LS, Hnida K, Neumann RS, et al. Efficient T cell-B cell collaboration guides autoantibody epitope bias and onset of celiac disease. Proc Natl Acad Sci U S A. 23 de julio de 2019;116(30):15134–9.
47. Thomazy VA, Vega F, Medeiros LJ, Davies PJ, Jones D. Phenotypic modulation of the stromal reticular network in normal and neoplastic lymph nodes: tissue transglutaminase reveals coordinate regulation of multiple cell types. Am J Pathol julio de. 2003;163(1):165–74.
48. Siegel M, Strnad P, Watts RE, Choi K, Jabri B, Omary MB, et al. Extracellular transglutaminase 2 is catalytically inactive, but is transiently activated upon tissue injury. PLoS One. 26 de marzo de 2008;3(3):e1861.
49. Zanoni G, Navone R, Lunardi C, Tridente G, Bason C, Sivori S, et al. In celiac disease, a subset of autoantibodies against transglutaminase binds toll-like receptor 4 and induces activation of monocytes. PLoS Med. septiembre de 2006;3(9):e358.
50. Diraimondo TR, Klöck C, Khosla C. Interferon-γ activates transglutaminase 2 via a phosphatidylinositol-3-kinase-dependent pathway: implications for celiac sprue therapy. J Pharmacol Exp Ther abril de. 2012;341(1):104–14.
51. Lebreton C, Ménard S, Abed J, Moura IC, Coppo R, Dugave C, et al. Interactions among secretory immunoglobulin A, CD71, and transglutaminase-2 affect permeability of intestinal epithelial cells to gliadin peptides. Gastroenterology septiembre de. 2012;143(3):698-707.e4.
52. Johansen BH, Vartdal F, Eriksen JA, Thorsby E, Sollid LM. Identification of a putative motif for binding of peptides to HLA-DQ2. Int Immunol febrero de. 1996;8(2):177–82.
53. Tollefsen S, Arentz-Hansen H, Fleckenstein B, Molberg O, Ráki M, Kwok WW, et al. HLA-DQ2 and -DQ8 signatures of gluten T cell epitopes in celiac disease. J Clin Invest agosto de. 2006;116(8):2226–36.
54. Sjöström H, Lundin KE, Molberg O, Körner R, McAdam SN, Anthonsen D, et al. Identification of a gliadin T-cell epitope in coeliac disease: general importance of gliadin deamidation for intestinal T-cell recognition. Scand J Immunol agosto de. 1998;48(2):111–5.
55. Kim C-Y, Quarsten H, Bergseng E, Khosla C, Sollid LM. Structural basis for HLA-DQ2-mediated presentation of gluten epitopes in celiac disease. Proc Natl Acad Sci U S A. 23 de marzo de 2004;101(12):4175–9.
56. Tye-Din JA, Stewart JA, Dromey JA, Beissbarth T, van Heel DA, Tatham A, et al. Comprehensive, quantitative mapping of T cell epitopes in gluten in celiac disease. Sci Transl Med. 21 de julio de 2010;2(41):41ra51.
57. Nilsen EM, Jahnsen FL, Lundin KE, Johansen FE, Fausa O, Sollid LM, et al. Gluten induces an intestinal cytokine response strongly dominated by interferon gamma in patients with celiac disease. Gastroenterology septiembre de. 1998;115(3):551–63.
58. Fina D, Sarra M, Caruso R, Del Vecchio BG, Pallone F, MacDonald TT, et al. Interleukin 21 contributes to the mucosal T helper cell type 1 response in coeliac disease. Gut julio de. 2008;57(7):887–92.

59. Bodd M, Ráki M, Tollefsen S, Fallang LE, Bergseng E, Lundin KEA, et al. HLA-DQ2-restricted gluten-reactive T cells produce IL-21 but not IL-17 or IL-22. Mucosal Immunol noviembre de. 2010;3(6):594–601.

60. DePaolo RW, Abadie V, Tang F, Fehlner-Peach H, Hall JA, Wang W, et al. Co-adjuvant effects of retinoic acid and IL-15 induce inflammatory immunity to dietary antigens. Nature. 10 de marzo de 2011;471(7337):220–4.

61. Salvati VM, MacDonald TT, Bajaj-Elliott M, Borrelli M, Staiano A, Auricchio S, et al. Interleukin 18 and associated markers of T helper cell type 1 activity in coeliac disease. Gut febrero de. 2002;50(2):186–90.

62. van Bergen J, Mulder CJ, Mearin ML, Koning F. Local communication among mucosal immune cells in patients with celiac disease. Gastroenterology mayo de. 2015;148(6):1187–94.

63. Korneychuk N, Meresse B, Cerf-Bensussan N. Lessons from rodent models in celiac disease. Mucosal Immunol enero de. 2015;8(1):18–28.

64. Kooy-Winkelaar YMC, Bouwer D, Janssen GMC, Thompson A, Brugman MH, Schmitz F, et al. CD4 T-cell cytokines synergize to induce proliferation of malignant and nonmalignant innate intraepithelial lymphocytes. Proc Natl Acad Sci U S A. 7 de febrero de 2017;114(6): E980–9.

65. Abadie V, Kim SM, Lejeune T, Palanski BA, Ernest JD, Tastet O, et al. IL-15, gluten and HLA-DQ8 drive tissue destruction in coeliac disease. Nature febrero de. 2020;578 (7796):600–4.

66. Christophersen A, Ráki M, Bergseng E, Lundin KE, Jahnsen J, Sollid LM, et al. Tetramer-visualized gluten-specific CD4+ T cells in blood as a potential diagnostic marker for coeliac disease without oral gluten challenge. United Eur Gastroenterol J agosto de. 2014;2(4):268–78.

67. Anderson RP, Degano P, Godkin AJ, Jewell DP, Hill AV. In vivo antigen challenge in celiac disease identifies a single transglutaminase-modified peptide as the dominant A-gliadin T-cell epitope. Nat Med marzo de. 2000;6(3):337–42.

68. Ráki M, Dahal-Koirala S, Yu H, Korponay-Szabó IR, Gyimesi J, Castillejo G, et al. Similar responses of intestinal T cells from untreated children and adults with celiac disease to deamidated gluten epitopes. Gastroenterology septiembre de. 2017;153(3):787-798.e4.

69. Risnes LF, Christophersen A, Dahal-Koirala S, Neumann RS, Sandve GK, Sarna VK, et al. Disease-driving CD4+ T cell clonotypes persist for decades in celiac disease. J Clin Invest. 1 de junio de 2018;128(6):2642–50.

70. Qiao S-W, Christophersen A, Lundin KEA, Sollid LM. Biased usage and preferred pairing of α- and β-chains of TCRs specific for an immunodominant gluten epitope in coeliac disease. Int Immunol enero de. 2014;26(1):13–9.

71. Dahal-Koirala S, Ciacchi L, Petersen J, Risnes LF, Neumann RS, Christophersen A, et al. Discriminative T-cell receptor recognition of highly homologous HLA-DQ2-bound gluten epitopes. J Biol Chem. 18 de enero de 2019;294(3):941–52.

72. Christophersen A, Risnes LF, Dahal-Koirala S, Sollid LM. Therapeutic and diagnostic implications of T cell scarring in celiac disease and beyond. Trends in Molecular Medicine. 1 de octubre de 2019;25(10):836–52.

73. Di Niro R, Mesin L, Zheng N-Y, Stamnaes J, Morrissey M, Lee J-H, et al. High abundance of plasma cells secreting transglutaminase 2-specific IgA autoantibodies with limited somatic hypermutation in celiac disease intestinal lesions. Nat Med. 26 de febrero de 2012;18(3):441–5.

74. Høydahl LS, Richter L, Frick R, Snir O, Gunnarsen KS, Landsverk OJB, et al. Plasma cells are the most abundant gluten peptide MHC-expressing cells in inflamed intestinal tissues from patients with celiac disease. Gastroenterology abril de. 2019;156(5):1428-1439.e10.

75. Salmi TT, Collin P, Järvinen O, Haimila K, Partanen J, Laurila K, et al. Immunoglobulin A autoantibodies against transglutaminase 2 in the small intestinal mucosa predict forthcoming coeliac disease. Aliment Pharmacol Ther. 1 de agosto de 2006;24(3):541–52.

76. Mesin L, Sollid LM, Di Niro R. The intestinal B-cell response in celiac disease. Front Immunol. 2012;3:313.

77. du Pré MF, Blazevski J, Dewan AE, Stamnaes J, Kanduri C, Sandve GK, et al. B cell tolerance and antibody production to the celiac disease autoantigen transglutaminase 2. J Exp Med. 3 de febrero de 2020;217(2).

78. Stamnaes J, Iversen R, du Pré MF, Chen X, Sollid LM. Enhanced B-Cell Receptor Recognition of the Autoantigen Transglutaminase 2 by Efficient Catalytic Self-Multimerization. PLoS One. 2015;10(8):e0134922.

79. Chirdo FG, Auricchio S, Troncone R, Barone MV. The gliadin p31–43 peptide: Inducer of multiple proinflammatory effects. Int Rev Cell Mol Biol. 2021;358:165–205.

80. Meresse B, Chen Z, Ciszewski C, Tretiakova M, Bhagat G, Krausz TN, et al. Coordinated induction by IL15 of a TCR-independent NKG2D signaling pathway converts CTL into lymphokine-activated killer cells in celiac disease. Immunity septiembre de. 2004;21(3):357–66.

81. Meresse B, Curran SA, Ciszewski C, Orbelyan G, Setty M, Bhagat G, et al. Reprogramming of CTLs into natural killer-like cells in celiac disease. J Exp Med. 15 de mayo de 2006;203 (5):1343–55.

82. Hüe S, Mention J-J, Monteiro RC, Zhang S, Cellier C, Schmitz J, et al. A direct role for NKG2D/MICA interaction in villous atrophy during celiac disease. Immunity septiembre de. 2004;21(3):367–77.

83. Anderson RP. Innate and adaptive immunity in celiac disease. Curr Opin Gastroenterol noviembre de. 2020;36(6):470–8.

84. Villella VR, Venerando A, Cozza G, Esposito S, Ferrari E, Monzani R, et al. A pathogenic role for cystic fibrosis transmembrane conductance regulator in celiac disease. EMBO J. 15 de enero de 2019;38(2).

85. Barone MV, Troncone R, Auricchio S. Gliadin peptides as triggers of the proliferative and stress/innate immune response of the celiac small intestinal mucosa. Int J Mol Sci. 7 de noviembre de 2014;15(11):20518–37.

86. Nanayakkara M, Lania G, Maglio M, Auricchio R, De Musis C, Discepolo V, et al. P31–43, an undigested gliadin peptide, mimics and enhances the innate immune response to viruses and interferes with endocytic trafficking: a role in celiac disease. Sci Rep. 17 de julio de 2018;8(1):10821.

87. Maiuri L, Ciacci C, Ricciardelli I, Vacca L, Raia V, Auricchio S, et al. Association between innate response to gliadin and activation of pathogenic T cells in coeliac disease. Lancet. 5 de julio de 2003;362(9377):30–7.

88. Stepniak D, Koning F. Celiac disease–sandwiched between innate and adaptive immunity. Hum Immunol junio de. 2006;67(6):460–8.

89. Jabri B, Kasarda DD, Green PHR. Innate and adaptive immunity: the yin and yang of celiac disease. Immunol Rev agosto de. 2005;206:219–31.

90. Barrat FJ, Crow MK, Ivashkiv LB. Interferon target-gene expression and epigenomic signatures in health and disease. Nat Immunol diciembre de. 2019;20(12):1574–83.

91. Abadie V, Kim SM, Lejeune T, Palanski BA, Ernest JD, Tastet O, et al. IL-15, gluten and HLA-DQ8 drive tissue destruction in coeliac disease. Nature. febrero de 2020;578 (7796):600-4.

92. Bethune MT, Siegel M, Howles-Banerji S, Khosla C. Interferon-gamma released by gluten-stimulated celiac disease-specific intestinal T cells enhances the transepithelial flux of gluten peptides. J Pharmacol Exp Ther mayo de. 2009;329(2):657–68.

93. Mazumdar K, Alvarez X, Borda JT, Dufour J, Martin E, Bethune MT, et al. Visualization of transepithelial passage of the immunogenic 33-residue peptide from alpha-2 gliadin in gluten-sensitive macaques. PLoS One. 19 de abril de 2010;5(4):e10228.

94. Bondar C, Araya RE, Guzman L, Rua EC, Chopita N, Chirdo FG. Role of CXCR3/CXCL10 axis in immune cell recruitment into the small intestine in celiac disease. PLoS One. 2014;9 (2):e89068.

95. Fernández-Bañares F, Carrasco A, Martín A, Esteve M. Systematic review and meta-analysis: accuracy of both gamma delta+ intraepithelial lymphocytes and coeliac lymphogram evaluated by flow cytometry for coeliac disease diagnosis. Nutrients. 23 de agosto de 2019;11 (9).
96. Setty M, Discepolo V, Abadie V, Kamhawi S, Mayassi T, Kent A, et al. Distinct and synergistic contributions of epithelial stress and adaptive immunity to functions of intraepithelial killer cells and active celiac disease. Gastroenterology septiembre de. 2015;149(3):681-691.e10.
97. Jabri B, Sollid LM. Tissue-mediated control of immunopathology in coeliac disease. Nat Rev Immunol diciembre de. 2009;9(12):858–70.

Autoimmunity and Celiac Disease

Stefano Guandalini

1 Introduction

Celiac disease (CD) is an autoimmune disorder, and it has been known for a long time to carry a strong association with other autoimmune diseases (AID), as recently confirmed in a large scale study in Israel [1]. AID originate from an aberrant immune response in distinguishing between self and non-self-antigens. More than 80 AID are presently known [2], and the link of CD with many other AID is well known: they include (see Table 1) endocrine disorders (Type 1 Diabetes—T1D -, Autoimmune Thyroid Diseases—AITD—such as Hashimoto and Graves-Basedow's disease, Addison's disease); but also rheumatologic diseases such as Systemic Lupus Erythematosus, rheumatoid arthritis, Juvenile Rheumatoid Arthritis, Sjogren's diseases, in addition to Psoriasis, Alopecia areata and Autoimmune Hepatitis. Genome-wide association studies (GWAS) have recently shown that genetic risk factors are shared by about half of all autoimmune diseases [3]. The most relevant group of genes that are found in common between CD and other autoimmune conditions is the family of Human Leukocyte Antigens (HLA) [4]. In fact, CD only develops in subjects who possess the HLA antigens DQ2 (DQA1*0501-DQB1*0201) and/or DQ8 (DQA1*0301- DQB1*0302), and these genes, tightly linked with DR3 and DR4, are the major common genetic predisposition with T1D and AITD, the two endocrine autoimmune disorders most commonly associated with CD. Of interest, while HLA-DR3 and DR4 are positively associated with many autoimmune diseases, the universal allele HLA-DRB1*0701 appears to be protective [5]. Some of the risk factors that appear to increase the possibility of developing an associated autoimmune disorders in CD patients have been shown [6] to be: female gender, family history for AIDs, anti-

S. Guandalini (✉)
Department of Pediatrics, Section of Gastroenterology, Hepatology and Nutrition, University of Chicago Medicine, Chicago, US
e-mail: sguandal@uchicago.edu

© The Author(s), under exclusive license to Springer Nature Switzerland AG 2022
J. Amil-Dias and I. Polanco (eds.), *Advances in Celiac Disease*,
https://doi.org/10.1007/978-3-030-82401-3_5

51

Table 1 Autoimmune disorders most commonly associated with Celiac Disease

System	Autoimmune disorder
Endocrinological	Type 1 Diabetes
	Autoimmune Thyroid Diseases
	Hashimoto's thyroiditis
	Grave-Basedow's disease
	Addison Disease
Rheumatological	Rheumatoid Arthritis
	Idiopathic Arthritis
	SLE
	Sjogren's syndrome
Dermatological	Dermatitis Herpetiformis
	Psoriasis
	Vitiligo
	Alopecia areata
Gastrointestinal	Autoimmune Hepatitis

gliadin IgG positivity, vitamin D deficiency, antinuclear antibody positivity >1/80 titre and having any musculoskeletal disease.

While the protective role of a gluten-free diet (GFD) in preventing the development of associated autoimmune disorders in CD patients is still debated, a GFD has been proposed in patients not only with endocrine autoimmune diseases, but also in other conditions such as Multiple Sclerosis and Psoriasis (reviewed in [7]).

However, the issue of whether the onset of autoimmune disorders in CD patients is favoured by the ingestion of gluten (either before or after diagnosis) remains controversial. An increased prevalence of autoimmune disorders was found in parallel with the increasing age at diagnosis of CD [8], suggesting that prolonged exposure to gluten may indeed favour the onset of autoimmune conditions. Further studies on this issue have revealed conflicting results, so that the issue is still unresolved [9].

2   Pathogenetic Links

It has become increasingly clear from epidemiological as well as animal studies that a major factor in inducing the development of AID is the occurrence of viral infections. Although the mechanisms linking viruses to the aberrant immune response found in AID are far from clear, it appears that viral-induced autoimmunity can be activated through multiple mechanisms including (see Fig. 1) molecular mimicry, epitope spreading, bystander activation, and immortalization of infected B cells [10].

While these supposed mechanisms may well be at work across all types of AID, much more is known in the specific case of CD. Infective agents have been investigated as environmental factors that contribute to trigger CD. GWAS

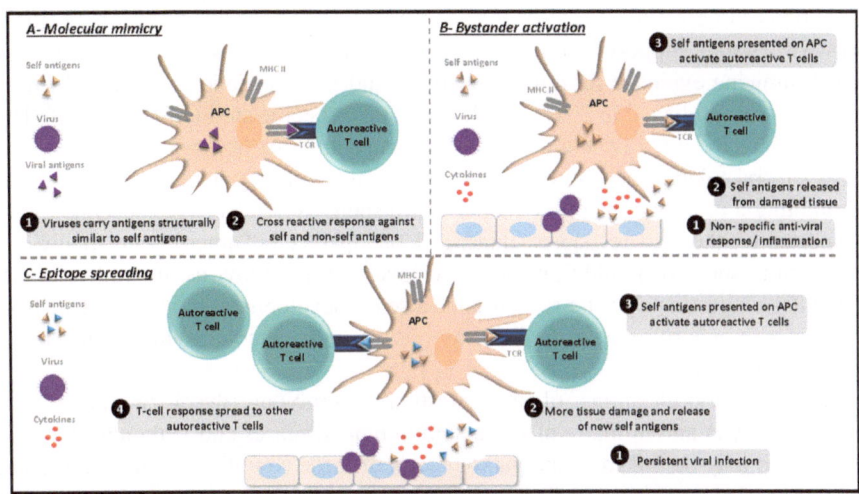

Fig. 1 How viruses can trigger development of autoimmunity (from Ref. [10]). **a** Molecular mimicry model: (1) Viruses carry epitopes structurally similar to self-epitopes. (2) Presentation of viral epitopes by antigen presenting cells (APCs) activate autoreactive T cells that bind to both, self and non-self-antigens, and induce tissue damage. **b** Bystander activation model: (1) Non-specific and over reactive antiviral immune responses lead to the liberation of self-antigens and release of inflammatory cytokines from the damaged tissue. (2) Self-antigen is taken up and presented by APCs. (3) Autoreactive T cells activated by APCs, leading to tissue destruction. **c** Epitope spreading model: (1) Persistent viral infection. (2) Continued tissue damage and release of new self-antigens. (3) Self-antigens are taken up and presented by APCs. (4) Nonspecific activation of more autoreactive T cells leading to autoimmunity

identified polymorphisms in genes involved in the response to viral infections to be associated with CD. Higher rates of summer births were described in children with CD, suggesting that exposure of 4-6-month-old infants to winter-linked viral infections such as rotavirus may play a role. The first correlation between viruses and CD dates back to the 1980s when a homology between alpha gliadin and a protein of the human adenovirus [11] had been described and related to increased frequency of adenovirus infection in CD patients [12] compared to controls. Increased anti-rotavirus antibody titres have also been associated with a moderate, but significantly increased risk of CD in HLA-susceptible children [13]; subsequently [14] anti-rotavirus antibodies have been found related with CD but not with Type 1 Diabetes (T1D) onset, despite the common genetic background and similar pathogenic mechanisms [15]. Over the past decade, several publications have explored the association between viral infections and CD development. Overall, they reported that a high number of infections in the first months of life is associated with an increased risk of later CD development [16]. Respiratory infections in particular [17, 18] appear to be the ones most commonly associated with CD. Importantly, the risk variance explained by respiratory infections was higher than that explained by sex or HLA [19]. And recently, a clear role for Enterovirus infections in early life has also been reported [20] Altogether these observations provide a strong support for the role of viral infections in CD development. The first work showing evidence for a mechanism behind the association of viral infections and CD was published in 2017 and showed that a specific strain of Reovirus (T1L), despite being cleared by the host, could disrupt immune homeostasis at intestinal sites of oral tolerance by suppressing peripheral regulatory T cell conversion and promoting a proinflammatory T-helper cell response to orally ingested gluten [21]. Importantly, the two responses depended on distinct but interplaying pathways. This study showed in addition increased levels of anti-reovirus antibodies in CD patients, suggesting that the risk of later disease development is independent from the severity or virulence of the infection, as Reovirus causes asymptomatic infections in humans.

Of note, modifications of the intestinal microbiota (dysbiosis) have been reported in CD and in many of these autoimmune diseases and is thought that this imbalance may be linked to the pathogenesis of these conditions. Thus, after some evidence of potential benefits of intervention with either bacterial strains or faecal microbiota transplantation (FMT) in animal models, clinical trials are already on their way in humans (reviewed in [22], with the aim of re-establishing a healthy microbiota and possibly reverse their clinical impact.

3   Associated Endocrinological Diseases

3.1   Type 1 Diabetes

While CD involves an autoimmune aggression to the small intestinal mucosa, T1D is characterized by an autoimmune attack to insulin-producing pancreatic β islet cells. The coexistence of these 2 conditions is well known. In reality, patients with T1D have a high prevalence not only of CD, but of a number of other autoimmune diseases, including AITD, gastric autoimmunity leading to pernicious anaemia, vitiligo, and adrenal gland insufficiency. The association of T1D with celiac disease was first reported in a 5-year-old girl in1969 by Walker-Smith and Grigor: "This is an association we have not previously observed, and the relationship between the two conditions, if any, is uncertain." [23]. Subsequently, a number of observations from various parts of the world have confirmed the strong association, with prevalence rates varying between 1.6 and 16.4%. Typically, when the two conditions coexist, development of overt T1D precedes that of CD; however, a large prospective study on a cohort of at risk children followed from 3 until 48 months of age with repeated determinations of serum levels of specific autoantibodies (Islet cell Antibodies—IAs—and tissue transglutaminase 2 antibodies—tTG) showed [24] that IAs preceded tTG development more often (67%) than vice versa (27%). This study also showed that having IAs significantly increased the risk of developing tTG. However, "co-occurrence significantly exceeded the expected rate. IAs usually, but not always, appeared earlier than tTG. IAs preceding tTG was associated with increasing risk of tTG" [24]. About two-thirds of children diagnosed with CD after T1D onset have elevated levels of celiac antibodies at the time of T1D diagnosis or within the first 24 months; however, an additional 40% of patients develop CD a few years after T1D onset [25], and even adults with a long history of T1D show a progressively higher prevalence of CD [26]. Thus, it is recommended that T1D children be repeatedly tested for CD. Of note, it has been shown that the presence or absence of GI symptoms in children with T1D has no predictable value for biopsy-confirmed CD or not [27]. In fact, many diabetic children who come to be diagnosed with CD have minimal or no symptoms. Testing children with T1D for CD is recommended by the American Diabetes Association, the International Society for Pediatric and Adolescent Diabetes, the British Society of Paediatric Gastroenterology, Hepatology and Nutrition, the European Society for Pediatric Gastroenterology, Hepatology and Nutrition (ESPGHAN) in their 2012 as well as 2020 guidelines, and the North American Society for Pediatric Gastroenterology, Hepatology and Nutrition [28–33]. However, patients with T1D can have elevated titres of tTG independently of CD (though usually at low to moderate titres), and such antibodies are known to fluctuate and even normalize spontaneously in a large number (about one-third) of patients with T1D who remain on gluten [34, 35]. In addition, even persistently positive tTG (especially in low titres) are frequently found in patients with T1D who have normal intestinal histology after endoscopy and biopsies. These cases are quite common and end up by being either considered a "false positive" tTG or are given the dubious diagnosis of potential celiac disease. For these reasons, it is recommended to proceed with the confirmatory biopsy only when their titre is >3 times the upper limit of normal, although recently a higher limit has been proposed [36] in order to avoid performing needless biopsies, raising some perplexity [37]. What is clear is that once diagnosed with CD, children with T1D need to follow a strict GFD. Such diet, in spite of being super-imposed to the restrictions of T1D, does not seem to negatively influence the quality of life and may be beneficial to patients with T1D and celiac disease, regardless of symptoms [38]. A study in a large series of such patients showed that those who achieved normalization of their tTG titres had better control of their glycated haemoglobin and growth. Data from prospective randomized clinical trials in this regard concluded there could be a decrease in hypoglycaemic episodes and better glycaemic control in patients with T1D with minimally symptomatic celiac disease following a GFD [39, 40]. A recent interventional trial also suggests that there might be a beneficial effect of GFD in patients with T1D, even if they do not have celiac disease [41].

3.2 Autoimmune Thyroid Diseases

Autoimmune thyroid diseases (both Hashimoto's thyroiditis and Graves-Basedow disease) are more frequent in CD than in controls [42, 43], and while the association had been reported for about a decade, it was not until 1981 that this was thought to be non-coincidental [44].

Hashimoto thyroiditis, like CD, is more common in females and is characterized by anti-thyroglobulin and/or anti-thyroid peroxidase antibodies. The disease is associated with a high thyroid stimulating hormone (TSH) and is considered the most common cause of adult-onset hypothyroidism. In these patients, GFD does not appear to prevent the progression of the autoimmune process [45]. Interestingly, symptoms of AITD and CD often overlap. In fact, fatigue, weight loss, constipation (or diarrhoea) and arthralgia or seronegative arthritis can be found in both. The latter indeed has recently been found to occur significantly more often in patients with AITD who are also celiac [46]. The prevalence of CD in patients with AITD has been the subject of many studies in various parts of the world and can be considered between 3 and 7% [47]. The reverse is also true: patients with CD have a higher prevalence of AITD than the general population, with this prevalence varying between 5 and 13% [48], increasing with advancing age, and up to 30% [49]. In spite of a recently reported apparent decrease, over the past 50 years, of the prevalence of AITD in celiac patients [50], it is evident that screening for CD all children and teenagers with AITD currently remains recommended [31].

3.3 Addison Disease

This condition too has been reported to have an increased prevalence in patients with CD, although mostly adults [51]. However, a Swedish national registry study [52] showed that both children and adults with CD had a significant association with Addison's disease and thus reiterated the recommendation that cases with adrenal insufficiency be screened for CD.

4 Rheumatological: Rheumatoid Arthritis and Idiopathic Arthritis

Musculoskeletal manifestations, including arthralgia and arthritis, are seen in roughly 5–10% of new celiac paediatric patients [53]. In 2017 [54], the prevalence of early joint involvement in children with CD was investigated with the use of musculoskeletal ultrasound, superior to conventional radiology in detecting a wide array of early inflammatory and structural abnormalities in joints. This test showed the presence of at least one abnormality in the joints of 32% of newly diagnosed celiac patients as compared to only 3% of those on the GFD for at least six months. The lower rates of joint involvement found in children on a GFD for more than six months is suggestive that the GFD may improve the joint abnormalities associated with CD.

The association of CD with rheumatological disorders has been known for decades, and screening for CD patients with rheumatological isolated manifestations has been recommended [43]. In addition to sharing some crucial HLA-related genes, CD and Rheumatoid Arthritis (RA) present similarities in their epidemiology, gender prevalence, pathogenetic aspects and clinical manifestations.

In fact, both conditions appear to be on the rise, affect about 1% of the general population, have a higher prevalence in females, can be favoured by dysbiosis and viral infections and are sustained by both innate and adaptive immunity. From the clinical standpoint, both are often associated with additional autoimmune disorders such as AITD, T1D, psoriasis, Sjogren's syndrome. It should also be noted that CD is well known to present, among its extra-intestinal manifestations, joint involvement in the form of arthralgia or frank arthritis. As these manifestations tend to regress on a GFD, this treatment has been also proposed for RA and Idiopathic and Idiopathic arthritis *in the absence of CD*, albeit with conflicting results.

5 Dermatological

Dermatitis Herpetiformis is the best-known skin condition considered to be a cutaneous manifestation of celiac disease [55]. But other dermatological conditions are also reported to have a higher prevalence in CD. In fact, a recent large epidemiological investigation [56] showed increased risks other autoimmune skin disorders such as psoriasis [57, 58], as confirmed by a recent meta-analysis [59], vitiligo [60] and alopecia areata [60]. Although the benefit of following a GFD on these skin disorders is still not completely clear, there is some evidence that this diet may be beneficial in treating psoriasis [57].

6 Autoimmune Hepatitis

A condition whose prevalence is on the rise, Autoimmune Hepatitis (AH) affects about 30 individuals (mostly adults) per 100,000 [61]. However, its prevalence in individuals with CD is estimated to be between 2 and 6% in children [62, 63] and about 4% in adults [64]. On the other hand, a larger proportion of patients with AH, around 15%, has been reported in a paediatric series [65] The response of AH to the GFD has been reported sporadically, and with conflicting outcomes [63, 66, 67].

7 Conclusions

CD shares a common genetic make-up with a number of other autoimmune diseases, leading to a higher prevalence of them in celiac patients, as well as to a higher prevalence of CD in patients affected by other autoimmune disorders, especially of the endocrinological type. While in most cases of coexisting CD and other AID, AID are diagnosed first, the reverse is also well known. The role of gluten intake *before* the diagnosis of CD in favouring the onset of associated AID is unclear, although there is some evidence in this regard. Equally, still uncertain is the role of the GFD in ameliorating—in addition of CD—the accompanying autoimmune condition.

In any case, there is universal agreement that patients with AID should be screened at the first opportunity for CD, and that this test should be periodically repeated. The same philosophy applies to patients with CD, in order to detect as soon as possible the onset of a new AID and begin appropriate treatment.

References

1. Assa A, Frenkel-Nir Y, Tzur D, Katz LH, Shamir R. Large population study shows that adolescents with celiac disease have an increased risk of multiple autoimmune and nonautoimmune comorbidities. Acta Paediatr. 2017;106(6):967–72.
2. Ercolini AM, Miller SD. The role of infections in autoimmune disease. Clin Exp Immunol. 2009;155(1):1–15.
3. Jonkers IH, Wijmenga C. Context-specific effects of genetic variants associated with autoimmune disease. Hum Mol Genet. 2017;26(R2):R185–92.
4. Bodis G, Toth V, Schwarting A. Role of Human Leukocyte Antigens (HLA) in autoimmune diseases. Rheumatol Ther. 2018;5(1):5–20.
5. Zakharova MY, Belyanina TA, Sokolov AV, Kiselev IS, Mamedov AE. The contribution of major histocompatibility complex II genes to an association with autoimmune diseases. Acta Naturae. 2019;11(4):4–12.
6. Demirezer Bolat A, Akin FE, Tahtaci M, Tayfur Yurekli O, Koseoglu H, Erten S, et al. Risk factors for polyautoimmunity among patients with celiac disease: a cross-sectional survey. Digestion. 2015;92(4):185–91.
7. Passali M, Josefsen K, Frederiksen JL, Antvorskov JC. Current evidence on the efficacy of gluten-free diets in multiple sclerosis, psoriasis, type 1 diabetes and autoimmune thyroid diseases. Nutrients. 2020;12(8).
8. Ventura A, Magazzu G, Greco L. Duration of exposure to gluten and risk for autoimmune disorders in patients with celiac disease. SIGEP Study Group for Autoimmune Disorders in Celiac Disease. Gastroenterology. 1999;117(2):297–303.
9. Elli L, Discepolo V, Bardella MT, Guandalini S. Does gluten intake influence the development of celiac disease-associated complications? J Clin Gastroenterol. 2014;48(1):13–20.
10. Smatti MK, Cyprian FS, Nasrallah GK, Al Thani AA, Almishal RO, Yassine HM. Viruses and autoimmunity: a review on the potential interaction and molecular mechanisms. Viruses. 2019;11(8).

11. Stene LC, Honeyman MC, Hoffenberg EJ, Haas JE, Sokol RJ, Emery L, et al. Rotavirus infection frequency and risk of celiac disease autoimmunity in early childhood: a longitudinal study. Am J Gastroenterol. 2006;101(10):2333–40.
12. Atkinson MA, Chervonsky A. Does the gut microbiota have a role in type 1 diabetes? Early evidence from humans and animal models of the disease. Diabetologia. 2012;55(11):2868–77.
13. Mathis D, Benoist C. Microbiota and autoimmune disease: the hosted self. Cell Host Microbe. 2011;10(4):297–301.
14. Dolcino M, Zanoni G, Bason C, Tinazzi E, Boccola E, Valletta E, et al. A subset of anti-rotavirus antibodies directed against the viral protein VP7 predicts the onset of celiac disease and induces typical features of the disease in the intestinal epithelial cell line T84. Immunol Res. 2013;56(2–3):465–76.
15. Goodwin G. Type 1 diabetes mellitus and celiac disease: distinct autoimmune disorders that share common pathogenic mechanisms. Horm Res Paediatr. 2019;92(5):285–92.
16. Marild K, Kahrs CR, Tapia G, Stene LC, Stordal K. Infections and risk of celiac disease in childhood: a prospective nationwide cohort study. Am J Gastroenterol. 2015;110(10):1475–84.
17. Karhus LL, Gunnes N, Stordal K, Bakken IJ, Tapia G, Stene LC, et al. Influenza and risk of later celiac disease: a cohort study of 2.6 million people. Scand J Gastroenterol. 2018;53 (1):15–23.
18. Tjernberg AR, Ludvigsson JF. Children with celiac disease are more likely to have attended hospital for prior respiratory syncytial virus infection. Dig Dis Sci. 2014;59(7):1502–8.
19. Auricchio R, Cielo D, de Falco R, Galatola M, Bruno V, Malamisura B, et al. Respiratory infections and the risk of celiac disease. Pediatrics. 2017;140(4).
20. Kahrs CR, Chuda K, Tapia G, Stene LC, Marild K, Rasmussen T, et al. Enterovirus as trigger of coeliac disease: nested case-control study within prospective birth cohort. BMJ. 2019;364: l231.
21. Bouziat R, Hinterleitner R, Brown JJ, Stencel-Baerenwald JE, Ikizler M, Mayassi T, et al. Reovirus infection triggers inflammatory responses to dietary antigens and development of celiac disease. Science. 2017;356(6333):44–50.
22. Marietta E, Mangalam AK, Taneja V, Murray JA. Intestinal dysbiosis in, and enteral bacterial therapies for, systemic autoimmune diseases. Front Immunol. 2020;11:573079.
23. Walker-Smith JA, Grigor W. Coeliac disease in a diabetic child. Lancet. 1969;1(7603):1021.
24. Hagopian W, Lee HS, Liu E, Rewers M, She JX, Ziegler AG, et al. Co-occurrence of type 1 diabetes and celiac disease autoimmunity. Pediatrics. 2017;140(5).
25. Barera G, Bonfanti R, Viscardi M, Bazzigaluppi E, Calori G, Meschi F, et al. Occurrence of celiac disease after onset of type 1 diabetes: a 6-year prospective longitudinal study. Pediatrics. 2002;109(5):833–8.
26. Scaramuzza AE, Mantegazza C, Bosetti A, Zuccotti GV. Type 1 diabetes and celiac disease: the effects of gluten free diet on metabolic control. World J Diab. 2013;4(4):130–4.
27. Bybrant MC, Ortqvist E, Lantz S, Grahnquist L. High prevalence of celiac disease in Swedish children and adolescents with type 1 diabetes and the relation to the Swedish epidemic of celiac disease: a cohort study. Scand J Gastroenterol. 2014;49(1):52–8.
28. Couper JJ, Haller MJ, Ziegler AG, Knip M, Ludvigsson J, Craig ME, et al. ISPAD clinical practice consensus guidelines 2014. Phases of type 1 diabetes in children and adolescents. Pediatr Diabetes. 2014;15 Suppl 20:18–25.
29. Hill ID, Dirks MH, Liptak GS, Colletti RB, Fasano A, Guandalini S, et al. Guideline for the diagnosis and treatment of celiac disease in children: recommendations of the North American Society for pediatric gastroenterology, hepatology and nutrition. J Pediatr Gastroenterol Nutr. 2005;40(1):1–19.
30. Hill ID, Fasano A, Guandalini S, Hoffenberg E, Levy J, Reilly N, et al. NASPGHAN clinical report on the diagnosis and treatment of gluten-related disorders. J Pediatr Gastroenterol Nutr. 2016;63(1):156–65.

31. Husby S, Koletzko S, Korponay-Szabo I, Kurppa K, Mearin ML, Ribes-Koninckx C, et al. European Society paediatric gastroenterology, hepatology and nutrition guidelines for diagnosing coeliac disease 2020. J Pediatr Gastroenterol Nutr. 2020;70(1):141–56.
32. Husby S, Koletzko S, Korponay-Szabo IR, Mearin ML, Phillips A, Shamir R, et al. European Society for pediatric gastroenterology, hepatology, and nutrition guidelines for the diagnosis of coeliac disease. J Pediatr Gastroenterol Nutr. 2012;54(1):136–60.
33. Murch S, Jenkins H, Auth M, Bremner R, Butt A, France S, et al. Joint BSPGHAN and coeliac UK guidelines for the diagnosis and management of coeliac disease in children. Arch Dis Child. 2013;98(10):806–11.
34. Castellaneta S, Piccinno E, Oliva M, Cristofori F, Vendemiale M, Ortolani F, et al. High rate of spontaneous normalization of celiac serology in a cohort of 446 children with type 1 diabetes: a prospective study. Diabetes Care. 2015;38(5):760–6.
35. Waisbourd-Zinman O, Hojsak I, Rosenbach Y, Mozer-Glassberg Y, Shalitin S, Phillip M, et al. Spontaneous normalization of anti-tissue transglutaminase antibody levels is common in children with type 1 diabetes mellitus. Dig Dis Sci. 2012;57(5):1314–20.
36. Wessels M, Velthuis A, van Lochem E, Duijndam E, Hoorweg-Nijman G, de Kruijff I, et al. Raising the cut-off level of anti-tissue transglutaminase antibodies to detect celiac disease reduces the number of small bowel biopsies in children with type 1 diabetes: A Retrospective Study. J Pediatr. 2020;223:87–92 e1.
37. Guandalini S. Type 1 diabetes and celiac disease: can (and should) we raise the cut-off of tissue transglutaminase immunoglobulin a to decide whether to biopsy? J Pediatr. 2020;223:8–10.
38. Nunes-Silva JG, Nunes VS, Schwartz RP, Mlss Trecco S, Evazian D, Correa-Giannella ML, et al. Impact of type 1 diabetes mellitus and celiac disease on nutrition and quality of life. Nutr Diab. 2017;7(1):e239.
39. Kaur P, Agarwala A, Makharia G, Bhatnagar S, Tandon N. Effect of gluten-free diet on metabolic control and anthropometric parameters in type 1 diabetes with subclinical celiac disease: a randomized controlled trial. Endocr Pract. 2020;26(6):660–7.
40. Nagl K, Bollow E, Liptay S, Rosenbauer J, Koletzko S, Pappa A, et al. Lower HbA1c in patients with type 1 diabetes and celiac disease who reached celiac-specific antibody-negativity-A multicenter DPV analysis. Pediatr Diab. 2019;20(8):1100–9.
41. Neuman V, Pruhova S, Kulich M, Kolouskova S, Vosahlo J, Romanova M, et al. Gluten-free diet in children with recent-onset type 1 diabetes: A 12-month intervention trial. Diab Obes Metab. 2020;22(5):866–72.
42. Minelli R, Gaiani F, Kayali S, Di Mario F, Fornaroli F, Leandro G, et al. Thyroid and celiac disease in pediatric age: a literature review. Acta Biomed. 2018;89(9-S):11–6.
43. Ghozzi M, Sakly W, Mankai A, Bouajina E, Bahri F, Nouira R, et al. Screening for celiac disease, by endomysial antibodies, in patients with unexplained articular manifestations. Rheumatol Int. 2014;34(5):637–42.
44. Troutman ME, Efrusy ME, Bennett GD, Kniaz JL, Dobbins WO 3rd. Simultaneous occurrence of adult celiac disease and lymphocytic thyroiditis. J Clin Gastroenterol. 1981;3(3):281–5.
45. Metso S, Hyytia-Ilmonen H, Kaukinen K, Huhtala H, Jaatinen P, Salmi J, et al. Gluten-free diet and autoimmune thyroiditis in patients with celiac disease. A prospective controlled study. Scand J Gastroenterol. 2012;47(1):43–8.
46. Lupoli GA, Tasso M, Costa L, Caso F, Scarpa R, Del Puente A, et al. Coeliac disease is a risk factor for the development of seronegative arthritis in patients with autoimmune thyroid disease. Rheumatology (Oxford). 2020.
47. Ch'ng CL, Jones MK, Kingham JG. Celiac disease and autoimmune thyroid disease. Clin Med Res. 2007;5(3):184–92.
48. Freeman HJ. Endocrine manifestations in celiac disease. World J Gastroenterol. 2016;22(38):8472–9.

49. Carta MG, Hardoy MC, Boi MF, Mariotti S, Carpiniello B, Usai P. Association between panic disorder, major depressive disorder and celiac disease: a possible role of thyroid autoimmunity. J Psychosom Res. 2002;53(3):789–93.
50. Castro PD, Harkin G, Hussey M, Christopher B, Kiat C, Chin JL, et al. Prevalence of coexisting autoimmune thyroidal diseases in coeliac disease is decreasing. United Eur Gastroenterol J. 2020;8(2):148–56.
51. O'Leary C, Walsh CH, Wieneke P, O'Regan P, Buckley B, O'Halloran DJ, et al. Coeliac disease and autoimmune Addison's disease: a clinical pitfall. QJM. 2002;95(2):79–82.
52. Elfstrom P, Montgomery SM, Kampe O, Ekbom A, Ludvigsson JF. Risk of primary adrenal insufficiency in patients with celiac disease. J Clin Endocrinol Metab. 2007;92(9):3595–8.
53. Jericho H, Sansotta N, Guandalini S. Extraintestinal manifestations of celiac disease: effectiveness of the gluten-free diet. J Pediatr Gastroenterol Nutr. 2017;65(1):75–9.
54. Garg K, Agarwal P, Gupta RK, Sitaraman S. Joint involvement in children with celiac disease. Indian Pediatr. 2017;54(11):946–8.
55. Reunala T, Hervonen K, Salmi T. Dermatitis herpetiformis: an update on diagnosis and management. Am J Clin Dermatol. 2021.
56. Lebwohl B, Soderling J, Roelstraete B, Lebwohl MG, Green PH, Ludvigsson JF. Risk of skin disorders in patients with celiac disease: a population-based cohort study. J Am Acad Dermatol. 2020.
57. De Bastiani R, Gabrielli M, Lora L, Napoli L, Tosetti C, Pirrotta E, et al. Association between coeliac disease and psoriasis: Italian primary care multicentre study. Dermatology. 2015;230 (2):156–60.
58. Ungprasert P, Wijarnpreecha K, Kittanamongkolchai W. Psoriasis and risk of celiac disease: a systematic review and meta-analysis. Indian J Dermatol. 2017;62(1):41–6.
59. Acharya P, Mathur M. Association between psoriasis and celiac disease: a systematic review and meta-analysis. J Am Acad Dermatol. 2020;82(6):1376–85.
60. Rodrigo L, Beteta-Gorriti V, Alvarez N, Gomez de Castro C, de Dios A, Palacios L, et al. Cutaneous and mucosal manifestations associated with celiac disease. Nutrients. 2018;10(7).
61. Tunio NA, Mansoor E, Sheriff MZ, Cooper GS, Sclair SN, Cohen SM. Epidemiology of Autoimmune Hepatitis (AIH) in the United States between 2014 and 2019: a population-based national study. J Clin Gastroenterol. 2020.
62. Vajro P, Paolella G, Maggiore G, Giordano G. Pediatric celiac disease, cryptogenic hypertransaminasemia, and autoimmune hepatitis. J Pediatr Gastroenterol Nutr. 2013;56 (6):663–70.
63. Di Biase AR, Colecchia A, Scaioli E, Berri R, Viola L, Vestito A, et al. Autoimmune liver diseases in a paediatric population with coeliac disease—a 10-year single-centre experience. Aliment Pharmacol Ther. 2010;31(2):253–60.
64. Mirzaagha F, Azali SH, Islami F, Zamani F, Khalilipour E, Khatibian M, et al. Coeliac disease in autoimmune liver disease: a cross-sectional study and a systematic review. Dig Liver Dis. 2010;42(9):620–3.
65. Caprai S, Vajro P, Ventura A, Sciveres M, Maggiore G, Disease SSGfALDiC. Autoimmune liver disease associated with celiac disease in childhood: a multicenter study. Clin Gastroenterol Hepatol. 2008;6(7):803–6.
66. Mounajjed T, Oxentenko A, Shmidt E, Smyrk T. The liver in celiac disease: clinical manifestations, histologic features, and response to gluten-free diet in 30 patients. Am J Clin Pathol. 2011;136(1):128–37.
67. Nastasio S, Sciveres M, Riva S, Filippeschi IP, Vajro P, Maggiore G. Celiac disease-associated autoimmune hepatitis in childhood: long-term response to treatment. J Pediatr Gastroenterol Nutr. 2013;56(6):671–4.

Value and Use of Serologic Markers of Celiac Disease

C. Ribes-Koninckx, M. Roca, and E. Donat

Abbreviations

AGA	Antigliadin antibodies
ARA	Antireticulin antibodies
TGA-IgA	IgA antibodies against type-2 (tissue) transglutaminase
TGA-IgG	IgG antibodies against type-2 (tissue) transglutaminase
EMA-IgA	IgA endomysial antibodies
DGP	Deamidated gliadin peptides
ELISA	Enzyme Immuno assay
IF	Immunofluorescence
DGPA-IgG	IgG Deamidated gliadin peptides antibodies
DGPA-IgA	IgA Deamidated gliadin peptides antibodies
SM	Serological marker
SBB	Small bowel biopsy
TG2	Tissue Transglutaminase

1 Introduction

Identification of SM of CD has no doubt modified our perception of the disease compelling to revisiting both disease definition and diagnostic approach, mainly over the last twenty years principally [1–6].

Antigliadin antibodies (AGA) both IgG and IgA [7, 8] and antireticulin antibodies (ARA) [9] were the first SM available and this since the early 1980s. Measurement of AGA by ELISA methods consolidated its widespread use while

C. Ribes-Koninckx (✉) · E. Donat
Pediatric Gastroenterology Unit, La Fe University Hospital, Valencia, Spain
e-mail: ribes_car@gva.es

C. Ribes-Koninckx · M. Roca · E. Donat
Celiac Disease and Digestive Immunopathology Unit, Instituto de Investigación Sanitaria La Fe, Valencia, Spain

© The Author(s), under exclusive license to Springer Nature Switzerland AG 2022
J. Amil-Dias and I. Polanco (eds.), *Advances in Celiac Disease*,
https://doi.org/10.1007/978-3-030-82401-3_6

ARA decayed overtime partially due to the need of immunofluorescence (IF) methods for detection [9]. In the late eighties a new marker was identified, the anti-endomysium antibodies (EMA), which turned out to display a higher diagnostic performance than the previous AGA and ARA [10–12]. Its widespread use in the nineties firmly demonstrated that initial EMA positivity in patients with small intestinal villous atrophy was, in children, as predictive of CD diagnosis as traditional gluten challenge and full Interlaken criteria [1, 13]. Thanks to availability of TGA and EMA the number of SBB required for diagnosis were reduced in the new 1990 ESPGHAN CD diagnostic criteria for the paediatric population [2].

In 1997 the identification of tissue transglutaminase (TG2) as the autoantigen in CD led to IgA and IgG antibodies detection specific for TG2 (TGA) [14]. The production of TGA was shown to be related to dietary exposure to gluten and to small bowel mucosal atrophy. Moreover easy to use automated methods helped promoting their wide use as SM for CD diagnosis after the year 2000 [15, 16].

Subsequent studies on the immunopathogenesis of CD demonstrated that selective deamination of gliadin by TG2 helps the gliadin fragments bind to the antigen-presenting cells which is a fundamental step in the immunological response that leads to CD [17]. Serologic assays based on deamidated gliadin peptides (DGP) were then developed detecting antibodies against DGP (DGPA); when comparing the DGPA tests performance with the previous SM, DGPA display a higher diagnostic accuracy than the traditional AGA test [18], but lower than TGA and EMA [19, 20].

Lately easy to run rapid visual tests, based on various SM and different SM combinations have been developed, which can be run at the patients bed side and display a high efficiency [21]. Results need however to be confirmed by conventional laboratorial based tests but can be useful as a preliminary diagnostic screening tool [22].

Availability of these high efficient SM from the 80's onwards have completely modified the approach to CD diagnosis, specially but not exclusively in the paediatric population [2–4, 6].

In this chapter we perform a thorough revision of the definition of the different SM, the methods used for detection and their accuracy for CD diagnosis.

2   Anti-endomysium Antibodies

In 1983, Chorzelsky et al. [23] detected for the first time IgA class EMA using sections of monkey oesophagus by indirect IF, in the serum of patients with dermatitis herpetiformis (DH) and CD. Ever since, EMA have been used for the diagnosis of CD due to their high sensitivity and specificity, replacing other less reliable tests such as AGA or ARA.

Enzyme tissue transglutaminase (TG2) was identified as the target antigen of EMA in 1997 [14], therefore EMA recognize the same antigen as TGA antibodies and they only differ in terms of detection method. Immunoenzyme assays (ELISA) used to detect TGA-IgA showed high sensitivity, so have gradually replaced the EMA test in the serological diagnosis of CD.

Methods for EMA Detection

EMA described in CD are IgA antibodies directed against the intermyofibril substance of the smooth muscle [23]. They are detected by indirect IF method on sections of monkey oesophagus or human umbilical cord as substrate. In short, a tissue section is incubated with serum from the patient under study. After washing to remove unbound Immunoglobulin to the tissue, an anti-human Immunoglobulin antibody labelled with a fluorochrome, usually fluorescein isothiocyanate (FITC), is added. Mounting medium is used before carefully covering with a coverslip. Slides should be blindly examined to identify the cellular or tissue location by means of the observation under the fluorescence or confocal microscope of antibodies recognized by the patient. These substrates allow the identification of the classic honeycomb pattern identifiable in *muscularis mucosae* (Fig. 1).

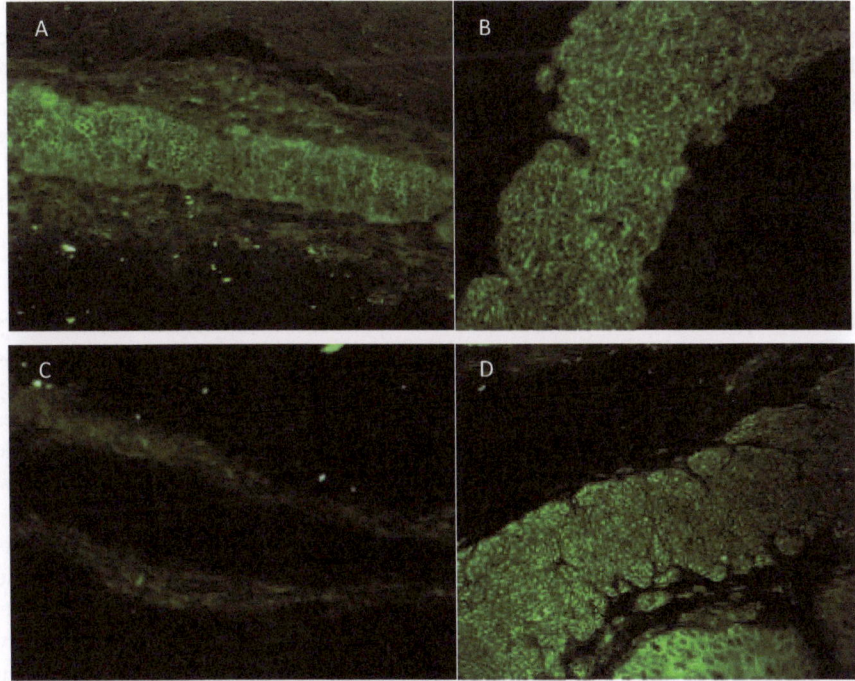

Fig. 1 IgA anti-endomysial antibodies by indirect immunofluorescence (IF) on section of monkey oesophagus. **a** Positive IF pattern of IgA-EMA staining the connective tissue structures that surround individual muscle fibrils (20x). **b** Positive classic honeycomb pattern (40x). **c** Negative pattern, the lack of fluorescence is remarkable (20x). **d** False positive pattern, fluorescence is found inside cells (20x)

Routine serial dilutions of patient's serum are not necessary for clinical purposes due to cost, time, and sacrifice of laboratory animals. Analysis of the sera at an initial dilution of 1:5 is recommended, but in case of doubtful positivity, the serum can be diluted or concentrated further, depending on whether there is an overlap with other autoantibody stains or weak stains, in order to draw conclusions. Regarding the optimal dilution of the fluorochrome-labelled antibody, the concentration should be adjusted for each antibody, according to the instructions of the manufacturer.

In patients with IgA deficiency, EMA-IgG can be performed, although according to the few studies published, it seems that the sensitivity of EMA-IgG is lower than of EMA-IgA [24]. However, a study in patients with IgA deficiency showed that EMA-IgG was very sensitive and specific [25]. CD is 5–20 times more common in patients with IgA deficiency compared to the general population, in these cases, EMA-IgG autoantibody tests are highly efficient in detecting celiac disease in IgA deficient patients.

To obtain accurate results of the fluorescence pattern, since there is a certain degree of subjectivity in interpreting the images, high-quality biological materials as well as expertise in result interpretation of the assay are required; therefore interpretation errors and added costs are found. Moreover, the IF can only be partially automated, so it should only be used in settings with appropriate expertise. Limitations would be, the fact that it cannot be completely automatized, and so are the workload and subjective reading of the results.

If typical pattern high positive results are found, the reading is easy, but in case of a low EMA-IgA level or atypical patterns, it can be more difficult. Despite the human umbilical cord is a good alternative to monkey oesophagus, the staining intensity is considerably weaker and the interpretation can be a real challenge.

Efficiency of EMA for CD Diagnosis

In mixed populations of children and adults, EMA-IgA tests have shown a very high specificity (~ 100%) in studies using monkey oesophagus. However, those studies had some variation in sensitivities; one study reported a very low sensitivity (75%), while in others, the sensitivity ranged from 86 to 98% [26]. In studies assessed EMA-IgA using human umbilical cord in a mixed-age population, the pooled sensitivity was 93% (95% CI, 88.1–95.4%), while the specificity was 100% (95% CI, 97.5–100%).

In adult patients, in different studies the sensitivity reported of EMA test is slightly above 86% and the specificity close to 100% [27–30]. The diagnostic performance of the EMA-IgA using monkey oesophagus as substrate in adults showed a pooled sensitivity of 97.4% (95% CI, 95.7–98.5), and a pooled specificity of 99.6% (95% CI, 98.8–99.9). The specificity of the EMA-IgA using human umbilical cord in adults was reported as 100% [26, 31]; however, there was greater variability in the sensitivity, ranging from 87 to 100%. The pooled sensitivity and specificity of this test were 90.2% (95% CI, 86.3–92.5) and 99.6% (95% CI, 98.4–99.9), respectively [26].

In children, IgA EMA using monkey oesophagus as substrate showed a pooled sensitivity and specificity of 96.1% (95% CI, 94.5–97.3) and 97.4% (95% CI, 96.3–98.2), respectively [26]. Studies that assessed EMA-IgA using human umbilical cord performance in children reported some variability in specificity [26] close to 100%. The pooled sensitivity in children was 96.9% (95% CI, 93.5–98.6).

In paediatric patients, a systematic meta-analysis of the diagnostic accuracy of CD antibody tests, the EMA-IgA sensitivity was ≥ 90%, pooled specificity 98.2% (ranged 94.7–100%), and the positive likelihood ratio is 31.8 [19]. This evidence report on CD serology estimates that EMA has a higher reliability for the diagnosis of CD that reveals almost an absolute specificity.

An international prospective study concluded that children can be accurately diagnosed with celiac disease without biopsy analysis, based on level of TGA-IgA tenfold or more the ULN, positive results from the EMA tests of 2 blood samples, and the presence of one symptom with a PPV of 99.75 (95% CI, 98.61–99.99). The inclusion of HLA analyses did not increase accuracy [32].

In a recent retrospective study in children, EMA-IgA and TG2-IgA, reached similar sensitivities (98% and 99%), while EMA had a higher specificity (99%) than anti-TG2 (93%). The results support the use of EMA to increase CD diagnostic accuracy in a non-biopsy approach, especially in asymptomatic children [33].

In a study including children and adults, EMA positivity has been observed as a very strong predictor of subsequent CD diagnosis irrespective of the initial titres or initial clinical presentation and it is a very strong predictor of forthcoming CD also in subjects with initially normal villi [34]. In multicentre studies, it has been found that inexperienced personnel in reading IF preparations can incorrectly evaluate EMA serological markers. Additionally, it has been found that EMA test specificity, detected through routine diagnostic analysis in a study performed over a long period, was considerably lower than expected, due to the degree of subjectivity in interpreting the results [35]. These statements support the importance of the evaluation in an expert laboratory by skilled technicians.

In general, EMA tests are more specific and TG2 more sensitive. Specificity is greater in EMA-IgA than in TG2-IgA, since EMA only recognize the TG2 epitopes related to CD, usually extracellular TG2 combined with fibronectin.

EMA is currently considered the most specific laboratory test for the diagnosis of CD due to its high sensitivity and specificity, even though there is no general agreement regarding its use. Furthermore, the EMA-IgA test is highly specific, but less sensitive than TG2-IgA, and should therefore preferably be used as a confirmation test.

3   Anti-tissue Transglutaminase Antibodies

In 1997 Dieterich and colleagues [14] identified the enzyme TG2 as the target antigen of EMA. They demonstrated that CD patient's serum with high EMA-IgA levels tested negative when preadsorbed with TG2, showing that TGA recognize the same antigen as EMA.

TG2 is a ubiquitous calcium-dependent enzyme. Eight different isoenzyme forms have been described, depending to their location in the tissues, like type TG2 of intestinal origin, type 3 (TG3 present) in the skin, which is the target of autoantibodies in dermatitis herpetiformis or type 6 (TG6), which targets the central nervous system and have been identified in patients with ataxia. TG2 plays a significant biological role, catalyzing the connection between glutamine and lysine in different proteins as well as in the conversion of glutamine into glutamic acid. It is involved in tissue repair and also in the removal of cell detritus after cell death and apoptosis [36]. In normal subjects, TG2 has been detected in all layers of the small intestinal wall.

TGA are present in different organs and can be detected not only in the blood, saliva or intestinal mucosa (TGA-IgA deposits) but also in liver and other tissues.

In CD patients an inappropriate immune response to gluten ingestion leads to mucosal damage and the release and activation of TG. Gluten and glutamine may be the target of the enzyme, which can bind it to other proteins including transglutaminase itself. TG deaminates gliadin peptides increases their affinity for HLA-DQ8 and DQ2 receptors [37] and activate lymphocytes T CD4 that afterwards stimulate B lymphocytes for the production of TGA antibodies of IgA and IgG class.

Methods for TGA Measurement

The first commercial immunoenzymatic (ELISA) assay were based on guinea pig liver TG and showed a very good diagnostic accuracy. However, after the later introduction of extractive or human recombinant TG2 (rhTG) as the antigen, higher sensitivity and specificity for CD diagnosis was obtained. Human erythrocytes are one of the most widely used sources of TG2, and human recombinant TG2 is obtained with eukaryotic expression systems or baculoviruses. The antigens obtained by these procedures show high stability and maintain the conformational epitopes of the protein unchanged, thus providing excellent analytical performance.

The use of TGA monoclonal antibodies demonstrates that the target region of the TGA antibodies in CD patients is located in the core of the molecule corresponding to a peptide not exceeding 237 aminoacids and that the epitopes are conformational, as they require the presence of the C and N terminal domains to maintain stability and immunogenicity [38].

The TGA can be detected by different methods: ELISA, Fluorometric Enzyme Immunoassay (FEIA) or Chemiluminescence Immunoassay (CLIA) and by radio binding assay (RBA) [39]

The tests are generally quantitative; as there are no international standards the values are expressed using different units based on calibration curves from each manufacturer. A standardization, harmonizing the results obtained in different laboratories with the various testing methods, is still needed. Thus cut-off level for each method and commercial kit needs to be identified on the basis of receiver operating characteristic (ROC) curve analysis.

Considering the recent guidelines, which attach considerable importance to the results obtained from the assay of TGA-IgA in diagnosing CD without biopsy, the importance of standardizing the results obtainable with the various commercial kits is all the more obvious. Although the measurement TGA-IgA antibodies is not standardized, most commercially available tests are highly accurate, especially at high values [40]. However, there is evidence of variability between different tests or different laboratories using the same test when it comes to moderate TGA levels.

A study comparing five commercial kits and RBA performed on the sera of children at risk for CD highlighted significant differences in the responses, affecting the interpretation of the results and the diagnosis of CD [41]. An European Workshop stressed the lack of reference materials and procedures [42] and also the American Gastroenterological Association Institute underlined the need for significant international collaboration to improve and harmonize the results of TGA testing [43].

Laboratories must be extremely rigorous in their internal quality control measures, accurately calculating the calibration curve, which should include the value of 10 times the ULN.

Due to increasing quality of current available assays, which are well-suited for automation and high-throughput testing, as well as to lower price of the assays, methods based on recombinant human enzyme has consolidated TGA in the last years almost a standard in CD diagnosis

Efficiency of TGA for CD Diagnosis

A meta-analysis of studies investigating the diagnostic accuracy of ELISA tests showed that the SROC curve indicated the absence of heterogeneity, and the superiority of recombinant human TGA (rh-TGA) and purified human TGA (ph-TGA) compared to guinea pig-TGA (gp-TGA). The sensitivities (all individual assays) for rh-TGA, ph-TGA, and gp-TGA were 94%, 94%, and 91%, respectively, and the specificities 95%, 94%, and 89%, respectively [44].

The sensitivity in adults of the TGA- IgA assays, using rh-TG, was above 95% and specificity in the range of 92–100% [4, 45]. The higher the value of the test, the greater the likelihood of a true positive result.

In paediatric patients a systematic meta-analysis of the diagnostic accuracy of CD antibody tests, covering the years 2004 to 2009, showed that the sensitivity was around 90% and specificity around 95% in most studies [19].

A recent systematic review with meta-analysis [46] in asymptomatic patients (children and adults) found a sensitivity of 92.8% (95% confidence interval [CI], 90.3–94.8%) and a specificity of 97.9% (95% CI, 96.4–98.8%).

In a study comparing 10 commercial ELISA kits a high level of accuracy was reported for all the methods examined. The sensitivity of TGA-IgA ranged from 91 to 97% and the specificity between 93 and 100%, using the producer's cut-off. The diagnostic accuracy of all the kits can be improved further by adjusting the cut-off through ROC-curve analysis [47].

On suspicion of CD the initial test in the diagnosis evaluation (in the diagnostic approach) should be the TGA-IgA test on account of its high sensitivity and specificity as well as its wide availability and the use of an automated and objective method. In previous years TGA-IgA was consider to have lower sensitivity in infants under 2 years of age but latter studies did not confirmed this [48].

In patients with IgA deficiency, whose risk of developing CD is higher, TGA-IgG is used. TGA-IgG achieved a performance inferior to TGA-IgA assays. Studies investigating the diagnostic accuracy of TGA-IgG have mainly been conducted using ELISA methods, in a mixed population which also included patients with selective IgA deficiency. The sensitivity of the test as reported in the various studies and comparing different commercial kits, ranges from 67.6 to 100%, and the specificity from 80 to 100% [19, 49, 50].

The relationship between TGA-IgA and the degree of the histological lesion has been evaluated, this showing sensitivity is significantly lower in cases with milder histological damage as for the EMA test; it drops to 67% in patients with partial mucosal atrophy and fall to only 7.69% in patients with Marsh 1 lesions, both in adults and in children [51]. Different groups of researchers [15, 52–54] have attempted to determine whether high TGA-IgA levels can justify avoiding diagnostic biopsies, especially in paediatric patients. Considering the differences between the various commercial methods, Hill et al. [55] report that a TGA-IgA equal or above 10 times the cut-off level for the specific test can detect 100% of patients with intestinal atrophy.

In contrast no studies have been conducted to establish the levels of TGA-IgG that can reliably predict the presence of enteropathy. There is also not enough evidence in children with type1 diabetes in whom a spontaneous normalization of CD serology at moderate titres [56] has been described.

False positive results for TGA have been occasionally reported, as in some diseases positive TGA-IgA can occur in the absence of CD and this, usually at low values and in a limited percentage (2–3%) of patients. This might be the case in autoimmune diseases, liver diseases, other foods sensitizations or infections [57–59] like giardiasis. On the other hand, false negative results can be expected in patients on immunosuppressive therapy and in dermatitis herpetiformis. Measurement in haemolyzed samples may also yield a false decrease in antibody levels [60]. Serological testing must be performed while the patient is consuming gluten regularly, as antibody levels decrease after initiation of a low-gluten or a gluten-free diet.

4   Anti-deamidated Gliadin Peptides Antibodies

For the traditional AGA test based on native gliadin, a wide range of sensitivity and specificity has been reported related mainly to the detection kit used and largely to the quality (specificity) of the gliadin employed in the ELISA test. Sensitivity for AGA-IgA ranges from 60.9 to 96.0% and specificity from 79.4 to 93.8%; worse performance, mainly for specificity, is reported for AGA-IgG in paediatric patients [19, 24]. Overall results were better for children than for adults [24]. In fact for many years AGA-IgA were considered extremely useful in the diagnostic approach in children as they were the only available marker and after the discovery of EMA, combined testing for AGA plus EMA was the most widely used laboratory approach, thus amending the lack of specificity of AGA [2]. After the year 2000, a higher sensitivity and specificity for TGA than for AGA was definitely confirmed and the use of the latter was no longer endorsed [3, 6, 61].

However new knowledge on the immunopathogenesis of CD, reported the observation that selective deamination of gliadin by TG2 is a crucial step in the immunologic pathway. This event changes glutamine into glutamic acid, thus enhancing the affinity of gliadin fragments for the antigen-presenting cells [17]. The HLA-DQ/gliadin/tTG complex induces a response by the immunocompetent cells, with production of TGA and also antibodies against DGP (DGPA), both IgA and IgG class; these can be detected circulating in serum of patients with active CD on a gluten containing diet [62, 63].

Thereafter several studies promptly showed that comparing the traditional AGA test with DGPA this latter has a higher diagnostic accuracy than AGA, both in terms of sensitivity and of specificity [18, 64].

Methods for DGPA Detection

DGPA were first determined by immune assay methods (ELISA methods) and promptly easy to run automated serologic assays using a pool of deamidated gliadin peptides as the antigen became commercially available for their detection.

Accuracy for CD Diagnosis

Contrary to AGA and EMA or TGA, preliminary studies found that accuracy of DGPA- IgG to be superior to DGPA-IgA for CD diagnosis, and this on account of a lower specificity for the latter. So, sensitivity of DGPA IgA in adults ranges from 83.6 to 98.3% with a specificity between 90.3 and 99.1% [65]. For DGPA- IgG the sensitivity is between 84.4 and 96.7% with a specificity of 98.5–100% [65]

Data reported in paediatric patients showed a sensitivity for DGPA-IgA in the range of 80.7–95.1% and a specificity between 86.3 and 93.1%; the positive likelihood ratios (LR+) and negative likelihood ratios (LR-) were from 6.9 to 12.7 and from 0.06 to 0.21 respectively. The diagnostic odds ratios (DOR) were between

56 and 93 which is lower than for TGA and EMA. The sensitivity in DGPA-IgG tests ranged between 80.1 and 98.6%, the specificity between 86.0 and 96.9%, the LR+ between 6.8 and 25.8, the LR_ between 0.02 and 0.21 and the DOR between 115 and 948 [19, 65]. In early studies patients with more severe mucosal damage had higher DGPA levels, but other studies have not confirmed a strict relationship. Also, although in general the levels of DGPA (IgG and IgA) increase in proportion to the degree of mucosal damage, a cut-off point to differentiate between patients with and without atrophy has not been identified [65, 66].

Although it is overall considered that performance of DGP-IgG is inferior to TGA, nevertheless there are still discrepancies on their use. Some authors affirm that as TGA tends to appear later in life than AGA, normally after the age of 1–2.5 years, DGPA should be used mainly in children under 2 years of age [67]. However this has not been confirmed in more recent studies [68] and it has even been shown that AGA can appear early in life and disappear without CD developing [48]. Thus, the role of DGPA in the diagnosis in children younger than 2–3 years still requires further assessment in large prospective studies, especially in comparison with TGA or EMA detection [62, 69].

There is also some debate on DGP being an earlier marker of mild histological lesion than TGA. Kurppa et al. reported that the sensitivity of DGPA was superior to TGA and comparable to EMA in patients having early-stage celiac disease with normal villous morphology. The authors conclude that DGPA seems to offer a promising tool for case-finding and follow-up in this entity [70].

Other authors however consider DGPA to have a diagnostic accuracy comparable to, or slightly lower than TGA-IgA [18]. So, it has been shown that DGPA-IgG are positive in the majority of patients negative for TGA, such as young children and patients of any age with selective IgA deficiency [50, 71].

Further studies confirmed these data as well as the fact that in this population DGPA-IgG maintained a high specificity and nowadays DGPA-IgG is considered to add real value in the diagnostic approach of CD in IgA deficiency [3].

5   POC Tests

In the last decade POC tests for CD have been developed and are commercially available worldwide. These are immunochromatographic rapid visual methods which are performed with whole blood/serum and use a strip coated with the antigen [21, 22, 72]. After the blood/serum diffuses down the strip, if antibodies are present in the patient's sample, antigen–antibody complexes are detected by labelled anti-human IgA and/or IgG antibodies, thus showing a series of coloured lines after a few minutes [21, 22, 72].

The first POC tests developed were based on AGA-IgA, but the number of published studies is too low to draw sounded conclusions about sensitivity and specificity, although both might be slightly above 95% [19, 21, 22, 72].

TGA-IgA based POC tests were developed later on with overall reported pooled sensitivities of 96.4% and pooled specificities of 97.7%; pooled LR+ was 40.6, LR- was 0.04, and DOR was 1343 [19, 22, 73], however, TGA-IgA or EMA perform better [3, 19, 22].

The main advantage of POCs is that they are easy to perform, do not require a laboratory or experienced laboratory staff, and results are rapidly available. Also because of stability for most of these methods, strips can be sent to a central lab for a centralized lecture or checking of the results.

Therefore, POCTs have the potential to increase CD diagnosis rates worldwide, facilitate early diagnosis, at a reduced cost. They are especially useful in primary care or in settings with inadequate infrastructure. Although they have also been used for screening in the general population, it is a matter of debate whether they could be performed by lay people if properly instructed [3, 19, 22]. Anyhow, results of POCs need to be confirmed by conventional tests performed in skilled laboratories by expert professionals, before a CD diagnosis can be established(Table 1).

6 Conclusions

Serological markers, specifically EMA-IgA, TGA-IgA and DGPA/ IgG have a high accuracy for CD diagnosis, particularly in the paediatric population, provided that IgA deficiency has been ruled out for IgA class SM and the patient is consuming a gluten-containing diet. The best performance for sensitivity is reported for TGA-IgA, which together with availability of automated methods, makes it the most popular SM used worldwide. EMA is superior in terms of specificity but detection by non-automated IF method requires a well-equipped laboratory and, above all, skilled personnel. One of the limitations of ELISA methods is the need for standardization, thus high quality commercial kits and quality controls of the laboratories are mandatory. POC tests, being rapid, cheap and easy to use, could be used in the initial approach of CD diagnosis but results needs to be confirmed by conventional assays.

Table 1 Summary of performance and characteristics of the CD serological markers

	Adults		Children		Detection methods	Pros	Cons	Use
	Specificity (range)	Sensitivity (range)	Specificity (range)	Sensitivity (range)				
EMA-IgA	>90% (86-100)	>99% (90-100)	≥90% (82-100)	98.2 % (94-100)	Manual IF methods	The most specific test	Time consuming and high cost plus subjective interpretation	Confirmatory test
TGA-IgA	≥95% (90-100)	≥97% (92-100)	≥95% (73.9-100)	≥95% (77.8-100)	Automatized ELISA detection method	The most sensitive test	No standardization	As initial screening test and follow-up
TGA-IgG	≥70% (67.6-100)	≥90% (80-100)	(12.6-99.3)	≥94%(86.3-100)	Automatized ELISA detection method	Usually positive in IgA deficient CD patients	Variability of commercial detection kits' precision	For IgA deficient patients
DPGA-IgG	(84.4-96)	(98.5-100)	≥95% (80.1-98.6)	≥90% (86.0-96.9)	Automatized ELISA detection method	Usually positive in TGA-IgA negative CD children	Lower accuracy than TGA-IgA	For children below 3 years of age and IgA deficient patients
DPGA-IgA	(83.6-98.3)	(90.3-99.1)	≥90% (80.7 - 95.1)	≥90% (86.3-93.1)	Automatized ELISA detection method	No benefit	Lower accuracy than TGA-IgA and DGP-IgG	Not recommended

EMA, anti-endomysial antibodies; TGA, antibodies against type-2 (tissue) transglutaminase; DPGA, deamidated gliadin peptides antibodies; IF, immunofluorescence staining; ELISA, enzyme linked immunosorbent assay; CD, celiac disease

References

1. GW Meeuwisse 1970 Diagnostic criteria in celiac disease Acta Paediatr Scand 59 909 911
2. JA Walker-Smith S Guandalini J Schmitz 1990 Revised criteria for diagnosis of coeliac disease Arch Dis Child 65 909 911
3. S Husby S Koletzko IR Korponay-Szabó 2012 European society for pediatric gastroenterology, hepatology, and nutrition guidelines for the diagnosis of coeliac disease J Pediatr Gastroenterol Nutr 54 136 160
4. A Rubio-Tapia ID Hill CP Kelly 2013 ACG clinical guidelines: diagnosis and management of celiac disease Am J Gastroenterol 108 656 676
5. S Husby JA Murray DA Katzka 2019 AGA clinical practice update on diagnosis and monitoring of celiac disease—changing utility of serology and histologic measures: expert review Gastroenterology 156 885 889
6. S Husby S Koletzko I Korponay-Szabó 2020 European Society paediatric gastroenterology, hepatology and nutrition guidelines for diagnosing coeliac disease 2020 J Pediatr Gastroenterol Nutr 70 141 156
7. CR Koninckx JP Giliams I Polanco 1984 IgA antigliadin antibodies in celiac and inflammatory bowel disease J Pediatr Gastroenterol Nutr 3 676 682
8. R Troncone A Ferguson 1991 Anti-gliadin antibodies J Pediatr Gastroenterol Nutr 12 150 158
9. M Mäki O Hällström T Vesikari 1984 Evaluation of a serum IgA-class reticulin antibody test for the detection of childhood celiac disease J Pediatr 105 901 905
10. A Lecea De C Ribes-Koninckx I Polanco 1996 Serological screening (antigliadin and antiendomysium antibodies) for non-overt coeliac disease in children of short stature Acta Paediatr Int J Paediatr Suppl 85 54 55
11. E Grodzinsky G Jansson T Skogh 1995 Anti-endomysium and anti-gliadin antibodies as serological markers for coeliac disease in childhood: a clinical study to develop a practical routine Acta Pædiatrica 84 294 298
12. TM Rossi CH Albini V Kumar 1993 Incidence of celiac disease identified by the presence of serum endomysial antibodies in children with chronic diarrhea, short stature, or insulin-dependent diabetes mellitus J Pediatr 123 262 264
13. IR Korponay-Szabó JB Kovács A Czinner 1999 High prevalence of silent celiac disease in preschool children screened with IgA/IgG antiendomysium antibodies J Pediatr Gastroenterol Nutr 28 26 30
14. W Dieterich T Ehnis M Bauer 1997 Identification of tissue transglutaminase as the autoantigen of celiac disease Nat Med 3 797 801
15. CC Barker C Mitton G Jevon 2005 Can tissue transglutaminase antibody titers replace small-bowel biopsy to diagnose celiac disease in select pediatric populations? Pediatrics 115 1341 1346
16. I Dahlbom IR Korponay-Szabó JB Kovács 2010 Prediction of clinical and mucosal severity of coeliac disease and dermatitis herpetiformis by quantification of IgA/IgG serum antibodies to tissue transglutaminase J Pediatr Gastroenterol Nutr 50 140 146
17. Ø Molberg SN Mcadam R Körner 1998 Tissue transglutaminase selectively modifies gliadin peptides that are recognized by gut-derived T cells in celiac disease Nat Med 4 713 717
18. E Schwertz F Kahlenberg U Sack 2004 Serologic assay based on gliadin-related nonapeptides as a highly sensitive and specific diagnostic aid in celiac disease Clin Chem 50 2370 2375
19. K Giersiepen M Lelgemann N Stuhldreher 2012 Accuracy of diagnostic antibody tests for coeliac disease in children: summary of an evidence report J Pediatr Gastroenterol Nutr 54 229 241
20. MJ Gould H Brill MA Marcon 2019 In screening for celiac disease, deamidated gliadin rarely predicts disease when tissue transglutaminase is normal J Pediatr Gastroenterol Nutr 68 20 25
21. S Ferre-López C Ribes-Koninckx C Genzor 2004 Immunochromatographic sticks for tissue transglutaminase and antigliadin antibody screening in celiac disease Clin Gastroenterol Hepatol 2 480 484

22. P Singh A Arora TA Strand 2019 Diagnostic accuracy of point of care tests for diagnosing celiac disease: a systematic review and meta-analysis J Clin Gastroenterol 53 535 542
23. TP Chorzelski J Sulej H Tchorzewska 1983 IgA class endomysium antibodies in dermatitis herpetiformis and coeliac disease Ann N Y Acad Sci 420 325 334
24. Rostom A, Dubé C, Cranney A, et al. The diagnostic accuracy of serologic tests for celiac disease: a systematic review. Gastroenterology; 128. Epub ahead of print 2005. https://doi.org/10.1053/j.gastro.2005.02.028.
25. IR Korponay-Szabó I Dahlbom K Laurila 2003 Elevation of IgG antibodies against tissue transglutaminase as a diagnostic tool for coeliac disease in selective IgA deficiency Gut 52 1567 1571
26. Rostom A, Dubé C, Cranney A, et al. Celiac disease. Evidence report/technology assessment (Summary) 2004; 1–6.
27. Mascart-Lemone F, Lambrechts A. Serology of coeliac disease: early diagnosis and therapeutic impact. In: Acta gastro-enterologica belgica;1995. p. 388–96.
28. A Picarelli M Tola Di L Sabbatella 2001 Identification of a new coeliac disease subgroup: antiendomysial and anti-transglutaminase antibodies of IgG class in the absence of selective IgA deficiency J Intern Med 249 181 188
29. N Tesei E Sugai H Vázquez 2003 Antibodies to human recombinant tissue transglutaminase may detect coeliac disease patients undiagnosed by endomysial antibodies Aliment Pharmacol Ther 17 1415 1423
30. V Baldas A Ventura T Not 2000 Development of a novel rapid non-invasive screening test for coeliac disease Gut 47 628 631
31. P Collin K Kaukinen H Vogelsang 2005 Antiendomysial and antihuman recombinant tissue transglutaminase antibodies in the diagnosis of coeliac disease: a biopsy-proven European multicentre study Eur J Gastroenterol Hepatol 17 85 91
32. Werkstetter KJ, Korponay-Szabó IR, Popp A, et al. Accuracy in diagnosis of celiac disease without biopsies in clinical practice. Gastroenterology; 153. Epub ahead of print 2017. https://doi.org/10.1053/j.gastro.2017.06.002.
33. M Roca E Donat N Marco-Maestud 2019 Efficacy study of anti-endomysium antibodies for celiac disease diagnosis: a retrospective study in a Spanish pediatric population J Clin Med 8 2179
34. K Kurppa T Räsänen P Collin 2012 Endomysial antibodies predict celiac disease irrespective of the titers or clinical presentation World J Gastroenterol 18 2511 2516
35. A Mubarak VM Wolters SAM Gerritsen 2011 A biopsy is not always necessary to diagnose celiac disease J Pediatr Gastroenterol Nutr 52 554 557
36. L Lorand RM Graham 2003 Transglutaminases: crosslinking enzymes with pleiotropic functions Nat Rev Mol Cell Biol 4 140 156
37. LM Sollid B Jabri 2011 Celiac disease and transglutaminase 2: a model for posttranslational modification of antigens and HLA association in the pathogenesis of autoimmune disorders Curr Opin Immunol 23 732 738
38. D Sblattero F Florian E Azzoni 2002 The analysis of the fine specificity of celiac disease antibodies using tissue transglutaminase fragments Eur J Biochem 269 5175 5181
39. M Bonamico C Tiberti A Picarelli 2001 Radioimmunoassay to detect antitransglutaminase autoantibodies is the most sensitive and specific screening method for celiac disease Am J Gastroenterol 96 1536 1540
40. P Vermeersch K Geboes G Mariën 2013 Defining thresholds of antibody levels improves diagnosis of celiac disease Clin Gastroenterol Hepatol 11 398 403
41. E Liu M Li F Bao 2005 Need for quantitative assessment of transglutaminase autoantibodies for celiac disease in screening-identified children J Pediatr 146 494 499
42. M Stern 2000 Comparative evaluation of serologic tests for celiac disease: a European initiative toward standardization J Pediatr Gastroenterol Nutr 31 513 519
43. M Li L Yu C Tiberti 2009 A report on the international transglutaminase autoantibody workshop for celiac disease Am J Gastroenterol 104 154 163

44. E Zintzaras AE Germenis 2006 Performance of antibodies against tissue transglutaminase for the diagnosis of celiac disease: meta-analysis Clin Vaccine Immunol 13 187 192
45. A Carroccio G Vitale L Prima Di 2002 Comparison of anti-transglutaminase ELISAs and an anti-endomysial antibody assay in the diagnosis of celiac disease: a prospective study Clin Chem 48 1546 1550
46. R Chou C Bougatsos I Blazina 2017 Screening for celiac disease: evidence report and systematic review for the US preventive services task force JAMA—J Am Med Assoc 317 1258 1268
47. B Meensel Van M Hiele I Hoffman 2004 Diagnostic accuracy of ten second-generation (human) tissue transglutaminase antibody assays in celiac disease Clin Chem 50 2125 2135
48. SL Vriezinga R Auricchio E Bravi 2014 Randomized feeding intervention in infants at high risk for celiac disease N Engl J Med 371 1304 1315
49. F Cataldo D Lio V Marino 2000 IgG1 antiendomysium and IgG antitissue transglutaminase (anti-tTG) antibodies in coeliac patients with selective IgA deficiency Gut 47 366 369
50. Villalta D, Alessio MG, Tampoia M, et al. Diagnostic accuracy of IgA anti-tissue transglutaminase antibody assays in celiac disease patients with selective IgA deficiency. In: Annals of the New York academy of sciences. Blackwell Publishing Inc.;2007. p. 212–20.
51. A Tursi G Brandimarte GM Giorgetti 2003 Prevalence of antitissue transglutaminase antibodies in different degrees of intestinal damage in celiac disease J Clin Gastroenterol 36 219 221
52. Diamanti A, Colistro F, Calce A, et al. Clinical value of immunoglobulin A antitransglutaminase assay in the diagnosis of celiac disease. Pediatrics; 118. Epub ahead of print December 2006. https://doi.org/10.1542/peds.2006-0604.
53. Vivas S, Ruiz De Morales JG, Riestra S, et al. Duodenal biopsy may be avoided when high transglutaminase antibody titers are present. World J Gastroenterol 2009; 15: 4775–80.
54. MR Donaldson LS Book KM Leiferman 2008 Strongly positive tissue transglutaminase antibodies are associated with Marsh 3 histopathology in adult and pediatric celiac disease J Clin Gastroenterol 42 256 260
55. PG Hill GKT Holmes 2008 Coeliac disease: a biopsy is not always necessary for diagnosis Aliment Pharmacol Ther 27 572 577
56. S Castellaneta E Piccinno M Oliva 2015 High rate of spontaneous normalization of celiac serology in a cohort of 446 children with type 1 diabetes: a prospective study Diabetes Care 38 760 766
57. F Ferrara S Quaglia I Caputo 2010 Anti-transglutaminase antibodies in non-coeliac children suffering from infectious diseases Clin Exp Immunol 159 217 223
58. HJ Freeman 2004 Strongly positive tissue transglutaminase antibody assays without celiac disease Can J Gastroenterol 18 25 28
59. Garcia-Peris M, Donat Aliaga E, Roca Llorens M, et al. Anti-tissue transglutaminase antibodies not related to gluten intake. An Pediatr. Epub ahead of print 2018. https://doi.org/10.1016/j.anpedi.2018.01.013.
60. Wolf J, Haendel N, Remmler J, et al. Hemolysis and IgA-antibodies against tissue transglutaminase: When are antibody test results no longer reliable? J Clin Lab Anal; 32. Epub ahead of print 1 May 2018. https://doi.org/10.1002/jcla.22360.
61. ID Hill MH Dirks GS Liptak 2005 Guideline for the diagnosis and treatment of celiac disease in children: Recommendations of the North American Society for pediatric gastroenterology, hepatology and nutrition J Pediatr Gastroenterol Nutr 40 1 19
62. D Agardh 2007 Antibodies against synthetic deamidated gliadin peptides and tissue transglutaminase for the identification of childhood celiac disease Clin Gastroenterol Hepatol 5 1276 1281
63. E Sugai H Vázquez F Nachman 2006 Accuracy of testing for antibodies to synthetic gliadin-related peptides in celiac disease Clin Gastroenterol Hepatol 4 1112 1117
64. Brusca I. Overview of biomarkers for diagnosis and monitoring of celiac disease. In: Adv Clin Chem. Academic Press Inc.;2015. p. 1–55.

65. I Brusca A Carroccio E Tonutti 2012 The old and new tests for celiac disease: which is the best test combination to diagnose celiac disease in pediatric patients? Clin Chem Lab Med 50 111 117

66. D Basso G Guariso P Fogar 2009 Antibodies against synthetic deamidated gliadin peptides for celiac disease diagnosis and follow-up in children Clin Chem 55 150 157

67. S Simell A Kupila S Hoppu 2005 Natural history of transglutaminase autoantibodies and mucosal changes in children carrying HLA-conferred celiac disease susceptibility Scand J Gastroenterol 40 1182 1191

68. O Olen AH Gudjónsdóttir L Browaldh 2012 Antibodies against deamidated gliadin peptides and tissue transglutaminase for diagnosis of pediatric celiac disease J Pediatr Gastroenterol Nutr 55 695 700

69. E Liu M Li L Emery 2007 Natural history of antibodies to deamidated gliadin peptides and transglutaminase in early childhood celiac disease J Pediatr Gastroenterol Nutr 45 293 300

70. K Kurppa K Lindfors P Collin 2011 Antibodies against deamidated gliadin peptides in early-stage celiac disease J Clin Gastroenterol 45 673 678

71. L Mozo J Gómez E Escanlar 2012 Diagnostic value of anti-deamidated gliadin peptide IgG antibodies for celiac disease in children and IgA-deficient patients J Pediatr Gastroenterol Nutr 55 50 55

72. LCB Baviera ED Aliaga L Ortigosa 2007 Celiac disease screening by immunochromato-graphic visual assays: results of a multicenter study J Pediatr Gastroenterol Nutr 45 546 550

73. IR Korponay-Szabó K Szabados J Pusztai 2007 Population screening for coeliac disease in primary care by district nurses using a rapid antibody test: diagnostic accuracy and feasibility study Br Med J 335 1244 1247

Histopathological Assessment of Celiac Disease

Villanacci Vincenzo⬤, Simoncelli Gloria, Monica Melissa, Caputo Alessandro, and Del Sordo Rachele

1 Introduction

Celiac disease (CD) is an immune-mediated enteropathy triggered by dietary gluten in genetically predisposed individuals [1, 2]. CD is among the most common autoimmune disorders, with a worldwide prevalence of about 0.5–1%. Wide variability in prevalence is observed in different areas of the world due to different prevalence of the predisposing genes, gluten consumption and other factors [3]. CD can be diagnosed at any age, but two peaks of incidence have been described: one in the first 2 years of life (after the introduction of gluten in the diet), and the other in adolescence and early adulthood [2]. CD is twice as common in females than in males (male:female ratio 1:2) [3].

Diagnosis of CD requires the integration of clinical, serological, histological, and genetic data. All these factors (except genetics) are altered by a gluten-free diet, and thus they must be evaluated while the patient is still ingesting gluten. The symptoms of CD can be intestinal (also known as "classical") and extra-intestinal. Due to the heterogeneity and non-specificity of symptoms, CD still represents an under-recognized and under-diagnosed condition, especially in adults, and has been compared to a chameleon [1, 4]. Diagnostic delays ranging from 4 to 13 years have been reported by some authors [4–9].

V. Vincenzo (✉) · S. Gloria · M. Melissa
Institute of Pathology Spedali Civili, Piazzale Spedali Civili 1, 25123 Brescia, Italy

C. Alessandro
University Hospital "San Giovanni di Dio e Ruggi D'Aragona", Salerno, Italy

D. S. Rachele
Department of Medicine and Surgery, Section of Anatomic Pathology and Histology, Medical School, University of Perugia, Perugia, Italy

© The Author(s), under exclusive license to Springer Nature Switzerland AG 2022   79
J. Amil-Dias and I. Polanco (eds.), *Advances in Celiac Disease*,
https://doi.org/10.1007/978-3-030-82401-3_7

2   The Duodenal Biopsy: Technical Considerations

Duodenal biopsy with histopathological analysis was, until some years ago, the gold standard procedure for the diagnosis of CD. It is the most invasive procedure in the diagnostic workup of CD, yet it is unavoidable in most cases [10]. Current guidelines underscore the centrality of duodenal biopsy in all adults with suspected CD, even when all other data are in favour of CD (serology, genetics, and symptoms) [10, 11]. In the paediatric population, on the other hand, the European Society for Paediatric Gastroenterology Hepatology and Nutrition (ESPGHAN) guidelines contemplate the possibility of diagnosing CD in a child with typical symptoms and compatible serology without histological evaluation [12]. The North American Society for Pediatric Gastroenterology, Hepatology and Nutrition (NASPGHAN) still recommends duodenal biopsy even in the paediatric population [13].

To obtain the best and most representative biopsies, close collaboration is required between the endoscopist, the endoscopy-room nurse, the pathology laboratory technician and the pathologist [14].

2.1   Site and Number of Biopsies

Biopsies should sample the second duodenal portion as well as the bulb (a crucial site in paediatric patients) [15–19]. Two biopsies from the bulb and four biopsies from the second duodenal portion are recommended for maximum diagnostic yield [11, 20].

2.2   Orientation of the Biopsy

Proper orientation of the biopsy sample on the final glass slide is of paramount importance to assess histological features of pathological duodenal mucosa, such as villous blunting.

We use and recommend a strip of acetate cellulose filter to ensure an adequate orientation of the biopsies. In brief, each bioptic sample is placed with the luminal surface pointing upwards on the strip of paper. Samples are placed on a straight line to allow precise microtome sectioning. One end of the filter paper is marked, identifying the beginning of the biopsy sequence. Consequently, the same strip can be used for the samples from the bulb and the second duodenal portion.

In the histology lab, the whole strip will be processed (without detaching the samples) and embedded in paraffin on its side (Fig. 1). The nitrocellulose polymers do not react with reagents used for fixation, tissue processing or staining, but are porous enough to closely adhere to the bioptic samples. This, coupled with the fact that samples have been placed on a straight line by the endoscopist, will allow the laboratory technician to cut sections that include all the bioptic samples.

One haematoxylin-and-eosin–stained section will be enough to assess all the morphological elements required for the histopathological diagnosis of CD. We recommend that one section be stained with CD3 immunohistochemistry to aid in counting intraepithelial T lymphocytes [21].

When properly performed, this technique allows not only a significant diagnostic improvement due to preserved orientation, but also considerable time and money savings due to the reduced number of tissue blocks, glass slides and specimen handling required in the histology lab.

3   The Duodenal Biopsy: Histological Considerations

3.1   Normal Duodenal Mucosa

For each compartment, the following characteristics are considered normal:

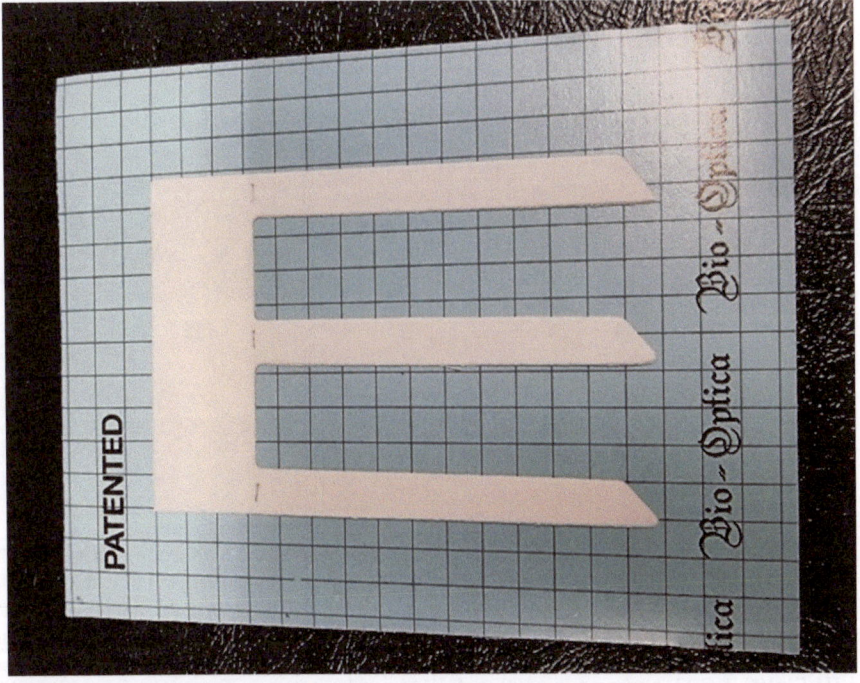

Fig. 1 Acetate cellulose filter (endokit)

Villi should be tall and slender, and their height should be at least three times the depth of the crypts (villus-crypt ratio > 3:1).

Intra-epithelial T lymphocytes (IELs) are present in normal subjects, but in smaller amounts than in CD patients. An IEL count greater than 25 per 100 epithelial cells is considered pathological [22, 23]. Furthermore, IELs tend to be more abundant along the sides of the villi and more sparse on the tips depending also on the orientation of the biopsies [14].

Crypts are made up of epithelial cells, endocrine cells, Paneth cells and stem cells. Due to their constant functional proliferation, one mitotic figure per crypt is expected. Epithelial turnover and thus crypt mitoses are greatly increased in CD patients.

Inflammatory cells normally present in small amounts in the lamina propria include plasma cells, eosinophils, histiocytes, mast cells and lymphocytes. A normal eosinophil count is fewer than 5 per high-power field (400x). Neutrophils are generally absent in the healthy duodenal mucosa (Fig. 2a, b).

3.2   Pathological Duodenal Mucosa

Histological diagnosis of CD is based on the recognition of three elementary lesions which together help in defining the diagnosis and grading the severity of CD:

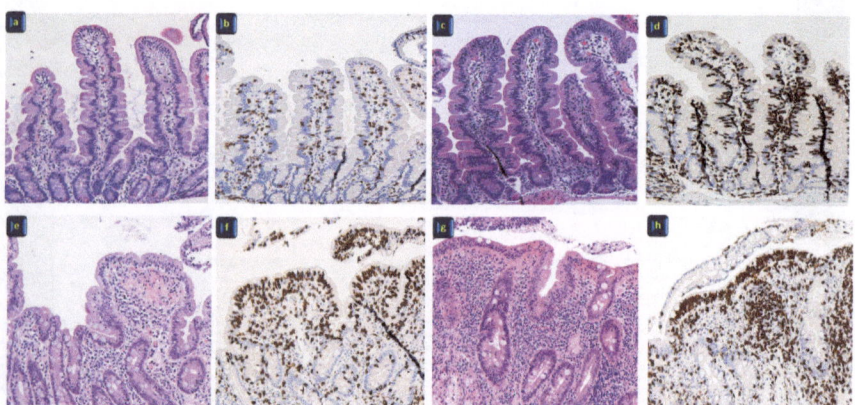

Fig. 2 a, b Normal duodenal mucosa: villus/crypt ratio over 3:1 and number of intraepithelial T lymphocytes <25 per 100 epithelial cells. **a** H&E x100, **b** CD3 immunostain x100. **c, d Type 1— Grade A lesion**: normal villi but with a pathological increase of intraepithelial T lymphocytes >25 per 100 epithelial cells. **c** H&E x100, **d** CD3 immunostain x100. **e, f** Mild to moderate villous atrophy **Type 3A-3B—Grade B1** with pathological increase of intraepithelial T lymphocytes **e** H&E x100, **f** CD3 immunostain x100. **g, h** Severe villous atrophy **Type 3C—Grade B2** with pathological increase of intraepithelial T lymphocytes **g** H&E x100, **h** CD3 immunostain x100

- **Intraepithelial lymphocytosis**: an IEL count greater than 25 per 100 enterocytes;
- **Crypt hyperplasia**: more than one mitotic figure per crypt, usually accompanied by a decrease in mucosecretive activity;
- **Villous atrophy**: a villus/crypt ratio lower than 3:1, i.e. a shortening or absence of villi.

These features are not, by themselves, pathognomonic of CD: each one of them can be seen in a number of other conditions. Therefore, the diagnosis of CD is based on the integration of histologic features with clinical, serological, and sometimes genetic data.

3.3 Histopathological Classification Schemes

The Marsh [24] classification is universally recognized and has been extensively validated. It classifies CD on the basis of presence or absence of elementary lesions, as follows:

- **Type 1** or **infiltrative lesion**: only intraepithelial lymphocytosis;
- **Type 2** or **hyperplastic lesion**: intraepithelial lymphocytosis and crypt hyperplasia;
- **Type 3** or **destructive lesion**: intraepithelial lymphocytosis, crypt hyperplasia, and villous atrophy.

Oberhuber [25] modified the Marsh classification by splitting the type 3 lesion in three subgroups based on the extent of villous atrophy:

- **Type 3a**: **mild** villous atrophy;
- **Type 3b: moderate** villous atrophy;
- **Type 3c: total** villous atrophy.

Further classifications aimed to achieve the simplification of the diagnostic criteria and reduction of the number of categories in order to increase the inter-rater agreement, facilitate the clinician's work and improve the communication between pathologist and clinician.

Corazza and Villanacci [26, 27] classify CD in only three categories:

- **Type A (Non-Atrophic)**: intraepithelial lymphocytosis, with or without crypt hyperplasia, but without villous atrophy;
- **Type B (Atrophic)**: intraepithelial lymphocytosis, crypt hyperplasia and villous atrophy, further subdivided into
- **Type B1**: low-moderate villous atrophy (villi still recognizable, but villus/crypt ratio < 3:1);
- **Type B2**: complete villous atrophy (villi no longer identifiable) (Fig. 2c–h).

Recently, an even more simplified classification with only two entities was proposed by Villanacci [28]:

- **Type A (Non-Atrophic)**: intraepithelial lymphocytosis, with or without crypt hyperplasia, but without villous atrophy;
- **Type B (Atrophic)**: intraepithelial lymphocytosis, crypt hyperplasia and villous atrophy, without further subdivisions.

These classifications schemes are summarized in Table 1.

3.4   Role of Immunohistochemistry

The IEL count is one of the key points in the histopathological diagnosis of CD, with more than 25 IELs per 100 epithelial cells representing a pathological count. While haematoxylin-and-eosin–stained slides allow for a reasonable IEL count, using an immunohistochemical stain for CD3 can be invaluable to aid the pathologist in obtaining a more accurate IEL count, especially on the first diagnostic biopsies (Fig. 2b–h) [29–32].

3.5   Non-celiac Gluten Sensitivity

Non-celiac gluten sensitivity (NCGS) is "a clinical entity induced by the ingestion of gluten leading to intestinal and/or extraintestinal symptoms that improve once the gluten-containing foodstuff is removed from the diet, and CD and wheat allergy have been excluded" [33]. The histologic characteristics of NCGS are still under investigation, ranging from normal histology to a slight intraepithelial lymphocytosis. Some authors described a normal number of T lymphocytes but a peculiar

Table 1 Comparison of the Marsh, Corazza-Villanacci, and Villanacci classification schemes

IELs (%)	Crypts	V:C ratio	Marsh mod. Oberhuber	Corazza-Villanacci	Villanacci
>25	Normal	Preserved	Type I Infiltrative	Type A Non atrophic	Type A Non atrophic
>25	Hyperplastic	Preserved	Type II Hyperplastic		
>25	Hyperplastic	Mild atrophy	Type III A Destructive	Type B1 Partial atrophy	Type B Atrophic
>25	Hyperplastic	Moderate atrophy	Type III B Destructive		
>25	Hyperplastic	Severe atrophy	Type III C Destructive	Type B2 Total atrophy	

disposition of these cells in small clusters of 3–4 elements in the superficial epithelium, or linearly in the deeper part of the mucosa, together with an increased number of eosinophils (>5/HPF) in the lamina propria. Further studies are needed to assess these findings as specific for NCGS (Fig. 3a–h) [34].

4 Differential Diagnosis

The histopathologic features of CD are not pathognomonic, but shared with numerous other disorders. A duodenal mucosa with all the characteristics of CD can at most be compatible with CD, thus requiring integration with clinical, laboratory, endoscopic and genetic data. Table 2 summarizes the main conditions that can share the histopathologic features of CD and thus should be considered in the differential diagnosis.

4.1 Pitfalls in the Determination of the IEL Count and Villous Architecture

Before discussing conditions that can mimic CD by replicating its histological features, a brief note is in order regarding the differential diagnosis of normal duodenal mucosa from CD. The pathologist must be aware of some pitfalls to avoid

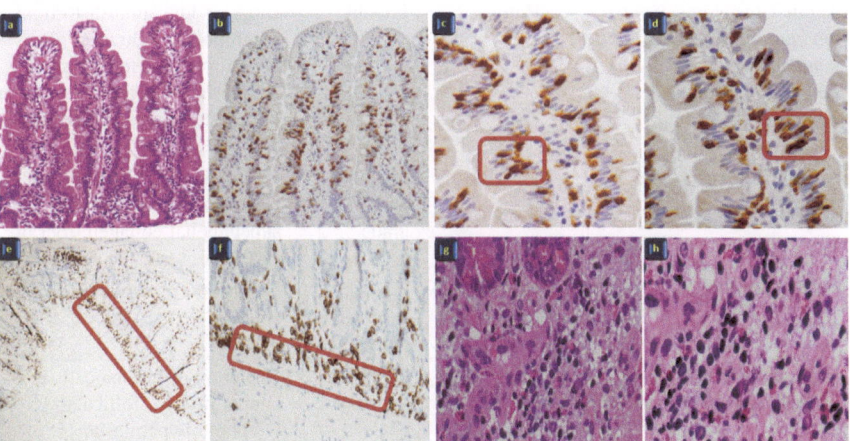

Fig. 3 **a, b** Normal villi; T lymphocytes <25 per 100 epithelial cells. **a** H&E x100, **b** CD3 immunostain x100. **c, d** Cluster of T Lymphocytes in the superficial epithelium. **c, d** CD3 immunostain x600 red rectangle. **e, f** Linear disposition of T lymphocytes in the deeper part of the mucosa. **e** CD3 immunostain x100, **f** CD3 immunostain x400 red rectangle. **g, h** Eosinophils in lamina propria H&E x400

Table 2 Major mimickers of celiac disease and histopathologic features useful for differential diagnosis

Mimicker	Increased IELs	Villous atrophy	Histopathologic tips for differential diagnosis
Infectious diseases			
Parasitic infestation	Rare (in children)	Rare (in children)	Identification of parasites (e.g. *Giardia*); increased eosinophils in lamina propria
HP-positive gastritis and peptic duodenitis	Possible	Possible, mild (if present)	Foveolar metaplasia of the duodenum; increased plasma cells in lamina propria; neutrophilic infiltration in lamina propria and epithelium; changes more prominent in the bulb; *HP* in gastric biopsies
Tropical sprue	Yes	Yes, usually low-grade	Extensive ileal involvement
Bacterial overgrowth	Yes	Possible	Mild lesions
Whipple disease	Rare	Yes	PAS-positive macrophages in the lamina propria
Viral gastroenteritis or post-infectious changes	Yes	Possible, variable grade	Mucosal recovery after infection resolution
Drugs			
NSAIDs	Possible	Rare, patchy, mild	Erosions, neutrophilic infiltration in the lamina propria
Antineoplastic and immune modulatory drugs (including immune checkpoint inhibitors)	Rare	Possible	Crypt architectural distortion; neutrophilic infiltration of lamina propria; foci of crypt apoptosis; involvement of other gastrointestinal sites (gastritis, colitis)
ARBs (Olmesartan and others)	Possible	Frequent, variable grade	Neutrophilic infiltration in lamina propria; deposition of subepithelial collagen, foci of crypt apoptosis
Collagenous sprue	Yes	Frequent, variable grade	Deposition of subepithelial collagen
Immunodeficiencies (including CVID)	Yes	Possible, variable grade	Depletion of plasma cells in the lamina propria, follicular lymphoid hyperplasia; concomitant giardiasis
Autoimmune enteropathy	Possible (celiac pattern)	Yes, variable grade	Neutrophilic infiltration in lamina propria; crypt apoptosis; reduction in goblet and Paneth

(continued)

Table 2 (continued)

Mimicker	Increased IELs	Villous atrophy	Histopathologic tips for differential diagnosis
			cells; diffuse involvement of other gastrointestinal tracts (gastritis, enteritis, colitis)
Crohn's disease and ulcerative colitis-associated duodenitis	Rare	Rare, patchy (if present)	Erosions/ulcerations, neutrophilic inflammation; crypt distortion; microgranulomas; basal plasmacytosis; ileal and colonic involvement
Eosinophilic gastroenteritis and food protein-sensitive enteropathies	Possible	Possible, usually not severe	Increased eosinophils in lamina propria; involvement of other gastrointestinal sites (enteritis and colitis)
Nutritional deficiencies (including pernicious anaemia)	Rare	Possible, usually not severe	Megaloblastic changes in epithelial cells in pernicious anaemia

Abbreviations: ARB: angiotension receptor blocker; CVID: common variable immunodeficiency; HP: Helicobacter pylori; IEL: intraepithelial lymphocyte; NSAID: non-steroidal anti-inflammatory drug; PAS: periodic acid–Schiff stain

false-positive (and rarely false-negative) determination of the histologic features defining CD.

4.1.1 Intraepithelial Lymphocyte Count

The thickness at which the histological section is cut can influence the IEL count by containing more or fewer lymphocytes. Standardization is of paramount importance to ensure that all sections are cut at equal thickness. Additionally, the IEL counts performed on slides immunohistochemically stained for CD3 will be higher than the corresponding H&E–stained slides. This is due to the fact that the lymphocyte plasma membrane is indistinguishable from enterocytes, so a lymphocyte is counted on a H&E slide only when its nucleus is included in the section and stained. IHC, on the other hand, labels the membrane of CD3+ cells, so even only a small anucleate section of lymphocyte membrane will be stained. Thus, when a T lymphocyte is cut so that only its membrane is included in the slide, then it will be visible in CD3–IHC slides and invisible in H&E.

Another caveat to be aware of is superficial lamina propria lymphocytes which may be erroneously counted as IELs due to their close relationship with the epithelium. Lymphocytes below the basement membrane should not be counted as IELs. IHC-stained slides pose a particular challenge in this regard because the detection reaction will make the stain extend slightly beyond the lymphocyte membrane, thus falsely appearing very close to enterocytes.

The degree of the increase in IELs should also be considered: while virtually all CD patients will show more than 25 IELs per 100 enterocytes, the vast majority of them will show greatly increased counts (i.e. >40/100). This is in contrast with most mimickers and borderline normal cases, where the IEL count will be borderline (20–30 IELs per 100 enterocytes).

The location of lymphocytes within the villus is also important. Normally, IELs are more abundant along the sides of the villus and fewer at the tip. Especially in imperfectly-oriented biopsies, the pathologist should be aware of this fact and he should try to identify the villus tips.

Finally, especially in duodenal bulb biopsies, the epithelium overlying lymphoid tissue patches can normally show an increased IEL count. Coupled with the fact that villi might even be blunted at this location, the epithelium overlying lymphoid patches should not be used to assess these features due to the risk of false positives.

4.1.2 Villous Blunting and Atrophy

The most important factor in the evaluation of villous architecture is orientation of the bioptic sample. However, even with impeccable pre-analytic procedures and religious use of acetate cellulose filters, some samples or some areas of the samples will not be oriented perfectly on the slide. Care should be taken not to overinterpret these areas as atrophic.

As mentioned earlier, some features of the lamina propria and submucosa (namely, lymphoid patches, gastric heterotopia and Brunner's glands nodules) may alter the architecture of the overlying villi to the point of mimicking atrophy. These areas should be avoided in the assessment of villous architecture [1].

Having discussed cases in which IELs and villous architecture are only falsely altered, let us now turn our attention to conditions that can actually cause alteration in these two features.

4.1.3 Differential Diagnosis with Other Pathological Conditions Other Than CD

True intraepithelial lymphocytosis without atrophy (i.e. a Corazza-Villanacci type A lesion) can be caused by a multitude of conditions other than CD (Table 2) [1, 35–37]. Most cases are, in fact, due to causes other than CD [3, 35, 38, 39]. However, atrophic changes can be present in some cases, such as with some drugs (olmesartan and other angiotensin receptor blockers, various immunomodulatory drugs), common variable immunodeficiency, autoimmune enteropathy, Whipple disease and tropical sprue [1, 40].

4.2 Drugs

Several medications, including nonsteroidal anti-inflammatory drugs (NSAIDs), immunomodulatory and antineoplastic drugs, can mimic CD histologically; however, villous atrophy is seldom described in these cases. In addition, the use of NSAIDs has been reported to cause mucosal erosions/ulcerations with an inflammatory infiltrate composed of plasma cells and neutrophils [41]. The use of checkpoint inhibitors or kinase inhibitors has been associated with crypt architectural distortion, neutrophilic infiltration, ischaemic changes, villous blunting, epithelial cell apoptosis in crypts and neutrophilic cryptitis [42, 43]. Angiotensin receptor blockers, and in particular Olmesartan, have been proven to cause partial or complete villous atrophy and intraepithelial lymphocytosis, thus mimicking CD histologically [44].

4.3 Infectious Diseases

Helicobacter pylori (HP) infection may cause an increased IEL count, generally without significant architectural changes [3]. Foveolar metaplasia and the presence of neutrophils may be of help in distinguishing HP-related peptic duodenitis from the microscopic alterations of CD [45].

Intestinal parasitosis may cause an increased IEL count as well as villous blunting. The pathologist should always search for parasites. Giardiasis, caused by Giardia lamblia, is one of the most common intestinal parasitoses. Giardia can be easily identified in duodenal biopsy samples as a pear-shaped organism with two paired nuclei, most often in the luminal debris overlying the epithelium. It does not usually cause significant histologic lesions, even though villous blunting, intraepithelial lymphocytosis, and crypt hyperplasia have been observed rarely in children [46]. It should be underscored that the presence of a parasite does not exclude a concomitant diagnosis of CD, especially if villous atrophy is marked [47].

Small intestinal bacterial overgrowth can also show villous blunting, intraepithelial lymphocytosis, crypt hyperplasia and increased chronic inflammation in the lamina propria [48].

4.4 Immune System Diseases

Food protein-sensitive enteropathies can also reproduce the histologic abnormalities of CD, but they tend to be transient and to respond to removal of the allergen from the diet. In duodenal biopsies from patients with pernicious anaemia, partial villous blunting and increased chronic inflammatory cells may be detected along with the more typical epithelial megaloblastic changes [49].

Collagenous sprue is a rare cause of malabsorption which is often misdiagnosed as CD; however, the identification of a thick subepithelial collagen band with inflammatory cells and capillaries entrapped may help the pathologist in reaching the correct diagnosis [50, 51]. A significant but variable fraction of cases is associated with CD and may be treated with a combination of gluten-free diet and immunosuppressive therapy (Fig. 4a, b).

Common variable immunodeficiency enteropathy (CVID) may mimic CD. However, two peculiar features may be found in duodenal samples of CVID patients which are usually absent in CD: depletion of lamina propria plasma cells and follicular lymphoid hyperplasia [52]. Furthermore, in CVID patients, pathologists should always search for Giardia lamblia accurately, as it was reported in 23% of cases by Malamut et al. [53]. In a minority of CVID patients, villous atrophy is gluten-sensitive [54].

Autoimmune enteropathy, a disease characterized by small intestinal mucosal atrophy and circulating autoantibodies towards enterocytes and/or goblet cells, may show an active enteritis pattern, characterized by expansion of the lamina propria by mixed inflammation with neutrophils, or a CD-like pattern [55–57]. Foci of apoptotic epithelial cells and reduction in goblet and Paneth cells may rarely be observed. Importantly, biopsies from other gastrointestinal sites often show histologic abnormalities and may aid in reaching the correct diagnosis. Lastly, it should be remembered that some forms of idiopathic villous atrophy (villous atrophy or sprue of unknown aetiology) may cause diagnostic challenges. Some of these patients have spontaneous histological recovery and show an excellent survival,

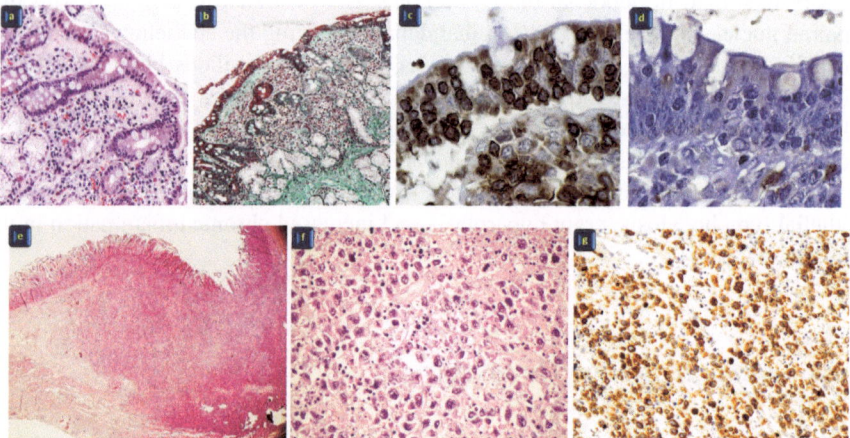

Fig. 4 a, b Collagenous Sprue; pathological increase in the thickness of the connective tissue band under the superficial epithelium >10 μm. **a** H&E x200, **b** Trichrome stain x100. **c, d** Refractory celiac disease; Pathological increase of T lymphocytes that are CD3 positive (**c**) and CD8 negative (**d**). **c, d** x400. **e–g** Enteropathy type T cell lymphoma, **e** H&E x10, **f** H&E x400, **g** CD3 immunostain x400

whereas others show persistent villous atrophy, with or without associated lymphoproliferative disorders [58].

Other immune system diseases that may mimic CD histologically include systemic lupus erythematosus, inflammatory bowel disease (IBD), lymphocytic colitis, eosinophilic gastroenteritis and HIV enteropathy [1, 3, 14].

4.5 Complications of CD

4.5.1 Refractory Celiac Disease (RCD)

CD is considered refractory when duodenal biopsies of a CD patient show persistent villous atrophy, along with malabsorption symptoms, despite a strict adherence to a gluten-free diet (GFD) for at least 12 months [59, 60]. Other causes of persistent villous atrophy must be excluded before diagnosing RCD. Endoscopic abnormalities such as mucosal erosions, ulcerations (ulcerative duodeno-jejunitis) or strictures may be observed.

RCD is a rare CD complication with variable incidence. A systematic review by Rowinski and Christensen [61] showed a cumulative incidence of 1–4% over 10 years and a prevalence of 0.31–0.38% in CD patients, while a study based on a cohort of celiac individuals in Austria reported an incidence over 25 years of 2.6% [62]. Globally, the incidence of RCD seems to have decreased during the last 20 years, probably because of an increase in CD awareness and a stricter adherence to GFD (also thanks to the increased availability of gluten-free products) [63, 64]. The mean age at diagnosis of RCD has been reported to be abound 63 years. Generally, the median time between the diagnosis of CD and the diagnosis of RCD is 21 months, although rare cases of RCD diagnosed at the time of first presentation have been described [65].

On the basis of clinical, histologic and molecular features, two types of RCD have been described.

- Type I RCD is characterized by intraepithelial lymphocytes with a normal immunophenotype (i.e. retained expression of surface CD3, CD8 and CD103) and lacking a monoclonal T cell receptor (TCR) gene rearrangement;
- Type II RCD, on the other hand, is characterized by an aberrant intraepithelial lymphocyte immunophenotype (i.e. >50% of intraepithelial T cells lacking CD8 by immunohistochemistry on formalin-fixed paraffin-embedded sections and/or >20–25% CD45+ T cells lacking surface CD3 on flow cytometry), and a monoclonal TCR gene rearrangement [61].

TCR gene rearrangement clonal analysis by multiple polymerase chain reaction may be efficiently performed also on formalin-fixed paraffin-embedded tissues. Because samples from the duodenal mucosa of healthy, CD or RCD type I patients may occasionally show TCR-β or TCR-γ clonality, the diagnosis and subtyping of

RCD should be only made by a gastroenterologist after an integrated evaluation of clinical information, histology, intraepithelial lymphocyte immunophenotype (by immunohistochemistry or flow cytometry) and clonal analysis [66, 67]. Flow cytometry seems to be better than CD8 immunohistochemistry in differential diagnosis between type I and II RCD. However, a recent study found that immunohistochemical expression of a NK biomarker, NKp46, on the T-cell surface, may help in distinguishing RCD type II (NKp46-positive) from RCD type I, usually showing no or few NKp46-positive T-cells [67]. Histology of RCD type I may be indistinguishable from untreated non-refractory CD; however, a collagenous sprue-type pattern and basal plasmocytosis have rarely been described in RCD type I. Subtyping RCD into type I or type II is very important, because the two diseases show a different prognosis, response to therapy and rate of development of lymphoproliferative malignancies. RCD type I has 5-year survival rates up to 95%, response rate to corticosteroids of 90% and odds of developing enteropathy-associated T-cell lymphoma (EATL) 5 year after RCD diagnosis lower than 14%; on the other hand, RCD type II has a 5-year survival of 58%, lower response rate to corticosteroids and higher rates of developing EATL [61, 68] (Fig. 4c, d).

4.6 Lymphoproliferative Malignancies

CD individuals, especially those with long-standing disease, have a relative risk of developing extra-nodal non-Hodgkin lymphoma approximately 3–4 times higher than the general population [67].

Enteropathy-associated T-cell lymphoma (EATL) is an aggressive malignancy complicating CD, most commonly involving the jejunum and ileum and characterized by markedly atypical malignant cells, densely infiltrating the epithelium (which typically shows severe villous atrophy) and extending in the lamina propria and below the muscularis mucosae. Neoplastic cells are positive for CD3 and CD103, negative for CD5 and CD4, express CD8 variably, contain cytotoxic granule-associated proteins and harbour a clonal rearrangement of TCRγ and/or TCRβ genes [68].

EATL should be distinguished from monomorphic epitheliotropic intestinal T-cell lymphoma (MEITL), which is composed of monomorphic, not significantly atypical, small- to medium-sized T cells, immunoreactive for CD3, CD8, CD56, CD103, and TIA1 and negative for CD5, CD4, and CD30. Similar to EATL, neoplastic T cells infiltrate the lamina propria and epithelium, causing villous atrophy. Although the latest WHO classification of lymphoid neoplasms denied any association of MEITL with CD, it was recently described in two CD patients [69]. Both EATL and MEITL have an ominous prognosis, with a reported 5-year survival rate lower than 20% [70] (Fig. 4e–g).

4.7 Small Bowel Carcinoma (SBC)

The incidence of SBC is increased in CD patients. It primarily affects the jejunum and occurs in patients with a median age of 53 years.

CD-associated SBCs harbour mismatch repair deficiency more frequently in comparison with Crohn's disease-associated or sporadic SBCs [71]. Accordingly, they often show a high number of tumour-infiltrating lymphocytes (TILs) and a subset also shows a medullary-type histology [72]. Importantly, they usually display a relatively indolent behaviour [71]. Recently, Giuffrida and colleagues found that as many as 35% of CD-associated SBCs are PD-L1–positive (combined positive score ≥ 1), paving the way for the usage of immunotherapy in these patients [73, 74].

4.7.1 Liver Complications

Altered liver function can be present in some CD patients that may have tests and/or develop a wide spectrum of liver diseases, encompassing cryptogenic hepatitis, steatohepatitis, cirrhosis, as well as liver autoimmune disorders that in many cases can be evidenced by pathological evaluation in liver biopsies [75].

References

1. Robert ME, Crowe SE, Burgart L, Yantiss RK, Lebwohl B, Greenson JK, et al. Statement on best practices in the use of pathology as a diagnostic tool for celiac disease. Am J Surg Pathol. 2018;42:e44–58. https://doi.org/10.1097/pas.0000000000001107.
2. Lindfors K, Ciacci C, Kurppa K, Lundin KE, Makharia GK, Mearin ML, et al. Coeliac disease. Nat Rev Dis Prim. 2019;5:1–18. https://doi.org/10.1038/s41572-018-0054-z.
3. Caio G, Volta U, Sapone A, Leffler DA, De Giorgio R, Catassi C, et al. Coeliac disease: a comprehensive current review. BMC Med. 2019;17:142. https://doi.org/10.1186/s12916-019-1380-z.
4. Ludvigsson JF, Rubio-Tapia A, van Dyke CT, Melton JL, Zinsmeister AR, Lahr BD, et al. Increasing incidence of coeliac disease in a North American population. Am J Gastroenterol. 2013;108:818–24. https://doi.org/10.1038/ajg.2013.60.
5. Mustalahti K, Catassi C, Reunanen A, Fabiani E, Heier M, McMillan S, et al. The prevalence of coeliac disease in Europe: results of a centralized, international mass screening project. Ann Med. 2010;42:587–95. https://doi.org/10.3109/07853890.2010.505931.
6. Sanders DS, Patel D, Stephenson TJ, Ward AM, McCloskey EV, Hadjivassiliou M, et al. A primary care cross-sectional study of undiagnosed adult coeliac disease. Eur J Gastroenterol Hepatol. 2003;15:407–13. https://doi.org/10.1097/00042737-200304000-00012.
7. Sanders DS, Hurlstone DP, Stokes RO, Rashid F, Milford-Ward A, Hadjivassiliou M, et al. Changing face of adult coeliac disease: experience of a single university hospital in South Yorkshire. Postgrad Med J. 2002;78:31–3. https://doi.org/10.1136/pmj.78.915.31.
8. West J, Fleming KM, Tata LJ, Card TR, Crooks CJ. Incidence and prevalence of celiac disease and dermatitis herpetiformis in the UK over two decades: population-based study. Am J Gastroenterol. 2014;109:757–68. https://doi.org/10.1038/ajg.2014.55.

9. Lo W, Sano K, Lebwohl B, Diamond B, Green PHR. Changing presentation of adult celiac disease. Dig Dis Sci. 2003;48:395–8. https://doi.org/10.1023/a:1021956200382.
10. Husby S, Murray JA, Katzka DA. AGA clinical practice update on diagnosis and monitoring of celiac disease—changing utility of serology and histologic measures: expert review. Gastroenterology. 2019;156:885–9. https://doi.org/10.1053/j.gastro.2018.12.010.
11. Rubio-Tapia A, Hill ID, Kelly CP, Calderwood AH, Murray JA. ACG clinical guidelines: diagnosis and management of celiac disease. Am J Gastroenterol. 2013;108:656–76. https://doi.org/10.1038/ajg.2013.79.
12. Husby S, Koletzko S, Korponay-Szabó IR, Mearin ML, Phillips A, Shamir R, et al. European Society for Pediatric Gastroenterology, Hepatology, and Nutrition guidelines for the diagnosis of coeliac disease. J Pediatr Gastroenterol Nutr. 2012;54:136–60. https://doi.org/10.1097/mpg.0b013e31821a23d0.
13. Hill ID, Dirks MH, Liptak GS, Colletti RB, Fasano A, Guandalini S, et al. Guideline for the diagnosis and treatment of celiac disease in children: recommendations of the North American Society for Pediatric Gastroenterology, Hepatology and Nutrition. J Pediatr Gastroenterol Nutr. 2005;40:1–19. https://doi.org/10.1097/00005176-200501000-00001.
14. Serra S, Jani PA. An approach to duodenal biopsies. J Clin Pathol. 2006;59:1133–50. https://doi.org/10.1136/jcp.2005.031260.
15. Pais WP, Duerksen DR, Pettigrew NM, Bernstein CN. How many duodenal biopsy specimens are required to make a diagnosis of celiac disease? Gastrointest Endosc. 2008;67:1082–7. https://doi.org/10.1016/j.gie.2007.10.015.
16. Latorre M, Lagana SM, Freedberg DE, Lewis SK, Lebwohl B, Bhagat G, et al. Endoscopic biopsy technique in the diagnosis of celiac disease: one bite or two? Gastrointest Endosc. 2015;81:1228–33. https://doi.org/10.1016/j.gie.2014.10.024.
17. Bonamico M, Thanasi E, Mariani P, Nenna R, Luparia RP, Barbera C, et al. Duodenal bulb biopsies in celiac disease: a multicenter study. J Pediatr Gastroenterol Nutr. 2008;47:618–22. https://doi.org/10.1097/mpg.0b013e3181677d6e.
18. Evans KE, Aziz I, Cross SS, Sahota GR, Hopper AD, Hadjivassiliou M, et al. A prospective study of duodenal bulb biopsy in newly diagnosed and established adult celiac disease. Am J Gastroenterol. 2011;106:1837–2742. https://doi.org/10.1038/ajg.2011.171.
19. Gonzalez S, Gupta A, Cheng J, Tennyson C, Lewis SK, Bhagat G, et al. Prospective study of the role of duodenal bulb biopsies in the diagnosis of celiac disease. Gastrointest Endosc. 2010;72:758–65. https://doi.org/10.1016/j.gie.2010.06.026.
20. De Leo L, Villanacci V, Ziberna F, Vatta S, Martelossi S, Di Leo G, et al. Immunohistologic analysis of the duodenal bulb: a new method for celiac disease diagnosis in children. Gastrointest Endosc. 2018;88:521–6. https://doi.org/10.1016/j.gie.2018.05.014.
21. Villanacci V, Ceppa P, Tavani E, Vindigni C, Volta U, Gruppo Italiano Patologi Apparato Digerente (GIPAD), et al. Coeliac disease: the histology report. Dig Liver Dis. 2011;43 Suppl 4:S385–95. https://doi.org/10.1016/S1590-8658(11)60594-X.
22. Hayat M, Cairns A, Dixon M, O'Mahony S. Quantitation of intraepithelial lymphocytes in human duodenum: what is normal? J Clin Pathol. 2002;55:393–4. https://doi.org/10.1136/jcp.55.5.393.
23. Veress B, Franzén L, Bodin L, Borch K. Duodenal intraepithelial lymphocyte-count revisited. Scand J Gastroenterol. 2004;39:138–44. https://doi.org/10.1080/00365520310007675.
24. Marsh MN. Grains of truth: evolutionary changes in small intestinal mucosa in response to environmental antigen challenge. Gut. 1990;31:111–4. https://doi.org/10.1136/gut.31.1.111.
25. Oberhuber G, Granditsch G, Vogelsang H. The histopathology of coeliac disease: time for a standardized report scheme for pathologists. Eur J Gastroenterol Hepatol. 1999;11:1185–94. https://doi.org/10.1097/00042737-199910000-00019.
26. Corazza GR, Villanacci V. Coeliac disease. J Clin Pathol. 2005;58:573–4. https://doi.org/10.1136/jcp.2004.023978.
27. Corazza GR, Villanacci V, Zambelli C, Milione M, Luinetti O, Vindigni C, et al. Comparison of the interobserver reproducibility with different histologic criteria used in celiac disease. Clin Gastroenterol Hepatol. 2007;5:838–43. https://doi.org/10.1016/j.cgh.2007.03.019.

28. Villanacci V. The histological classification of biopsy in celiac disease: time for a change? Dig Liver Dis. 2015;47:2–3. https://doi.org/10.1016/j.dld.2014.09.022.

29. Shidrawi RG, Przemioslo R, Davies DR, Tighe MR, Ciclitira PJ. Pitfalls in diagnosing coeliac disease. J Clin Pathol. 1994;47:693–4. https://doi.org/10.1136/jcp.47.8.693.

30. Eigner W, Wrba F, Chott A, Bashir K, Primas C, Eser A, et al. Early recognition of possible pitfalls in histological diagnosis of celiac disease. Scand J Gastroenterol. 2015;50:1088–93. https://doi.org/10.3109/00365521.2015.1017835.

31. Freeman HJ. Pearls and pitfalls in the diagnosis of adult celiac disease. Can J Gastroenterol. 2008;22:273–80. https://doi.org/10.1155/2008/905325.

32. Ravelli A, Villanacci V. Tricks of the trade: How to avoid histological pitfalls in celiac disease. Pathol Res Pract. 2012;208:197–202. https://doi.org/10.1016/j.prp.2012.01.008.

33. Fasano A, Sapone A, Zevallos V, Schuppan D. Nonceliac gluten sensitivity. Gastroenterology. 2015;148:1195–204. https://doi.org/10.1053/j.gastro.2014.12.049.

34. Zanini B, Villanacci V, Marullo M, Cadei M, Lanzarotto F, Bozzola A, et al. Duodenal histological features in suspected non-celiac gluten sensitivity: new insights into a still undefined condition. Virchows Arch. 2018;473:229–34. https://doi.org/10.1007/s00428-018-2346-9.

35. Kakar S, Nehra V, Murray JA, Dayharsh GA, Burgart LJ. Significance of intraepithelial lymphocytosis in small bowel biopsy samples with normal mucosal architecture. Am J Gastroenterol. 2003;98:2027–33. https://doi.org/10.1111/j.1572-0241.2003.07631.x.

36. Memeo L, Jhang J, Hibshoosh H, Green PH, Rotterdam H, Bhagat G. Duodenal intraepithelial lymphocytosis with normal villous architecture: common occurrence in H. pylori gastritis. Mod Pathol. 2005;18:1134–44. https://doi.org/10.1038/modpathol.3800404.

37. Yousef MM, Yantiss RK, Baker SP, Banner BF. Duodenal intraepithelial lymphocytes in inflammatory disorders of the esophagus and stomach. Clin Gastroenterol Hepatol. 2006;4:631–4. https://doi.org/10.1016/j.cgh.2005.12.028.

38. Brown I, Mino-Kenudson M, Deshpande V, Lauwers GY. Intraepithelial lymphocytosis in architecturally preserved proximal small intestinal mucosa: an increasing diagnostic problem with a wide differential diagnosis. Arch Pathol Lab Med. 2006;130:1020–5.

39. Biagi F, Bianchi PI, Campanella J, Badulli C, Martinetti M, Klersy C, et al. The prevalence and the causes of minimal intestinal lesions in patients complaining of symptoms suggestive of enteropathy: a follow-up study. J Clin Pathol. 2008;61:1116–8. https://doi.org/10.1136/jcp.2008.060145.

40. Dai Y, Zhang Q, Olofson AM, Jhala N, Liu X. Celiac disease: updates on pathology and differential diagnosis. Adv Anat Pathol. 2019;26:292–312. https://doi.org/10.1097/pap.0000000000000242.

41. Bjarnason I, Zanelli G, Smith T, Prouse P, Williams P, Smethurst P, et al. Nonsteroidal antiinflammatory drug-induced intestinal inflammation in humans. Gastroenterology. 1987;93:480–9. https://doi.org/10.1016/0016-5085(87)90909-7.

42. Louie CY, DiMaio MA, Matsukuma KE, Coutre SE, Berry GJ, Longacre TA. Idelalisib-associated enterocolitis: clinicopathologic features and distinction from other enterocolitides. Am J Surg Pathol. 2015;39:1653–60. https://doi.org/10.1097/pas.0000000000000525.

43. Gonzalez RS, Salaria SN, Bohannon CD, Huber AR, Feely MM, Shi C. PD-1 inhibitor gastroenterocolitis: case series and appraisal of 'immunomodulatory gastroenterocolitis.' Histopathology. 2017;70:558–67. https://doi.org/10.1111/his.13118.

44. Rubio-Tapia A, Herman ML, Ludvigsson JF, Kelly DG, Mangan TF, Wu TT, et al. Severe spruelike enteropathy associated with olmesartan. Mayo Clin Proc. 2012;87:732–8. https://doi.org/10.1016/j.mayocp.2012.06.003.

45. Ensari A. Gluten-sensitive enteropathy (celiac disease): controversies in diagnosis and classification. Arch Pathol Lab Med. 2010;134:826–36. https://doi.org/10.1043/1543-2165-134.6.826.

46. Koot BG, ten Kate FJ, Juffrie M, Rosalina I, Taminiau JJ, Benninga MA. Does Giardia lamblia cause villous atrophy in children?: a retrospective cohort study of the histological

abnormalities in giardiasis. J Pediatr Gastroenterol Nutr. 2009;49:304–8. https://doi.org/10. 1097/mpg.0b013e31818de3c4.

47. Hanevik K, Wik E, Langeland N, Hausken T. Transient elevation of anti-transglutaminase and anti-endomysium antibodies in Giardia infection. Scand J Gastroenterol. 2018;53:809–12. https://doi.org/10.1080/00365521.2018.1481522.

48. King CE, Toskes PP. Small intestine bacterial overgrowth. Gastroenterology. 1979;76:1035–55.https://doi.org/10.1016/s0016-5085(79)91337-4.

49. Bianchi A, Chipman DW, Dreskin A, Rosensweig NS. Nutritional folic acid deficiency with megaloblastic changes in the small-bowel epithelium. New Engl J Med. 1970;282:859–61. https://doi.org/10.1056/nejm197004092821510.

50. Maguire AA, Greenson JK, Lauwers GY, Ginsburg RE, Williams GT, Brown IS, et al. Collagenous sprue: a clinicopathologic study of 12 cases. Am J Surg Pathol. 2009;33:1440–9. https://doi.org/10.1097/pas.0b013e3181ae2545.

51. Vakiani E, Arguelles-Grande C, Mansukhani MM, Lewis SK, Rotterdam H, Green PH, et al. Collagenous sprue is not always associated with dismal outcomes: a clinicopathological study of 19 patients. Mod Pathol. 2009;23:12–26. https://doi.org/10.1038/modpathol.2009.151.

52. Daniels JA, Lederman HM, Maitra A, Montgomery EA. Gastrointestinal tract pathology in patients with common variable immunodeficiency (CVID): a clinicopathologic study and review. Am J Surg Pathol. 2007;31:1800–12. https://doi.org/10.1097/pas.0b013e3180cab60c.

53. Malamut G, Verkarre V, Suarez F, Viallard JF, Lascaux AS, Cosnes J, et al. The enteropathy associated with common variable immunodeficiency: the delineated frontiers with celiac disease. Am J Gastroenterol. 2010;105:2262–75. https://doi.org/10.1038/ajg.2010.214.

54. Biagi F, Bianchi PI, Zilli A, Marchese A, Luinetti O, Lougaris V, et al. The significance of duodenal mucosal atrophy in patients with common variable immunodeficiency. Am J Clin Pathol. 2012;138:185–9. https://doi.org/10.1309/ajcpeiilh2c0wfye.

55. Akram S, Murray JA, Pardi DS, Alexander GL, Schaffner JA, Russo PA, et al. Adult autoimmune enteropathy: Mayo Clinic Rochester experience. Clin Gastroenterol Hepatol. 2007;5:1282–90. https://doi.org/10.1016/j.cgh.2007.05.013.

56. Masia R, Peyton S, Lauwers GY, Brown I. Gastrointestinal biopsy findings of autoimmune enteropathy. Am J Surg Pathol. 2014;38:1319–29. https://doi.org/10.1097/pas. 0000000000000317.

57. Villanacci V, Lougaris V, Ravelli A, Buscarini E, Salviato T, Lionetti P, et al. Clinical manifestations and gastrointestinal pathology in 40 patients with autoimmune enteropathy. Clin Immunol. 2019;207:10–7. https://doi.org/10.1016/j.clim.2019.07.001.

58. Schiepatti A, Sanders DS, Aziz I, De Silvestri A, Goodwin J, Key T, et al. Clinical phenotype and mortality in patients with idiopathic small bowel villous atrophy: a dual-centre international study. Eur J Gastroenterol Hepatol. 2020;32:938–49. https://doi.org/10.1097/ meg.0000000000001726.

59. Malamut G, Afchain P, Verkarre V, Lecomte T, Amiot A, Damotte D, et al. Presentation and long-term follow-up of refractory celiac disease: comparison of type I with type II. Gastroenterology. 2009;136:81–90. https://doi.org/10.1053/j.gastro.2008.09.069.

60. Hujoel IA, Murray JA. Refractory celiac disease. Curr Gastroenterol Rep. 2020;22:18. https:// doi.org/10.1007/s11894-020-0756-8.

61. Rowinski SA, Christensen E. Epidemiologic and therapeutic aspects of refractory coeliac disease—a systematic review. Dan Med J. 2016;63:A5307.

62. Eigner W, Bashir K, Primas C, Kazemi-Shirazi L, Wrba F, Trauner M, et al. Dynamics of occurrence of refractory coeliac disease and associated complications over 25 years. Aliment Pharmacol Ther. 2017;45:364–72. https://doi.org/10.1111/apt.13867.

63. Hussein S, Gindin T, Lagana SM, Arguelles-Grande C, Krishnareddy S, Alobeid B, et al. Clonal T cell receptor gene rearrangements in coeliac disease: implications for diagnosing refractory coeliac disease. J Clin Pathol. 2018;71:825–31. https://doi.org/10.1136/jclinpath-2018-205023.

64. Celli R, Hui P, Triscott H, Bogardus S, Gibson J, Hwang M, et al. Clinical insignficance of monoclonal T-cell populations and duodenal intraepithelial T-cell phenotypes in celiac and

nonceliac patients. Am J Surg Pathol. 2019;43:151–60. https://doi.org/10.1097/PAS. 0000000000001172.

65. Cheminant M, Bruneau J, Malamut G, Sibon D, Guegan N, van Gils T, et al. NKp46 is a diagnostic biomarker and may be a therapeutic target in gastrointestinal T-cell lymphopro-liferative diseases: a CELAC study. Gut. 2019;68:1396–405. https://doi.org/10.1136/gutjnl-2018-317371.

66. Al-Toma A, Verbeek WH, Hadithi M, von Blomberg BM, Mulder CJ. Survival in refractory coeliac disease and enteropathy-associated T-cell lymphoma: retrospective evaluation of single-centre experience. Gut. 2007;56:1373–8. https://doi.org/10.1136/gut.2006.114512.

67. West J. Celiac disease and its complications: a time traveller's perspective. Gastroenterology. 2009;136:32–4. https://doi.org/10.1053/j.gastro.2008.11.026.

68. Swerdlow SH, Campo E, Harris NL, Jaffe ES, Pileri SA, Stein H, et al. WHO classification of tumours of haematopoietic and lymphoid tissues, vol. 2, Revised 4th ed. Lyon: IARC; 2017.

69. Lenti MV, Biagi F, Lucioni M, Di Sabatino A, Paulli M, Corazza GR. Two cases of monomorphic epitheliotropic intestinal T-cell lymphoma associated with coeliac disease. Scand J Gastroenterol. 2019;54:965–8. https://doi.org/10.1080/00365521.2019.1647455.

70. Nijeboer P, Malamut G, Mulder CJ, Cerf-Bensussan N, Sibon D, Bouma G, et al. Enteropathy-associated T-cell lymphoma: improving treatment strategies. Dig Dis. 2015;33:231–5. https://doi.org/10.1159/000369542.

71. Potter DD, Murray JA, Donohue JH, Burgart LJ, Nagorney DM, van Heerden JA, et al. The role of defective mismatch repair in small bowel adenocarcinoma in celiac disease. Cancer Res. 2004;64:7073–7. https://doi.org/10.1158/0008-5472.can-04-1096.

72. Vanoli A, Di Sabatino A, Furlan D, Klersy C, Grillo F, Fiocca R, et al. Small bowel carcinomas in coeliac or Crohn's disease: clinico-pathological, molecular, and prognostic features. A study from the small bowel cancer Italian consortium. J Crohns Colitis. 2017;11:942–53. https://doi.org/10.1093/ecco-jcc/jjx031.

73. Vanoli A, Di Sabatino A, Martino M, Klersy C, Grillo F, Mescoli C, et al. Small bowel carcinomas in celiac or Crohn's disease: distinctive histophenotypic, molecular and histogenetic patterns. Mod Pathol. 2017;30:1453–66. https://doi.org/10.1038/modpathol. 2017.40.

74. Giuffrida P, Arpa G, Grillo F, Klersy C, Sampietro G, Ardizzone S, et al. PD-L1 in small bowel adenocarcinoma is associated with etiology and tumor-infiltrating lymphocytes, in addition to microsatellite instability. Mod Pathol. 2020;33:1398–409. https://doi.org/10.1038/s41379-020-0497-0.

75. Majumdar K, Sakhuja P, Puri AS, Gaur K, Haider A, Gondal R. Coeliac disease and the liver: spectrum of liver histology, serology and treatment response at a tertiary referral centre. J Clin Pathol. 2018;71:412–9. https://doi.org/10.1136/jclinpath-2017-204647.

Value and Use of Genetic Test of Celiac Disease

Concepción Núñez and Mercedes Rubio

1 Introduction

Celiac disease (CD) is characterized by an auto-immune disease leading to en-teropathy, triggered by dietary gluten ingestion in genetically susceptible individ-uals. Gluten is the environmental factor necessary to develop the disease, but genetics also constitutes a key element outlining the set of individuals prone to present CD. Since the first studies describing a genetic association with the Human Leukocyte Antigen (HLA) complex, numerous genes have been associated with CD susceptibility, delineating a complex pattern of inheritance [1–5]. However, con-trasting with the small contribution of the described DNA polymorphisms, the HLA locus has a major role in the disease, with the classical HLA associated variants accounting for 22% of CD heritability [6].

A genetic test is included in the work-up of CD, but currently it is only based on HLA genotyping. New approaches are being explored and we expect that additional genetic variants can be included in the near future, refining the exclusion criteria and improving CD risk prediction of the test.

2 HLA Genetics in Celiac Disease

2.1 Background

The HLA locus lies on the short arm of chromosome 6 (6p21) and harbours hundreds of genes that are broadly grouped into three regions: class I, class II and class III. Besides being highly polygenic, this region possesses other notorious

C. Núñez (✉) · M. Rubio
Laboratory of Research in Genetics of Complex Diseases, Clínico San Carlos Hospital,
Health Research Institute of the Clínico San Carlos Hospital (IdISSC), 28040 Madrid, Spain

© The Author(s), under exclusive license to Springer Nature Switzerland AG 2022   99
J. Amil-Dias and I. Polanco (eds.), *Advances in Celiac Disease*,
https://doi.org/10.1007/978-3-030-82401-3_8

features, such as the extremely high polymorphism and the co-dominance inheritance. These properties ensure a high diversity of molecules that allow binding a huge variety of peptides with diverse preference depending on their biochemical features. Many HLA products are expressed on the surface of different cell types and are involved in self- and foreign-antigen recognition, orchestrating cellular and humoral immune responses.

HLA class II genes encode cell-surface glycoproteins that exhibit restricted cell expression, being predominantly present on the surface of antigen-presenting cells (APCs), such as macrophages, dendritic cells and B lymphocytes, as well as other cells under certain anomalous conditions. They bind peptides mainly derived from extracellular proteins to present to CD4$^+$ T cells and play an integral role in the adaptive immune responses and maintenance of self-tolerance. These genes have been associated with numerous autoimmune diseases since the early 1970s, including CD [1, 2]. Specifically, *HLA-DQA1* and *HLA-DQB1* were later pinpointed as the main predisposing genes to CD [7].

Despite the high polymorphism, HLA shows an extensive linkage disequilibrium, which makes that the inherited allele combinations (haplotypes) observed in human populations are much smaller than theoretical expectations. The pattern of linkage disequilibrium varies between populations, which must be considered when studying disease susceptibility. In CD, the analysis of the haplotypes and genotypes conformed by *HLA-DQA1* and *HLA-DQB1* variants is required to understand CD risk.

2.2 Role of HLA in Celiac Disease Pathogenesis

HLA-DQA1 and *HLA-DQB1* encode the α and β chains, respectively, that conform the heterodimeric HLA-DQ molecule (Fig. 1). The specific alleles present in these genes determine the particular HLA-DQ receptors present in each individual and the range of peptides to be bound. The HLA alleles conferring CD risk encode HLA-DQ heterodimers with peptide-binding grooves showing a high affinity for negatively charged peptides, as the ones originated after deamidation (conversion of glutamine residues to glutamic acid) by the transglutaminase type 2 enzyme of gluten peptides that survived gastrointestinal digestion [8, 9]. In CD patients, HLA-DQ-gluten complexes are recognized by CD4$^+$ T cells and elicit gluten-specific T cell responses [10, 11]. This gluten-induced activation of the immune system will also depend on the kinetic stability of those complexes. HLA-DQ receptors showing a higher ability to form stable complexes with gluten T-cell epitopes will contribute to higher CD risk [12, 13].

This central role in CD pathogenesis makes that certain HLA variants are considered necessary to develop CD, being translated meaningfully into clinical practice: the absence of permissive genetics allows excluding CD diagnosis and forms the rationale basis of the genetic test.

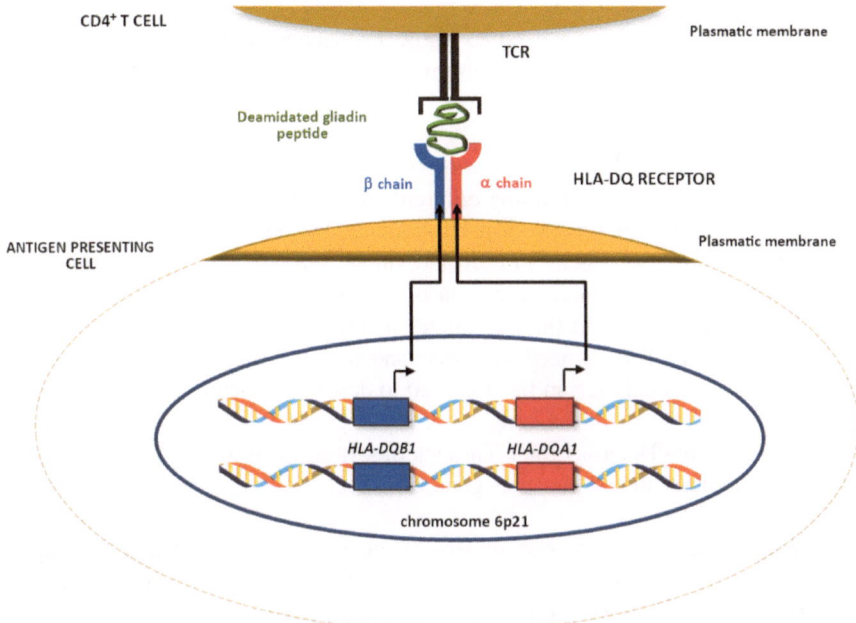

Fig. 1 The HLA-DQ receptor is conformed by two juxtaposed subunits, the α and the β chains, which are encoded by the *HLA-DQA1* and *HLA-DQB1* genes, respectively. This receptor is present on the surface of antigen presenting cells and can bind gluten-derived antigens to activate CD4⁺ T cells through TCR recognition, thus eliciting the immunological reaction leading to celiac disease. Transglutaminase type 2-mediated deamidation increases the affinity of HLA-DQ receptors by gluten peptides

2.3   HLA Genetic Risk Variants

HLA-DQA1 and *HLA-DQB1* allele combinations encoding the HLA-DQ receptors eliciting gluten-specific immune responses constitute the genetic risk variants for CD development (Fig. 2). For a better understanding of the genetic terms and the HLA nomenclature that is used below, see Boxes 1 and 2.

Box 1. Basic terms in genetics

A lasting understanding of the genetic test requires being familiar with some specific terms. We present them in a simplified form in order to facilitate comprehension.

Gene(s)/Locus (loci). Gene refers to a DNA segment that constitutes the functional and physical unit of inheritance. Locus is the physical site of the gene along the chromosome. There is a subtle distinction between them and they can be synonymous in some contexts.

In practical terms, we need to know that genes contain encoded information to make a protein (although this is not true for all genes) and they are passed from parents to offspring. We have two copies of each gene: each one inherited from each parent.

Allele. They are the variant forms of a gene. Some genes only have one variant and others have multiple different alleles, being this last situation the most commonly one observed in HLA genes. Despite this, only two alleles can be present at one gene in each individual. Different alleles may or may not have functional consequences on the encoded product.

Genotype. It refers to the two alleles inherited for a single gene or set of genes. This is what we need to understand in relation to the genetic test of CD, although genotype can be also used to describe the entire set of genes in an individual.

Homozygote/Heterozygote. These terms indicate if the individual have both inherited alleles in one gene identical (homozygote) or different (heterozygote).

Haplotype. Related to our subject of interest, we will consider the haplotype as the combination of alleles at different genes found in the same chromosome and inherited together from each parent.

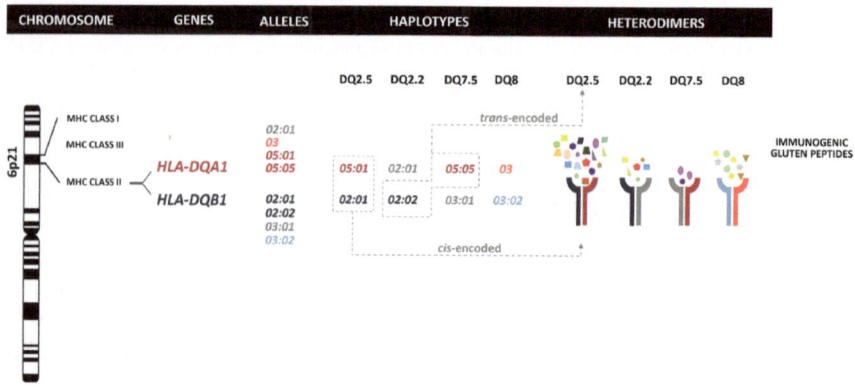

Fig. 2 HLA genetic risk for celiac disease (CD). *HLA-DQA1* and *HLA-DQB1*, on chromosome 6p21, constitute the main HLA predisposing genes to CD. Risk alleles are those encoding the heterodimeric receptors HLA-DQ2.5, HLA-DQ2.2, HLA-DQ7.5 and HLA-DQ8, which are mainly present in the haplotypes receiving the same name. One exception is HLA-DQ2.5 that can be formed by inheritance of the HLA-DQ2.5 haplotype (*cis* configuration) or the HLA-DQ2.2 and HLA-DQ7.5 haplotypes (*trans* configuration). Different HLA-DQ receptors can bind diverse repertoires of immunogenic gluten peptides

Cis/trans configuration. They indicate if the alleles present in different genes have been inherited on the same chromosome (*cis*) or on opposite chromosomes (*trans*). In practical terms, we will refer to *cis* inheritance for alleles constituting a parental haplotype (inherited from a single parent) and to *trans* inheritance for alleles received from both parents.

Polymorphism. It means "several forms" and refers to several alternative states of DNA. It can be defined at different scales. Genetic association studies commonly consider the smallest unit composing DNA: the nucleotide site. However, it can be used as a synonymous of allele.

Tagging SNP or tag SNP. It defines a SNP that proxies for a set of SNPs or other type of variants in linkage disequilibrium with itself, thus allowing to predict other variants in the region with small probability of error. A set of tagging SNPs are commonly used to uniquely define the variation that reside in a haplotype block, leading to talk about haplotype tagging SNPs.

Box 2. HLA nomenclature

The HLA complex harbour numerous genes, most of them characterized by being highly polymorphic. This makes that new alleles are constantly being reported and implies the need for a systematic nomenclature.

HLA allele names include the HLA prefix, followed by the hyphen-separated name of the specific gene and, since 2012, up to four sets of digits separated by colons (:) [63]. An asterisk (*) is used as a separator before the numerical allele designation. The first two digits describe the allele group or allele family, which often corresponds to the serological antigen carried by the allotype, although consistent linkage between allele names assigned on the basis of nucleotide sequences and the serological profiles of the encoded proteins is not always possible. The next two digits describe the specific allele. Longer names are not always necessary, since fifth to eight digits indicate synonymous nucleotide substitutions within the coding region or changes in non-coding regions, respectively. Specifically, last four digits are not necessary to understand HLA risk genetics to CD and the nomenclature of interest for us can be summarized as follows: HLA prefix-name of the particular HLA locus*two digits indicating the allele group: two digits indicating the specific HLA allele.

Alleles that differ in the first four digits show one or more nucleotide substitutions that change the amino-acid sequence of the encoded protein.

To differentiate HLA alleles of their coding proteins, it is recommended to use uppercase and italicized letters for gene symbols and a similar nomenclature but not italicized for protein symbols.

The strongest association with CD susceptibility has been described with the alleles of the family *HLA-DQA1*05* and *HLA-DQB1*02*, which encode the HLA-DQ2.5 receptor. These alleles can be inherited in *cis* configuration, by carrying the HLA-DQ2.5 haplotype, characterized by the alleles *HLA-DQA1*05:01* and *HLA-DQB1*02:01;* or in *trans* configuration, mainly by the presence of the alleles *HLA-DQA1*05:05* and *HLA-DQB1*02:02,* most commonly on haplotypes HLA-DQ7.5 and HLA-DQ2.2, respectively. The HLA-DQ receptors resultant from a *cis* or *trans* inheritance differ in residues outside the peptide-binding groove and confer similar risk to CD. HLA-DQ2.5 is the predominant receptor in CD, being present in around 86–95% of the patients [14].

The haplotype HLA-DQ8 is also associated with CD susceptibility. It is conformed by the alleles *HLA-DQA1*03* (being *HLA-DQA1*03:01* and *HLA-DQA1*03:02* the most frequent alleles in populations of European ancestry) and *HLA-DQB1*03:02*. It encodes the HLA-DQ8 receptor, which appears in approximately 4–8% of CD patients lacking HLA-DQ2.5.

In absence of the allele combinations encoding the HLA-DQ2.5 and HLA-DQ8 receptors, CD is mainly characterized by the presence of *HLA-DQB1*02:02*. This allele is most commonly found accompanied by *HLA-DQA1*02:01* in the haplotype HLA-DQ2.2, giving rise to the heterodimer that receives the same name (HLA-DQ2.2).

HLA-DQA1*05 has been described in almost all the patients lacking the aforementioned HLA genetic risk variants. It is commonly observed as part of the HLA-DQ7.5 haplotype, which contains the alleles *HLA-DQA1*05:05* and *HLA-DQB1*03:01* and encodes the HLA-DQ7.5 receptor. Several works question its involvement in CD due to its very low frequency, but its presence in some CD patients is well documented [15–17].

HLA-DQB1*02 and *HLA-DQA1*05* encode each one of the two chains required to form HLA-DQ2.5. In European populations, the frequency of carrying only one of these alleles has been reported as ranging from 3 to 10% of the total CD (excluding their co-expression with HLA-DQ2.5 or HLA-DQ8), most cases corresponding to *HLA-DQB1*02* carriers, who comprise 55 to 100% of non-HLA-DQ2.5/DQ8 CD patients [15, 16].

CD has also been reported in patients lacking all the above-described HLA variants, although this is an extremely rare event. In those cases, no clear association has been found with any previously unknown HLA risk allele, except for *HLA-DQA1*03-HLA-DQB1*03:03*, present in the haplotype HLA-DQ9.3 and described as a CD susceptibility factor in China, where it reaches considerable frequency [18].

2.4 Risk Gradient

Different risk to develop CD has been described depending on the individual HLA-DQ genotype, which is commonly explained according to a gene dosage model (Table 1). The highest risk is conferred by carrying the HLA-DQ2.5 heterodimer in presence of two copies of the *HLA-DQB1*02* allele [19–22]. This corresponds to subjects homozygous for the HLA-DQ2.5 haplotype or heterozygous HLA-DQ2.5/HLA-DQ2.2. The following risk grade appears in HLA-DQ2.5 subjects with one copy of *HLA-DQB1*02*, either by the inheritance of the HLA-DQ2.5 receptor encoded in *trans* or by the presence of one HLA-DQ2.5 haplotype. In this last group, some authors have described several grades of risk depending on the accompanying haplotype. When the HLA-DQ2.5 heterodimer is absent, moderate risk is conferred by the presence of HLA-DQ8 or HLA-DQ2.2. A dosage effect for these two haplotypes has been described in some populations, but not in others. Accordingly, patients carrying them in homozygosis or compound heterozygotes HLA-DQ8/HLA-DQ2.2 could show higher risk and similar to the ascribed to some HLA-DQ2.5 heterozygous subjects [16, 17, 23]. A final category includes the presence of the *HLA-DQA1*05* allele alone, with a non-significantly different frequency between CD and the general population, but being present in almost the totality of subjects lacking the other known HLA-risk alleles.

The discrepancies between studies involving the categories of intermediate risk are probably reflecting diversity in the studied populations, which may differ in the HLA frequencies. As a matter of fact, HLA frequencies show a north–south gradient in Europe. These differences can be especially relevant if affecting to putative risk factors additional to the described HLA-DQ alleles [6, 24, 25]. It must be also noted the high complexity of the HLA region, with haplotype interactions and complex regulation described in different autoimmune diseases including CD [26–28]. In addition, divergence between studies may be partially due to the characteristics of the studied group (children/adults, proportion of patients belonging to risk groups, rate of diagnosed/undiagnosed cases...) [29].

Several works have contributed to explain the underlying cause of the CD risk stratification based on HLA genetics. As previously stated in this manuscript, the resistance of gluten-derived peptides to proteolytic degradation as well as the ability of TG2 to modify these peptides influence the selection of gluten T-cell epitopes, but this selection ultimately relies on the HLA, considering both the affinity HLA-DQ-gluten sequence and the stability of that complex. HLA-DQ2.5 heterodimers bind a larger gluten peptide repertoire and show higher ability to form stable complexes than the other HLA-DQ receptors, which also differ regarding those properties [12, 13, 30–32].

Up to four different HLA-DQ receptors can be found in every subject since the coding products of the four potentially available *HLA-DQA1* and *HLA-DQB1* alleles can be combined in *cis* or in *trans* to form the HLA-DQ heterodimers. Accordingly, the individual *HLA-DQ* genotype (Table 1) will determine the APC surface density of "risk" heterodimers and the chance of developing CD. In addition

Table 1 Celiac disease risk stratification based on the individual *HLA-DQ* genotypes

HLA-DQA1-HLA-DBQ1 haplotypes	*DQA1*05-DQB1*02* (HLA-DQ2.5)	*DQA1*02-DQB1*02* (HLA-DQ2.2)	*DQA1*05-DQB1*03:01* (HLA-DQ7.5)	*DQA1*03-DQB1*03:02* (HLA-DQ8)	Others
DQA1*05-DQB1*02 (HLA-DQ2.5)	Very high	Very high	High	High	High
DQA1*02-DQB1*02 (HLA-DQ2.2)		Moderate/high	High[a]	Moderate/high	Moderate
DQA1*05-DQB1*03:01 (HLA-DQ7.5)			Low	Moderate	Low
DQA1*03-DQB1*03:02 (HLA-DQ8)				Moderate/high	Moderate
Others					None[b]

[a]HLA-DQ2.5 receptor encoded in *trans*

[b]*HLA-DQA1*03-DQB1*03:03* has been described as susceptibility factor to celiac disease in China

Different alleles of the *HLA-DQA1*05, HLA-DQA1*03* and *HLA-DBQ1*02* (simplified as *DQA1*05, DQA1*03* and *DQB1*02* in the Table) family can be found in patients with celiac disease, being *HLA-DQA1*05:01, HLA-DQA1*05:05, HLA-DQA1*03:01, HLA-DQA1*03:02, HLA-DQB1*02:01* and *HLA-DQB1*02:02* the most frequent alleles in populations of European ancestry

Moderate/High is used for genotypes with non-uniform data in the literature. Differences across populations are probably present

to this dosage model, recent studies have suggested that risk stratification could be explained according to a gene expression model [33]. They are based on the preferential expression of the *HLA-DQA1*05:01* and *HLA-DQB1*02:01* alleles with respect to non-risk alleles observed in APCs obtained from heterozygous subjects. A higher rate of transcription of the risk alleles resulted in higher expression of the α and β -DQ chains and gave a comparable amount of -DQ2.5 heterodimers on the cell surface of APCs in DQ2.5 homozygous or heterozygous subjects [34, 35]. Differences in gene expression of the risk alleles were also observed between CD patients and controls, being lower in this last group.

2.5 Genetic Test

2.5.1 Rationale Basis

The very unlikely development of CD in absence of the HLA described genetic variants constitutes the rationale basis of the genetic test of CD, which can be used for a lifelong exclusion of the disease. However, although present in almost the totality of CD patients, those variants also appear in around 30–40% of the general population, at least in those of European descent [14]. Therefore, the presence of those specific *HLA-DQ* allele combinations is considered necessary but not sufficient for CD onset. This confers HLA testing a very high negative predictive value, of almost 100%, but it cannot be used to confirm CD.

2.5.2 Performance

Full genotyping of *HLA-DQA1* and *HLA-DQB1* or determination of the alleles associated with CD risk (*HLA-DQA1*05, HLA-DQA1*03, HLA-DQB1*02* and *HLA-DQB1*03:02*) constitute the genetic test of CD. Traditionally, *HLA-DRB1* genotyping was included in the study due to the high linkage disequilibrium with the *HLA-DQ* loci. Nowadays, only *HLA-DQA1* and *HLA-DQB1* are studied since they are the genes involved in CD susceptibility.

HLA risk alleles can be determined using different technical methods [36, 37]. Those based on PCR coding sequence, either considering full *HLA-DQA1* and *HLA-DB1* genotyping or just by determining the specific associated variants, are preferred to single nucleotide polymorphism (SNP)-based approaches for individual patient management. Correlation of tagging SNPs and HLA genotypes may not reach 100% and may show variations depending on the linkage disequilibrium pattern of the studied population.

The genetic test can be performed using blood or saliva and it needs to be performed just once in the lifetime, as it is not affected by time or type of diet (normal or gluten-free diet).

2.6   Genetic Report

The genetic information obtained for each individual must be incorporated and properly interpreted into the genetic report of CD, taking special care in including:

1. Full genotyping of *HLA-DQA1* and *HLA-DQB1*, providing information of the four alleles present in each individual and preferably indicating the specific methodology used for genotyping. Alternatively, determination of only the four alleles involved in CD risk can be performed: *HLA-DQA1*05*, *HLA-DQA1*03*, *HLA-DQB1*02* and *HLA-DQB1*03:02*.
2. A reference to the four aforementioned HLA alleles specifically denoting their presence/absence. Note that CD risk exists when both or only one of the HLA-DQ2.5-encoding alleles are present, but both alleles encoding HLA-DQ8 must be present to confer CD risk.
3. The clinical interpretation of the genetic results. The presence or absence of the HLA-DQ2.5, HLA-DQ8 and HLA-DQ2.2 heterodimers must be indicated. When absent, the report should indicate the presence or absence of the *HLA-DQA1*05* allele. Additionally, it must be clearly stated if the individual genetics is considered permissive or non-permissive for CD development. Permissive genetics must include the presence of the haplotype HLA-DQ8 and the presence of both or one of the alleles encoding HLA-DQ2.5. Only excluding all these combinations, the negative predictive value reaches almost the 100%.
4. In the case of permissive genetics, a brief statement must underline that the genetic study of CD inform us about the likelihood of developing CD, the so-called genetic predisposition or genetic susceptibility. However, it cannot be used to confirm CD and consequently, it must not be considered that people with a positive genetic test have, or it will have, CD. It must be kept in mind the poor positive predictive value of the genetic test.
5. In the case of non-permissive genetics, a brief statement can be included to indicate the very low likelihood of CD (<1%).

The risk grade based on the particular HLA-DQ genotype can be added to the genetic report. It has not been translated into clinical practice according to current diagnostic guidelines, which only recommend the genetic test based on its high negative predictive value. However, it can result useful in specific scenarios (see next section). The number of risk categories and the genotypes included in each group is still a matter of debate. The highest risk has been uniformly established in the presence of HLA-DQ2.5 with double dose of *HLA-DQB1*02*. Similarly, small discrepancies have been described at the lowest risk categories following the gradient: single dosage of HLA-DQ8 > single dosage of HLA-DQ2.2 > HLA-DQ7.5. This contrasts with the different risks reported for subjects carrying HLA-DQ2.5 in heterozygosis depending on the accompanying haplotype and, in non-HLA-DQ2.5 carriers, for subjects with double dosage of HLA-DQ8, HLA-DQ2.2 or even for HLA-DQ8/DQ2.2 individuals [29]. To be properly used, risk classifications should be calculated in each population, but in absence of data, the grades showed in Table 1 could be used.

Some examples of genetic report models have been published [38, 39]. To avoid misinterpretation, it is convenient that a practitioner with experience interprets the genetic report providing genetic counselling.

2.7 Diagnostic Algorithms: Who Must Be Tested

The value of the genetic test derives from its very high negative predictive value. Therefore, clinical guidelines recommend to use it in the diagnostic work-up of cases requiring or obtaining a benefit of ruling out CD. This mainly involves two specific scenarios: individuals belonging to risk groups and individuals with uncertain diagnosis (Table 2).

2.7.1 Risk Groups

Asymptomatic first-degree relatives of CD patients or individuals with CD related conditions (other immune-related disorders or some syndromes such as Down's, Turner's or Williams´s syndromes) can be initially screened with the genetic test. The presence of permissive genetics will lead to periodic serological testing to

Table 2 Clinical scenarios to include the genetic test in the diagnostic work-up of celiac disease (CD) and further actions based on the observed HLA genetic results

Group of study	Permissive HLA	Non-permissive HLA
Asymptomatic children with increased CD risk: CD familiarity, immune-related diseases except type 1 diabetes[a], other CD-related conditions	CD screening: serological study and duodenal biopsy (if needed)	Do not start CD screening
Individuals with positive serology and normal histology	Periodic clinical and serological follow-up	Consider alternative diagnosis
Individuals with high clinical suspicion but seronegative for anti-TG2 antibodies	Consider duodenal biopsy	
Individuals with high clinical suspicion but a biopsy showing increased intestinal intraepithelial lymphocytes without the presence of atrophy	CD diagnosis support (not confirmatory)	
Individuals with a lack of response to the GFD		
Individuals following a GFD without a previous diagnosis	Gluten challenge and CD screening	

[a]Serologic screening to all patients based on cost-effectiveness
TG2: transglutaminase type 2; GFD: gluten-free diet

avoid a misdiagnosis of subclinical CD or at least to be alert to perform CD screening if developing symptoms. In contrast, non-permissive genetics will avoid CD screening at that moment and further in the lifetime. The use of the genetic test in risk groups has been formally stated for children [14], although it can be also useful in adults.

Children with type 1 diabetes (T1D) have been excluded from the group to perform CD screening based on the genetic test. T1D and CD share a strong HLA genetic background and all subjects with type 1 diabetes are recommended to be tested for CD serology based on cost-effectiveness [40, 41].

In risk groups, especially in first-degree relatives of CD patients, the proportion of permissive genetics reaches values that greatly exceed 40% (even values of 80–90% can be observed). Therefore, in these settings, it is especially important to use the genetic test to rule out CD, but never to confirm diagnosis.

2.7.2 Uncertain CD Diagnosis

Individuals with high clinical suspicion but seronegative CD or low grade enteropathy, as well as individuals showing a discrepancy serology-histology can benefit of being genetically tested. The absence of permissive genetics will suggest to consider an alternative diagnosis, while a permissive test will lead to continue the study by following a different approach depending on the specific scenario (Table 2). The genetic test is also recommended in subjects with lack of response to the gluten-free diet in order to avoid wrong diagnosis; and in those who adopt a gluten-free diet before diagnosis, especially if they are unwilling or unable to undertake gluten challenge. All these cases are more frequently observed in adults, who show lower sensitivity and specificity of serological tests and more often take action on their own initiative without consulting a specialist, but they can be also applied to children.

In addition to these cases, some studies suggest the use of HLA testing in other scenarios, but further research is needed. Risk stratification is not currently used in the work-up of CD. However, it could be useful to develop personalized programmes and establishing the interval of serological screening in children belonging to risk groups [23, 42, 43]. In first-degree relatives of CD patients younger than 10 years, HLA genotyping has been suggested to be used as the first-line screening test to identify cases with permissive genetics in order to repeat periodically the serological study and identify most cases of CD. After 10 years old, one serological testing in asymptomatic individuals would be enough due to the low likelihood of developing CD [44, 45].

The specific HLA genotype of a recently diagnosed patient may be also useful to obtain a comprehensive indication of the likelihood of finding new CD cases in the family and take the decision of starting CD screening.

3 Non-classical HLA Genetics in Celiac Disease

3.1 Genome-Wide Association Studies

For many years, multiple attempts to find additional genetic variants of susceptibility were performed trying to explain the high heritability of CD, to understand
the mechanisms underlying the disease and to support diagnosis. Unfortunately, the
linkage analysis in families with CD members and the case–control studies following candidate gene approaches were not able to identify susceptibility genes
beyond the HLA-DQ region [46]. Around 15 years ago, after the sequencing of the
human genome and the advent of technological progress, the genome-wide association studies (GWAS) raised as a new approach for genetic studies. A GWAS is a
hypothesis-free observational study where thousands of genetic polymorphisms
across the genome are analyzed in very high numbers of individuals to identify
associations between genetic variants and a disease or trait of interest. Most commonly, SNPs have been used as the type of variation because of their simplicity,
abundance and dispersion in the genome [47]. To perform GWAS, a large collection of DNA samples of well-phenotyped subjects is needed, hence the importance of the consortia. These studies represent a very useful approach for complex
diseases with a strong genetic component, thus CD was not by chance one of the
first diseases considered by GWAS [48].

The first GWAS in CD was published in 2007 [3]. van Heel et al. tested 310,605
SNPs in 778 CD cases and 1,422 controls in the UK population. As expected, they
found the strongest association around the HLA *locus*, but outside this region, only
the IL2-IL21 locus was significantly involved. This could explain less than 1% of
the increased familial risk to CD, suggesting that there were additional unidentified
susceptibility genes. In 2008 and 2009, two subsequent follow-up studies were
performed including additional European populations and combining their data
with the originally obtained in the GWAS, thus trying to find new associated
regions. In the first follow-up study, 1,020 of the most strongly previously associated non-HLA markers were genotyped in an additional set of 1,643 cases and
3,406 controls from the UK, Ireland and Netherlands. Seven previously unknown
risk regions, six of them harbouring genes controlling immune responses, were
found [49]. In the second follow-up study, 458 additional SNPs with more modest
original association were studied in the same UK, Irish and Dutch cohorts. The
SNPs with the strongest association although not reaching the stringent statistical
threshold required for GWAS significance were subsequently genotyped in an
independent Italian cohort and two novel CD risk regions were identified [50].

The first follow-up study including non-European populations was performed by
Garner et al. in 2009 [51]. They genotyped 975 out of the top 1,020 SNPs from the
GWAS in a sample of 906 CD cases of self-reported European descent from the
USA and 3,819 ethnically matched controls and found evidence for association of
one new genomic region.

In 2010, a second-generation GWAS was performed in order to identify additional risk variants, particularly of smaller effect size, using 4,533 CD individuals and 10,750 control subjects in European populations from UK, Italy, Finland and Netherlands. They also followed-up the most promising findings by genotyping 131 selected SNPs in a further 4,918 cases and 5,684 controls from 7 independent cohorts [4]. They found that the HLA *locus* and all the other 13 previously reported CD risk *loci* showed evidence for association and identified 13 novel risk regions. Interestingly, most variants mapped near genes with functions related to the immune system.

In 2011, using the Immunochip array, a new study was performed including 12,041 CD patients and 12,228 controls from six geographic regions (five European and one Indian collection) and 13 new CD risk *loci* were described [5]. This array contained around 200,000 SNPs to fine-map *loci* previously associated with 12 major immune-mediated diseases.

A new GWAS was performed in 2014, the first one studying CD patients from North America [52]. Association with CD risk was ratified to one *locus* with previously described suggestive association.

In 2015, a fine mapping of the MHC region taking advantage of high-density imputation studied almost 12,000 cases and 12,000 controls of European ancestry and identified five new independent variants in this complex region [6].

Several meta-analyses of Immunochip discoveries were also performed. In 2016, Coleman et al. replicated the Trynka et al. study [5] in an Irish population and performed a meta-analysis with the previous data being able to identify two further *loci* [53]. Also in 2016, Sharma et al. identified eight new regions as associated with CD using a large international cohort of children at-risk for CD and suggested the existence of country-specific additional associations [54]. In 2020, two cohorts of self-reported Argentinian origin were genotyped on the Immunochip and two new regions were identified after combining their data with those from several European populations (Netherlands, Spain, Italy, Ireland, and Poland) [55].

All these large-scale studies, mostly developed in European populations, have described 47 genomic regions associated with CD susceptibility independent of HLA-DQ, the great majority (42 regions) outside the MHC complex. Eight additional regions can be considered if including the study developed with the paediatric cohort at risk for CD [54]. All together explain around 50% of CD heritability.

3.2 Non-classical HLA Genetic Variants in Clinical Practice

Since the first GWAS published in 2007, hundreds of SNPs have been associated with CD susceptibility. However, it became evident the almost negligible effect and low predictive value of the individual described genetic variants. Most of them were located in non-coding regions of the genome and, even thought to play a predominant role in gene expression, it was unclear the genes and the cells or physiological scenarios under regulation [30]. This made difficult the translation of

GWAS findings into clinical interventions. However, with the advances in technology and the reduction of costs, it has become possible to obtain the genetic make-up of each individual. Combined with the development of bioinformatics tools, the cumulative effect of thousands of genetic variations on disease risk can be determined and appears as a promising instrument to be used in clinical practice. Thus, recent and previous attempts to create models of risk diagnosis using genetic associated variants could be closer to be implemented in the daily clinical routine once confirmed their accuracy.

In CD, several models have been suggested to add the genotyping of different subsets of the SNPs associated with CD susceptibility in the GWAS or related studies to the known HLA variants attempting to refine and improve CD risk prediction.

The first alternative genetic model was proposed in 2009 by Romanos et al. [56], who combined HLA information with genotype data of 10 SNPs from nine non-HLA genomic regions associated in the first GWAS in CD and its follow-up studies [3, 49, 50]. Regarding HLA, three categories based on the HLA-DQ2 dosage effect were considered using tagging SNPs: high (HLA-DQ2.5/HLA-DQ2.5 and HLA-DQ2.5/HLA-DQ2.2), intermediate (HLA-DQ2.5 or HLA-DQ2.2 heterozygous and HLA-DQ2.2/HLA-DQ2.2) and low risk (neither HLA-DQ2.5 nor HLA-DQ2.2). The authors found that the number of non-HLA risk alleles was higher in CD than in controls (OR = 6.2 for individuals carrying more than 13 alleles compared to those carrying 0–5 risk alleles). However, this only entailed a small percentage of patients (7.5%) moving from the intermediate-risk category to the high-risk one when compared with the HLA-based risk classification.

In 2014, those authors expanded the study calculating genetic risk scores (GRS) by using different variants [57]. GRS aggregate the effects of different genetic variants across the human genome to provide a single numerical value per individual to evaluate their risk to present a disease. The weighted effects of each variant are assigned depending on its published strength of association with the risk of the disease. Romanos et al. included three different subsets of 10, 26 or 57 SNPs in addition to the HLA data. HLA risk was also established in three categories, but HLA-DQ8 subjects were included in the category of intermediate instead of low risk. The best results were observed combining HLA with 57 non-HLA SNPs data, which resulted in reclassifying 11% of the individuals into a more accurate risk group.

HLA and non-HLA variants were also combined to calculate CD risk improvement in 157 families with at least one child affected with CD [58]. Only three non-HLA SNPs were included, which were selected out of a total of 10 SNPs associated in the second GWAS in CD [4] after confirming their involvement in the studied families. HLA risk was graded in five categories based on a previous work [43]: HLA-DQ2.5/HLA-DQ2.5 and HLA-DQ2.5/HLA-DQ2.2; HLA-DQ2.5 in *trans*; heterozygous HLA-DQ2.5 in *cis*; HLA-DQ2.2/HLA-DQ2.2, HLA-DQ2.2/HLA-DQ8 or HLA-DQ8/HLA-DQ8; other genotypes. By using a Bayesian approach, the authors showed that the addition of those three non-HLA SNPs moderately increased the sensitivity of CD-risk prediction based on HLA in all the considered HLA groups, but with larger effect in the groups with low HLA risk.

In 2014, Abraham et al. suggested a different approach in order to refine CD genetic risk [59]. The authors calculated a GRS based on the analysis of thousands of SNPs following supervised learning models. They did not limit themselves to use SNPs previously associated in GWAS considering that some SNPs with predictive ability could be ignored due to the stringent corrections applied in GWAS to overcome multiple-testing. Their model, finally reduced to 228 SNPs, showed a good predictive performance and better results than the ones obtained using the model previously proposed by Romanos et al. including 57 non-HLA SNPs [57].

On the following year, Abraham et al. constructed GRS specific to HLA-DQ2.5 individuals [60]. HLA variation was imputed using two different approaches: one based on SNPs and a second one via SNP2HLA, which included HLA variants additional to SNPs and enabled fine-mapping of the region. The authors found similar results with both approaches, which offered increased precision in diagnosis than three other predictive models: imputed HLA-DQ2.5 zygosity (HLA-DQ2.5 homozygous or heterozygous), 3-level HLA haplotype risk score obtained by genotyping (established according to Romanos et al. in 2014 [57]) and the model they have recently constructed with 228 SNPs [59]. The models they proposed allowed reducing the number of unnecessary follow-up tests, although it was accompanied by a slight loss in sensitivity (around 10% of HLA-DQ2.5 CD patients were lost).

Last year, two new works dealing with these issues were published [26, 61]. Sharp et al. hypothesized that a good description of HLA-DQ variants and their interactions would need small numbers of non-HLA-DQ SNPs to generate more discriminative GRS. They used data of SNPs associated with CD in GWAS or related studies [5, 6, 62] and combined them with SNPs tagging the four CD-related HLA-DQ haplotypes (HLA-DQ2.5, HLA-DQ8, HLA-DQ2.2 and HLA-DQ7.5) obtained by imputation. They started with 42 SNPs to generate GRS: 4 HLA-DQ, 5 HLA non-DQ and 33 non-HLA SNPs. Using backwards stepwise regression, they reduced that number to 23 SNPs obtaining almost the same discriminative ability. The authors concluded that their GRS was more discriminative than the model based on HLA stratification (considering 9 different genotypes which ignored the risk conferred by carrying DQ2.2 or DQ7.5 alone). Interestingly, the authors validated their model in two different settings: population level screening (UK biobank) and a high-risk population (paediatric clinic for CD). In the second work, Erlichster et al. paid special attention to HLA [26]. They performed a fine-grain stratification of CD risk based on the 15 different genotypes established from combining the four CD-related classical haplotypes. Interestingly, Erlichster et al. described two novel interactions. The risk conferred by the HLA-DQ2.5 risk haplotype could be modulated depending on the presence of HLA-DQ6.2 (*HLA-DQA1*01:02* and *HLA-DQB1*06:02*) and HLA-DQ7.3 (*HLA-DQA1*03:03* and *HLA-DQB1*03:01*), giving to higher or lower risk than expected, respectively. These interactions showed variability across the studied populations (United Kingdom, Finland, Netherlands and Italy). Incorporating the genotypes HLA-

DQ2.5/HLA-DQ6.2 and HLA-DQ2.5/HLA-DQ7.3 to the model, they obtained a CD-risk score that improved prediction considering only 15 genotypes and that gave equivalent results to the model proposed by Abraham et al. [59] that used hundreds of markers. Moreover, they found that CD-risk haplotypes could be tagged with six SNPs. Erlichster et al. concluded that HLA typing had been being undervalued to calculate CD risk and proposed a new model with only six HLA tagging SNPs that maintained the nearly 100% negative predictive value but increased positive predictive value in high-risk populations. Slightly better results were obtained adding 6 additional SNPs to their model, only 2 SNPs clearly independent of the HLA locus.

4 Present Status and Future Perspectives

Currently, CD diagnosis can be supported by a genetic test based on *HLA-DQA1* and *HLA-DQB1* genotyping. While offering a negative predictive value close to 100%, this test shows a low positive predictive value and specificity, which limits its use in the daily clinical practice to exclude CD. Based on the numerous variants associated with CD susceptibility in large-scale studies, more specific predictions of CD risk by combining HLA with the effect of several SNPs across the genome were expected. However, the several attempts of creating new genetic models in CD seem to indicate that the highest discriminative ability remains in the HLA complex. This is especially suggested by the study of Erlichster et al. [26], which is consistent with the previous observation by Abraham et al. when focusing on HLA-DQ2.5 subjects [60]. Most likely, a better predictive ability will be achieved emphasizing the study in the HLA region, although adding few non-HLA markers could contribute to refine risk.

Further research in this area is clearly needed. The complexity of the HLA region, especially the influence of low-risk variants, regulatory mechanisms and structural variation, and the variability present across populations should be addressed in future studies. In this respect, GRS can be useful since they are flexible and can be easily adapted to new findings. However, several pros and cons can be attributed to these models. Just mention that to be implemented in the near future, GRS must prove to be cost-effective but, most of all, to provide accurate information in different clinical settings and also in different populations. Most of the large-scale studies in CD have included only populations of White European ancestry. The final score for each individual (GRS) is obtained depending on the variants considered and their published weights. Therefore, differences between populations will compromise the predictive ability. For a general use, multi-ethnic data must be analyzed to guarantee the accuracy of GRS.

Finally, it must be considered that we are dealing with a complex disease. Models integrating information of the different players contributing to CD development will be essential to get risk refinement and to approach us to the long-awaited personalized medicine.

Acknowledgements Concepción Núñez received a grant from Instituto de Salud Carlos III-Fondo Europeo de Desarrollo Regional (FEDER) (PI18/0089).

References

1. Falchuk ZM, Rogentine GN, Strober W. Predominance of histocompatibility antigen HL-A8 in patients with gluten-sensitive enteropathy. J Clin Invest. 1972;51(6):1602–5.
2. Stokes PL, Asquith P, Holmes GK, Mackintosh P, Cooke WT. Histocompatibility antigens associated with adult coeliac disease. Lancet. 1972;2(7769):162–4.
3. van Heel DA, Franke L, Hunt KA, Gwilliam R, Zhernakova A, Inouye M, et al. A genome-wide association study for celiac disease identifies risk variants in the region harboring IL2 and IL21. Nat Genet. 2007;39(7):827–9.
4. Dubois PC, Trynka G, Franke L, Hunt KA, Romanos J, Curtotti A, et al. Multiple common variants for celiac disease influencing immune gene expression. Nat Genet. 2010;42(4):295–302.
5. Trynka G, Hunt KA, Bockett NA, Romanos J, Mistry V, Szperl A, et al. Dense genotyping identifies and localizes multiple common and rare variant association signals in celiac disease. Nat Genet. 2011;43(12):1193–201.
6. Gutierrez-Achury J, Zhernakova A, Pulit SL, Trynka G, Hunt KA, Romanos J, et al. Fine mapping in the MHC region accounts for 18% additional genetic risk for celiac disease. Nat Genet. 2015;47(6):577–8.
7. Sollid LM, Markussen G, Ek J, Gjerde H, Vartdal F, Thorsby E. Evidence for a primary association of celiac disease to a particular HLA-DQ alpha/beta heterodimer. J Exp Med. 1989;169(1):345–50.
8. Johansen BH, Vartdal F, Eriksen JA, Thorsby E, Sollid LM. Identification of a putative motif for binding of peptides to HLA-DQ2. Int Immunol. 1996;8(2):177–82.
9. van de Wal Y, Kooy Y, van Veelen P, Pena S, Mearin L, Papadopoulos G, et al. Selective deamidation by tissue transglutaminase strongly enhances gliadin-specific T cell reactivity. J Immunol. 1998;161(4):1585–8.
10. Lundin KE, Scott H, Hansen T, Paulsen G, Halstensen TS, Fausa O, et al. Gliadin-specific, HLA-DQ(alpha 1*0501,beta 1*0201) restricted T cells isolated from the small intestinal mucosa of celiac disease patients. J Exp Med. 1993;178(1):187–96.
11. Molberg O, Kett K, Scott H, Thorsby E, Sollid LM, Lundin KE. Gliadin specific, HLA DQ2-restricted T cells are commonly found in small intestinal biopsies from coeliac disease patients, but not from controls. Scand J Immunol. 1997;46(3):103–9.
12. Bodd M, Kim CY, Lundin KE, Sollid LM. T-cell response to gluten in patients with HLA-DQ2.2 reveals requirement of peptide-MHC stability in celiac disease. Gastroenterology. 2012;142(3):552–61.
13. Fallang LE, Bergseng E, Hotta K, Berg-Larsen A, Kim CY, Sollid LM. Differences in the risk of celiac disease associated with HLA-DQ2.5 or HLA-DQ2.2 are related to sustained gluten antigen presentation. Nat Immunol. 2009;10(10):1096–101.
14. Husby S, Koletzko S, Korponay-Szabo IR, Mearin ML, Phillips A, Shamir R, et al. European Society for Pediatric Gastroenterology, Hepatology, and Nutrition guidelines for the diagnosis of coeliac disease. J Pediatr Gastroenterol Nutr. 2012;54(1):136–60.
15. Fernandez-Banares F, Arau B, Dieli-Crimi R, Rosinach M, Nunez C, Esteve M. Systematic review and meta-analysis show 3% of patients with celiac disease in Spain to be negative for HLA-DQ2.5 and HLA-DQ8. Clin Gastroenterol Hepatol. 2017;15(4):594–6.
16. Karell K, Louka AS, Moodie SJ, Ascher H, Clot F, Greco L, et al. HLA types in celiac disease patients not carrying the DQA1*05-DQB1*02 (DQ2) heterodimer: results from the European Genetics Cluster on Celiac Disease. Hum Immunol. 2003;64(4):469–77.

17. Pietzak MM, Schofield TC, McGinniss MJ, Nakamura RM. Stratifying risk for celiac disease in a large at-risk United States population by using HLA alleles. Clin Gastroenterol Hepatol. 2009;7(9):966–71.

18. Wang H, Zhou G, Luo L, Crusius JB, Yuan A, Kou J, et al. Serological screening for celiac disease in adult Chinese patients with diarrhea predominant irritable bowel syndrome. Medicine (Baltimore). 2015;94(42):e1779.

19. Demarchi M, Carbonara A, Ansaldi N, Santini B, Barbera C, Borelli I, et al. HLA-DR3 and DR7 in coeliac disease: immunogenetic and clinical aspects. Gut. 1983;24(8):706–12.

20. Louka AS, Nilsson S, Olsson M, Talseth B, Lie BA, Ek J, et al. HLA in coeliac disease families: a novel test of risk modification by the "other" haplotype when at least one DQA1*05-DQB1*02 haplotype is carried. Tissue Antigens. 2002;60(2):147–54.

21. Ploski R, Ek J, Thorsby E, Sollid LM. On the HLA-DQ(alpha 1*0501, beta 1*0201)-associated susceptibility in celiac disease: a possible gene dosage effect of DQB1*0201. Tissue Antigens. 1993;41(4):173–7.

22. van Belzen MJ, Koeleman BP, Crusius JB, Meijer JW, Bardoel AF, Pearson PL, et al. Defining the contribution of the HLA region to cis DQ2-positive coeliac disease patients. Genes Immun. 2004;5(3):215–20.

23. Martinez-Ojinaga E, Molina M, Polanco I, Urcelay E, Nunez C. HLA-DQ distribution and risk assessment of celiac disease in a Spanish center. Rev Esp Enferm Dig. 2018;110 (7):421–6.

24. Medrano LM, Dema B, Lopez-Larios A, Maluenda C, Bodas A, Lopez-Palacios N, et al. HLA and celiac disease susceptibility: new genetic factors bring open questions about the HLA influence and gene-dosage effects. PLoS One. 2012;7(10):e48403.

25. Margaritte-Jeannin P, Babron MC, Bourgey M, Louka AS, Clot F, Percopo S, et al. HLA-DQ relative risks for coeliac disease in European populations: a study of the European Genetics Cluster on Coeliac Disease. Tissue Antigens. 2004;63(6):562–7.

26. Erlichster M, Bedo J, Skafidas E, Kwan P, Kowalczyk A, Goudey B. Improved HLA-based prediction of coeliac disease identifies two novel genetic interactions. Eur J Hum Genet. 2020;28(12):1743–52.

27. Handunnetthi L, Ramagopalan SV, Ebers GC, Knight JC. Regulation of major histocompatibility complex class II gene expression, genetic variation and disease. Genes Immun. 2010;11(2):99–112.

28. Lenz TL, Deutsch AJ, Han B, Hu X, Okada Y, Eyre S, et al. Widespread non-additive and interaction effects within HLA loci modulate the risk of autoimmune diseases. Nat Genet. 2015;47(9):1085–90.

29. Espino L, Nunez C. The HLA complex and coeliac disease. Int Rev Cell Mol Biol. 2021;358:47–83.

30. Dieli-Crimi R, Cenit MC, Nunez C. The genetics of celiac disease: a comprehensive review of clinical implications. J Autoimmun. 2015;64:26–41.

31. Vader W, Stepniak D, Kooy Y, Mearin L, Thompson A, van Rood JJ, et al. The HLA-DQ2 gene dose effect in celiac disease is directly related to the magnitude and breadth of gluten-specific T cell responses. Proc Natl Acad Sci U S A. 2003;100(21):12390–5.

32. Henderson KN, Tye-Din JA, Reid HH, Chen Z, Borg NA, Beissbarth T, et al. A structural and immunological basis for the role of human leukocyte antigen DQ8 in celiac disease. Immunity. 2007;27(1):23–34.

33. Del Pozzo G, Farina F, Picascia S, Laezza M, Vitale S, Gianfrani C. HLA class II genes in precision-based care of childhood diseases: what we can learn from celiac disease. Pediatr Res. 2021;89(2):307–12.

34. Gianfrani C, Pisapia L, Picascia S, Strazzullo M, Del Pozzo G. Expression level of risk genes of MHC class II is a susceptibility factor for autoimmunity: new insights. J Autoimmun. 2018;89:1–10.

35. Pisapia L, Camarca A, Picascia S, Bassi V, Barba P, Del Pozzo G, et al. HLA-DQ2.5 genes associated with celiac disease risk are preferentially expressed with respect to

non-predisposing HLA genes: implication for anti-gluten T cell response. J Autoimmun. 2016;70:63–72.

36. Monsuur AJ, de Bakker PI, Zhernakova A, Pinto D, Verduijn W, Romanos J, et al. Effective detection of human leukocyte antigen risk alleles in celiac disease using tag single nucleotide polymorphisms. PLoS One. 2008;3(5):e2270.

37. Kunkel M, Duke J, Ferriola D, Lind C, Monos D. Molecular methods for human leukocyte antigen typing: current practices and future directions. In: Manual of molecular and clinical laboratory immunology. 2016. p. 1069–90.

38. Nunez C, Garrote JA, Arranz E, Bilbao JR, Fernandez Banares F, Jimenez J, et al. Recommendations to report and interpret HLA genetic findings in coeliac disease. Rev Esp Enferm Dig. 2018;110(7):458–61.

39. Tye-Din JA, Cameron DJ, Daveson AJ, Day AS, Dellsperger P, Hogan C, et al. Appropriate clinical use of human leukocyte antigen typing for coeliac disease: an Australasian perspective. Intern Med J. 2015;45(4):441–50.

40. Husby S, Koletzko S, Korponay-Szabo I, Kurppa K, Mearin ML, Ribes-Koninckx C, et al. European Society Paediatric Gastroenterology, Hepatology and Nutrition guidelines for diagnosing coeliac disease 2020. J Pediatr Gastroenterol Nutr. 2020;70(1):141–56.

41. Mitchell RT, Sun A, Mayo A, Forgan M, Comrie A, Gillett PM. Coeliac screening in a Scottish cohort of children with type 1 diabetes mellitus: is DQ typing the way forward? Arch Dis Child. 2016;101(3):230–3.

42. Megiorni F, Mora B, Bonamico M, Barbato M, Nenna R, Maiella G, et al. HLA-DQ and risk gradient for celiac disease. Hum Immunol. 2009;70(1):55–9.

43. Bourgey M, Calcagno G, Tinto N, Gennarelli D, Margaritte-Jeannin P, Greco L, et al. HLA related genetic risk for coeliac disease. Gut. 2007;56(8):1054–9.

44. Lionetti E, Castellaneta S, Francavilla R, Pulvirenti A, Tonutti E, Amarri S, et al. Introduction of gluten, HLA status, and the risk of celiac disease in children. N Engl J Med. 2014;371 (14):1295–303.

45. Wessels MMS, de Rooij N, Roovers L, Verhage J, de Vries W, Mearin ML. Towards an individual screening strategy for first-degree relatives of celiac patients. Eur J Pediatr. 2018;177(11):1585–92.

46. Wolters VM, Wijmenga C. Genetic background of celiac disease and its clinical implications. Am J Gastroenterol. 2008;103(1):190–5.

47. Uitterlinden AG. An introduction to genome-wide association studies: GWAS for dummies. Semin Reprod Med. 2016;34(4):196–204.

48. Dehghan A. Genome-wide association studies. Methods Mol Biol. 2018;1793:37–49.

49. Hunt KA, Zhernakova A, Turner G, Heap GA, Franke L, Bruinenberg M, et al. Newly identified genetic risk variants for celiac disease related to the immune response. Nat Genet. 2008;40(4):395–402.

50. Trynka G, Zhernakova A, Romanos J, Franke L, Hunt KA, Turner G, et al. Coeliac disease-associated risk variants in TNFAIP3 and REL implicate altered NF-kappaB signalling. Gut. 2009;58(8):1078–83.

51. Garner CP, Murray JA, Ding YC, Tien Z, van Heel DA, Neuhausen SL. Replication of celiac disease UK genome-wide association study results in a US population. Hum Mol Genet. 2009;18(21):4219–25.

52. Garner C, Ahn R, Ding YC, Steele L, Stoven S, Green PH, et al. Genome-wide association study of celiac disease in North America confirms FRMD4B as new celiac locus. PLoS One. 2014;9(7):e101428.

53. Coleman C, Quinn EM, Ryan AW, Conroy J, Trimble V, Mahmud N, et al. Common polygenic variation in coeliac disease and confirmation of ZNF335 and NIFA as disease susceptibility loci. Eur J Hum Genet. 2016;24(2):291–7.

54. Sharma A, Liu X, Hadley D, Hagopian W, Liu E, Chen WM, et al. Identification of non-HLA genes associated with celiac disease and country-specific differences in a large, international pediatric cohort. PLoS One. 2016;11(3):e0152476.

55. Ricano-Ponce I, Gutierrez-Achury J, Costa AF, Deelen P, Kurilshikov A, Zorro MM, et al. Immunochip meta-analysis in European and Argentinian populations identifies two novel genetic loci associated with celiac disease. Eur J Hum Genet. 2020;28(3):313–23.
56. Romanos J, van Diemen CC, Nolte IM, Trynka G, Zhernakova A, Fu J, et al. Analysis of HLA and non-HLA alleles can identify individuals at high risk for celiac disease. Gastroenterology. 2009;137(3):834–40, 840.e1–3.
57. Romanos J, Rosen A, Kumar V, Trynka G, Franke L, Szperl A, et al. Improving coeliac disease risk prediction by testing non-HLA variants additional to HLA variants. Gut. 2014;63 (3):415–22.
58. Izzo V, Pinelli M, Tinto N, Esposito MV, Cola A, Sperandeo MP, et al. Improving the estimation of celiac disease sibling risk by non-HLA genes. PLoS One. 2011;6(11):e26920.
59. Abraham G, Tye-Din JA, Bhalala OG, Kowalczyk A, Zobel J, Inouye M. Accurate and robust genomic prediction of celiac disease using statistical learning. PLoS Genet. 2014;10(2): e1004137.
60. Abraham G, Rohmer A, Tye-Din JA, Inouye M. Genomic prediction of celiac disease targeting HLA-positive individuals. Genome Med. 2015;7(1):72.
61. Sharp SA, Jones SE, Kimmitt RA, Weedon MN, Halpin AM, Wood AR, et al. A single nucleotide polymorphism genetic risk score to aid diagnosis of coeliac disease: a pilot study in clinical care. Aliment Pharmacol Ther. 2020;52(7):1165–73.
62. Plaza-Izurieta L, Castellanos-Rubio A, Irastorza I, Fernandez-Jimenez N, Gutierrez G, Cegec, et al. Revisiting genome wide association studies (GWAS) in coeliac disease: replication study in Spanish population and expression analysis of candidate genes. J Med Genet. 2011;48(7):493–6.
63. Marsh SG, Albert ED, Bodmer WF, Bontrop RE, Dupont B, Erlich HA, et al. Nomenclature for factors of the HLA system, 2010. Tissue Antigens. 2010;75(4):291–455.

Gluten Free Diet

Paula Crespo-Escobar

1 Introduction

For regulatory purposes gluten is defined as 'the protein fraction from wheat, barley, rye, oats or their crossbred varieties and derivatives that is insoluble in water and 0.5 M NaCl [1]. However, there are a varied range of definitions for gluten in the literature that have been developed from different perspectives. Cereal grain proteins, including gluten, have been classically defined according to their solubility according Osborne classification. Osborne classified the storage proteins into groups on the basis of their extraction and solubility in water (albumins), dilute saline (globulins), alcohol ether mixtures (prolamins), and dilute acid or alkali (glutelins) [2].

In the case of wheat, the storage proteins from the gluten fraction are important because of their properties are largely responsible for the ability to use wheat flour to make bread and other products. The wheat grain can be divided into two main groups: gliadins and glutenins. Gliadins are subdivided into four groups on the basis of mobility at low pH in gel electrophoresis (α-, β-, ϒ-, ω-gliadins in order of decreasing mobility). Glutenins are divided into two groups according to their molecular weight: high molecular weight glutenin subunits (HMW-GS) and low molecular weight glutenin subunits (LMW-GS). We can also find homolog gluten genes in rye and barley [3–5].

Glutenins and gliadins are widely studied due to their contribution to the quality of the end-product of bakery and pasta goods, including the rheological characteristics of dough made from wheat flour. However, besides the importance of gluten proteins in food quality, the gluten has a direct impact on the human health

P. Crespo-Escobar (✉)
Department of Health Sciences, Universidad Europea Miguel de Cervantes, Valladolid, Spain

P. Crespo-Escobar
Nutrition Unit of Campo Grande Hospital, Valladolid, Spain

J. Amil-Dias and I. Polanco (eds.), *Advances in Celiac Disease*,
https://doi.org/10.1007/978-3-030-82401-3_9

by triggering wheat related food disorders such as celiac disease (CD). Indeed, currently, the only effective therapy for CD is a strict lifelong gluten-free diet (GFD). This means the elimination of products containing wheat, rye and barley [6].

Although conceptually simple, following a GFD presents significant challenges and many barriers to compliance that have impact on life quality of patients. It can be exceedingly difficult to completely avoid gluten-containing foods, and adherence to a GFD is estimated to be only 45–80% [7]. Comprehensive understanding of the factors associated with optimal GFD adherence is needed to develop strategies and resources to assist individuals with CD maintain a GFD.

On the other hand, a GFD has been the most popular elimination diet for more than a decade. The evidence indicates that the number of people on a GFD is constantly increasing, not only among people with gluten-related disorders, because it is associated with being healthier [8]. However, some epidemiological studies indicate nutritional imbalances for people following GFD. They refer both to macronutrients and micronutrients including minerals [9]. Therefore, it is extremely important to educate the patients in how to follow a GFD not only avoiding gluten, but also combining foods in a healthy way. The gluten-free diet must be safe and healthy.

2 Gluten-Free Diet

The GFD pattern is the only treatment for CD, gluten sensitivity and wheat allergies. However, to ensure accurate test results it is imperative that the patient does not initiate a GFD until a final diagnosis is obtained. Once the diagnosis is reached, then the GFD should be implemented.

The basis of GFD is to remove all food products that contain wheat, rye and barley. In Table 1, we can find the naturally gluten-free foods, foods and products that naturally have no gluten but they may have been contaminated during processing and finally, those gluten containing foods and products.

2.1 *Healthy Gluten-Free Diet and Nutritional Deficiencies*

Recently, GFD has been associated with being healthier. However, epidemiological studies indicate nutritional imbalances for people following GFD, referring both to macronutrients and micronutrients including minerals [9].

This is an important issue because affects directly to the nutritional status of CD patients. Furthermore, the nutritional condition depends on the length of time the disease is active, the extent of intestinal inflammation, the degree of malabsorption, and dietary intake [10].

Table 1 Sources of gluten

Gluten-containing grains and their derivatives	Foods that may contain gluten[a]	Naturally gluten-free foods[b]
• Wheat and derivatives: *triticale*, *durum*, emmer, semolina, spelt, farina, farro, graham, KAMUT® khorasan wheat, einkorn wheat • Wheat starch that has not been processed to remove the presence of gluten to below the limit 20 ppm • Rye • Barley • Malt in various forms including: malted barley flour, malted milk or milkshakes, malt extract, malt syrup, malt flavouring, malt vinegar • Brewer's Yeast All products with those ingredients are **not** allowed for CD people	• **Processed lunch meats, friednuts, sauces** • **Energy bars and granola bars**—some bars may contain wheat as an ingredient, and most use oats that are not gluten-free • **French fries:** risk in the batter containing wheat flour or cross-contact from fryers • **Potato chips** seasonings: could contain malt vinegar or wheat starch • **Candy and Candy bars** • **Ready to eat soups**—pay special attention to cream-based soups, which have flour as a thickener. Many soups also contain barley • **Multi-grain or artisan tortilla chips or tortillas** that are not entirely corn-based (specify in the ingredient list) may contain a wheat • **Salad dressings and marinades** • **Starch or dextrin** without specify the origin of starch: if found on meat product • **Brown rice syrup**—may be made with barley traces • **Meat substitutes** made with seitan such as vegetarian burgers, vegetarian sausage, imitation meat or fish derivatives (bacon, seafood) • **Soy sauce** without specify that is gluten-free • **Self-basting poultry and eggs served at restaurants** or bars–some restaurants put pancake batter in their scrambled eggs and omelettes	• Rice, cassava, corn (maize) • Soy • Tubercles: potato, tapioca, yucca • Beans • Sorghum, quinoa, millet, buckwheat groats, amaranth • Arrowroot • Teff • Chia • Gluten-free oats certificated • Nut flours • Natural nuts, seeds and oils • Fresh meat, fish and eggs • Dairy • Fresh fruits and vegetables • Legumes

[a]These foods must be verified by reading the label or checking with the manufacturer and/or kitchen staff

[b]Although these foods are naturally gluten free, it is necessary to check always the labels of manufacturing products

For that reason, it is extremely important to monitor nutritional status and diet at the time of diagnosis and during follow-up. Some studies have revealed that deficiencies in iron, calcium, zinc, vitamin B12, vitamin D and folate are by far the most common nutritional inadequacies claimed for newly diagnosed celiac patients, whereas macronutrient inadequacies are rarely identified at diagnosis. On the other hand, after a while following the gluten-free diet, the most important nutritional deficiencies are: iron, calcium, selenium, zinc, magnesium, vitamin B and B12, excess of fat and simple sugar intake and poor fibre consumption [11, 12].

We have considered that most of the vitamin and mineral deficiencies are consequence of the villous atrophy in the small intestine, because their absorption occurs mainly in different sections of the intestine. But after start the GFD and when the gut recovers, some of these deficiencies are solved.

However, the nutritional deficiencies after several years of GFD are due to an unhealthy diet in the most cases. In general, an inadequate macronutrient intake has been associated above all with the fact that CD patients are focusing on the avoidance of gluten and often leaving back the importance of nutritional quality of the choice.

The poor intake of fibre is associated with the necessity of the avoidance of several kinds of foods naturally rich in fibre (i.e. grain) and the low content of fibre of GF products that are usually made with starches and/or refined flours. As is well known, a consumption of adequate amounts of dietary fibre is related to potential health benefits such as prevention of obesity, diabetes, cardiovascular diseases and various cancers. So, it is important to encourage CD patients to review their fibre consumption [13].

Concerning micronutrients, deficiencies of some vitamins and minerals may persist and this required a particular attention to the quality of the GFD, especially those implicated in crucial metabolic functions. Calcium and vitamin D deficiencies can lead to osteopenia and osteoporosis and may cause growth problems and difficulties in peak bone mass achievement in young patients. But this deficiency also are important in elderly CD people because it results in a lowered mineral density and increased bone fracture risk [14, 15]. The zinc deficiency can affect protein synthesis and leads to growth arrest, whereas magnesium deficiency can compromise the metabolism of proteins, nucleic acids, glucose, fats, and transmembrane transportation. We should control blood levels of these minerals [12, 16]. Finally, iron deficiency anaemia and vitamin B12 deficiency, are among of the most common extra-intestinal manifestations of CD patients. Although in most patients this is reversed after starting a gluten-free diet, we should monitor regularly, especially in vegan and vegetarian CD patients [9, 10].

Lastly, regarding macronutrients there are disagreements among studies mainly related with protein, sugars and fat intake. But anyway, what is clear is that, following all this evidence, from the practical point of view the nutritional recommendation to CD patients for following a healthy and balance GFD should be:

1. **The basis of your diet should be naturally gluten-free foods**: fruits, vegetables, meat, fish, eggs, sugar free dairy, whole grains, gluten free starches and flours, legumes, nuts, seeds and oils.
2. Eat at least **3 daily portions of fruits**
3. Eat at least **2 daily portions of vegetables**, and try to choose one of them the raw version.
4. Choose **gluten-free whole grains**
5. **Pseudocereals** such as amaranth, buckwheat, and quinoa are good sources of vitamins, minerals, healthy fats and fibre.
6. **Use the Healthy Eating Plate** developed by the Harvard School of Public Health, as a guide for creating healthy, balanced meals, whether served at the table or packed in a lunch box
7. **Try to avoid specific gluten-free products with high amounts** of added sugar, refined flours and saturated fats.

In Table 2, we can see a general example of a healthy gluten-free menu, with the main gluten-free food groups distributed.

Table 2 Example of a healthy gluten free menu

	Breakfast	Lunch	Dinner
Monday	Overnight chia seed pudding with Greek yoghurt	Meat with vegetables and fruit (desert)	Fish with quinoa salad and fruit
Tuesday	Gluten-free cereals with milk and nuts	Legumes with vegetables and fruit (desert)	Egg with vegetables and fruit
Wednesday	Gluten-free toast with avocado and an egg	Fish with vegetables, potatoes and fruit (desert)	Meat with vegetables and fruit
Thursday	Gluten-free cereals with milk and nuts	Legumes with vegetables and fruit (desert)	Egg with salad and fruit
Friday	Fruit smoothie with gluten-free cereals	Meat with vegetables, potatoes and fruit (desert)	Fish with vegetables and fruit
Saturday	Porridge with nuts	Legumes with vegetables and fruit (desert)	Egg with vegetables and fruit
Sunday	Overnight chia seed pudding with Greek yoghurt	Rice with vegetables and fruit (desert)	Fish with quinoa salad and fruit

2.2 Oats and Gluten-Free Diet

The inclusion of oat into the GFD is still controversial. The main limitation to its use is the contamination of oat by wheat, barley, or rye. Indeed, gluten contamination of oat occurs frequently and commercially available oats are not suitable in a GFD for CD due to their routine contamination with gluten-containing cereals. Only gluten-free oat is acceptable as a foodstuff for celiac patients [17].

However, the main problem with the cultivation and processing of gluten-free oat is that oats requires sophisticated technology. The prevention of contamination of oat by wheat, rye, or barley includes to have separate fields with an appropriate distance and a natural barrier between the fields. Moreover, a field previously planted with gluten-free cereals cannot be used for oat for at least eight years. Oat fields must also be routinely inspected for contaminating cereal plants (containing prolamins immunogenic for celiac patients), and those plants have to be removed [18]. Gluten-free oat must meet the legislative criteria for gluten-free foodstuff, i.e., the content of gluten in the end-products must be less than 20 mg/kg.

In general, uncontaminated oats are safe for almost all patients with CD. A small percentage of patients with CD maybe sensitive to oats and develop symptoms or even mucosal damage. One of the most recent systematic analysis about oat safety for celiac disease patients, concludes that supplementing a GFD with oats can potentially diminish nutrient deficiency and may provide significant health and quality of life benefits as well. However, the debate regarding the safety of oats for CD patients' needs to be settled first. The authors specify that a large-scale clinical trial using the high-quality GF oats is needed to confirm the real safety of this cereal in CD patients [19].

Experts recommend to avoid the introduction of oat into a gluten-free diet for newly diagnosed celiac patients, since a strict adherence to a gluten-free and oat-free diet is required for newly diagnosed celiac patients [18, 20, 21].

It also important to highlight that the recent evidence indicates gluten-free oats only should be introduce in patients with clinical remission, without symptoms and negative serology and, during the introduction, celiac patients should be under medical supervision due to individual susceptibility too. Although the dietary oat (without contamination with gliadins) is tolerated by the majority of celiac patients, the individual sensitivity to oat cannot be excluded [20].

The current recommendation of the European Society for the Study of Celiac Disease regarding the introduction of oats in the GFD is that *"Oats are safely tolerated by the majority of CD patients; its introduction into the diet should be cautious and patients should be monitored for possible adverse reaction"*. (Strong recommendation, moderate level of evidence) [22].

2.3   Pseudocereals in Gluten-Free Diet

Pseudocereals (amaranth, quinoa and buckwheat) are composed mainly of albumins and globulins and contain very little or no storage prolamin proteins, which are toxic for CD patients; thus, they are good substitutes for cereal in GF foods.

Also, they have an interesting nutritional value as compared with wheat and different important GF flour. All pseudocereals have more calcium, magnesium and iron than wheat. As we have mentioned, these are one of the most compromised micronutrients the diet of GFD [23].

Amaranth has a nutritional value better than that of any other vegetable and much higher amounts of fibre, protein and minerals than any other GF grain as well as important amino acids such as lysine, arginine, tryptophan, and sulphur-containing amino acids Moreover, some food industries have used this ingredient to enrich cereal-based foods, including GF pasta [24].

Quinoa is a good complement for legumes (low amounts of methionine and cysteine) because its protein is rich in lysine, methionine, and cysteine. In addition, quinoa is a good source of Vitamin E and B-group vitamins and has high levels of calcium, iron, and phosphorous. It also has a suitable fatty acid composition and low amylase contents, this particularity is necessary to have a high shear in extrusion cooking, so quinoa could be used for several gluten-free products [24–26].

Buckwheat has a low glycaemic index which is beneficial for lowering blood pressure and control cholesterol levels. It has been demonstrated that the replacement of cornstarch with buckwheat flour in GF bread and GF crackers showed to have a positive effect on the texture and leads to products with acceptable sensory qualities. Buckwheat and quinoa breads have a higher volume than other kinds of GF breads [24, 27].

Apart from these pseudocereals it is important to highlight the nutritional value of other minor cereals used as alternative to gluten containing cereals: sorghum, teff, millet and wild rice. In recent analyses, these cereals have demonstrated to have a good nutritional profile, mainly in those nutrients which CD patients have deficiencies. Finally, the authors concluded that it is possible to use the combined mix of these flours in order to improve the nutritional value of cereal-based gluten-free products [28].

3   Gluten-Free Products

3.1   Gluten Free Products Regulation

The gluten-free products (GFP) are regulated by two European Commission Regulation: The Regulation (EU) N° 828/2014 which entered into force on 20 July 2016. This Regulation lays down harmonized requirements for the provision of information to consumers on the absence or reduced presence of gluten in food. More specifically, this legislation sets out the conditions under which foods may be labelled as "gluten-free" or "very-low gluten". This new regulation repealed the previous one: Commission Regulation (EC) No 41/2009.

The other is Regulation (EU) N° 1169/2011 which lays down rules requiring the mandatory labelling for all foods of ingredients such as gluten-containing ingredients, with a scientifically proven allergenic or intolerance effect. In order to ensure clarity and consistency, it was considered that all the rules applying to gluten should be set by the same piece of legislation and, for this reason, Regulation (EU) No 609/2013 established that Regulation (EU) No 1169/2011 should also be the framework for the rules related to information on the absence of gluten in food [29].

The Crossed Grain Trademark

The Crossed Grain Trademark is registered and protected across the European Union, Switzerland, Norway, Croatia, Montenegro, Serbia and Bosnia Herzegovina developed by the Association of European Celiac Societies (AOECS).

This trademark can be licenced only for multiple ingredient and/or processed products or when there is a high risk of contamination. For example, fresh fruit and vegetables cannot be licenced as they are naturally gluten-free; but fruit bars or buckwheat flour can be licenced as they have undergone a process which may hold a risk for gluten contamination.

This symbol is recognized by CD patients in Europe and is particularly important when the consumer is unsure on the gluten status of a product, or whilst travelling and unable to understand the language in which the label is displayed. It is a quality and safety guarantee as all producers wishing to use it must conform to high and safe standards of production.

The use of this trademark protected symbol is strictly monitored by AOECS and its Member societies [30].

3.2   Challenges of Gluten Free Products

Gluten Functionality and Replacement Strategies

One of the main challenges is the gluten functionality. The gluten protein fraction displays unique structure building properties that are used in food processing because gluten in wheat flour forms a three-dimensional protein network upon proper hydration and mixing. These network-forming properties are used in baking applications to form viscoelastic dough matrices. In addition, gluten functionality in food includes water binding and viscosity yielding, which make gluten a widely used food additive. For that reason, the replacement of gluten as a vital ingredient in numerous food products is not straightforward. Different ingredients and processing techniques have been investigated to date. However, the quality of gluten-free products is often not comparable to gluten-containing products [4, 31].

There are a lot of trials which have attempted to imitate the cohesiveness and elasticity of a gluten-containing dough using a wide range of alternative raw ingredients and/or additives. The most studied gluten-replacing combinations with acceptable quality effects have been: starches, gluten-free flours of cereals/pseudocereals, hydrocolloids, and proteins with enzymes and emulsifiers [31].

Cost

One of the main challenges of GFP is the cost. The GF commercial foods are more expensive as compare with their gluten-containing counterparts (GCC). One of the most recent studies concluded that the cost of GF products was significantly more expensive than their wheat-based counterparts for all ten product categories which include staple foods; (bread, cereals, pastas), snack foods; (crackers, pretzels, cookies), and convenience foods (waffles, pizza, macaroni, and cheesecake). The overall cost of GF products was 183% more expensive than their wheat-based counterparts. The largest difference between GF and wheat-based products was for crackers (snack food category) which were 270% more expensive [32]. Previous studies have shown similar results between 150 and 200% more expensive GFP as compared with their counterparts [33–35].

Although, the cost GF products has been declined over the past 10 years, it remains significantly higher than their wheat-based counterparts.

Nutritional Value of Gluten-Free Products

It is well-known that overall the nutritional profile of GFP is associated with a lower content of protein as compared to their (GCC). The lower protein content of GFP is the result of the ingredients used in the formulation such as cornstarch, corn flour and rice flour that naturally have a high carbohydrate and low-protein content. This is unlikely to be a problem because of the high protein intake in general population, and this deficiency can be easily covered by the consumption of other gluten-free products: eggs, meat, fish, legumes and dairy [36–39].

However, regarding other macronutrients such as fibre, saturated fatty acids and added sugars there are controversial results. Some studies found significant differences in these ingredients in all GFP and others only found differences in few food groups (bread and biscuits). But in general, the main studies agree on two main points: (1) In GFP different kinds of fats are frequently added to the GF dough so as to replace the texture given by the network that gluten forms and also to enhance flavour and acceptance. (2) The main ingredients used in the formulation of most GFP are corn flour and rice flour, ingredients which are made of up to 70–80% of amylopectin, a glucose polymer, which results in a high glycaemic index, related to the risk of metabolic syndrome [35–47].

Additionally, the main limitation of all these studies is that the authors only analyze products from their own countries and from specific brands available there, so their results cannot be generalized to other countries and we should be cautious with the interpretation. Other important limitation is most of the studies only investigate about macronutrients. Few data are currently available regarding the vitamin content of GF products, despite nutritional deficiencies emerging from analysis of the nutritional status of CD patients on a GFD. So, it is necessary that vitamin and mineral content in GF food products should be investigated in order to evaluate the necessity for fortification of GF products. The use of alternative ingredients, such as pseudocereals and legumes, should be also considered in order to improve the protein profile of GF products [12].

Anyway, it is recommendable to highlight the importance of basing the GFD in natural gluten-free foods and choose minimally processed products. And this is also recommendable for general population: not to abuse of processed products with high amounts of low-quality fats, added sugars and refined grains. And the recommendation is the same for the gluten-free and gluten-containing food industry: a reduction of fat, carbohydrate, sugars and sodium should become a priority for manufacturing products [48].

4   Cross Contamination

Cross contamination occurs when a gluten-free or food product is exposed by either direct or indirect contact with a gluten-containing ingredient or food—making it unsafe for people with celiac disease to eat.

However, it is almost impossible to maintain a diet with a zero-gluten content because gluten contamination is very common in food and may contain undetectable amounts of gluten proteins. Currently, the maximum level of gluten contamination (expressed as parts per million, ppm) that can be tolerated in products that are marketed for the treatment of CD is 20 ppm [1]. Nevertheless, the relation between the intestinal damage induced by trace intakes of gluten and the long-term complications of CD remains to be elucidated, but the most acceptable dangerous amount established is that as little as 50 mg gluten can damage the architecture of the small intestine in patients being treated for CD [49].

The cross contamination can be a source of stress and anxiety for people and, although a gluten-free diet has become easier to follow with the proliferation of gluten-free products and increasing options available, eating out and risk of contamination is still one of the main challenges of CD people.

For that reason, it is recommendable to provide specific and basic procedures for gluten-free food preparation to avoid the cross contamination at home and out of home [30].

Basic recommendations at home:

- Store gluten-free and gluten-containing foods separately and labelled clearly, especially if removed from original packaging.
- Have dedicated butters and spreads for gluten-free use
- Clean surfaces after preparing foods containing gluten as well as chopping boards, knives and other cooking utensils used in food preparation.
- Have a separate toaster or use a clean sandwich press/grill.

- Use separate water in a clean pot for cooking or re-heating gluten-free pasta. Use a separate colander for gluten-free pasta or drain it first.
- Do not dust meats or fish or cake tins with gluten-containing flour prior to cooking
- Wash hands before handling gluten-free food, especially after preparation of other food.
- Clean oil should be used when deep frying. If sharing with family, make sure the gluten-free item is fried first and then fry the gluten-containing items.

Basic recommendations out of home:

- Keep a few gluten-free snacks at work, in the car or handbag for any time of day
- Speaking to the restaurant or host prior to the event
- If appropriate, offer to help with the meal, either by bringing a gluten-free dish to share, or with preparation and serving.
- Discuss the menu and suggest gluten-free alternatives, such as brands of gluten-free sauce or stock
- If the holiday involves a flight, try to pre-arrange a gluten-free meal
- When eating out explain the situation to waiting staff and ask them if they can check the ingredients of dishes with the chef.

5 Adherence to Gluten Free Diet

Complete gluten withdrawal in patients diagnosed with CD has been shown to lead to normalization of intestinal atrophy, disappearance of the symptoms as well as improvement in the majority of related problems including osteoporosis and osteopenia, anaemia, risk of malignancy, gastrointestinal symptoms and in several studies, psychological well-being and quality of life [50–58]. However, despite the proven benefits of the GFD, it can be exceedingly difficult to completely avoid gluten-containing foods, and adherence to a GFD is estimated to be only 45–80% [7]. Generally, better dietary adherence is achieved (in 90–95%of cases), on average in the paediatric population, or in those people whose disease is diagnosed in early childhood [13]. For that reason, it is necessary to have a comprehensive understanding of the factors associated with optimal GFD adherence to develop strategies and resources to assist individuals with CD to maintain a GFD.

Physicians and dietician involved in the management of CD should insist strongly to their patients that compliance with the GFD is fundamental and is the cornerstone of the success of this treatment. They need to explain this concept convincingly to the patients, as well as the main features of the GFD, with the greatest possible clarity and simplicity at the time of diagnosis: strict adherence to the diet includes careful monitoring of ingredients, food preparation, and reading of labels will help avoid any potential cross contact and unintentional ingestion of gluten.

It is important to consider the possible factors that contribute to gluten-free diet adherence. One of the most exhaustive studies conclude that up to thirteen factors can compromise the adherence to GFD [7]. Those significantly associated with improved adherence including: understanding of the gluten-free diet, membership of a celiac disease advocacy group, and perceived ability to maintain adherence despite travel or changes in mood or stress.

This study support others results confirm by recent studies. On the one hand, that patient associations or support groups can provide help with trying to achieve proper compliance with the diet. These associations offer detailed information about the importance of a strict GFD and answer all questions related to the characteristics of gluten-free foods and different recipes. They also organize regular meetings, during which patients can share information about CD with other patients and thereby improve compliance with their diet [59].

On the other hand, the nutritional education and the educational interventions are crucial to improve the adherence. When patients feel confident about their self-efficacy to follow the gluten-free diet, it is easier for them to adhere to the GFD.

How to Measure the Adherence to GFD

Compliance with the GFD can be evaluated through different approaches, and various health professionals, may participate or collaborate to carry this out, in line with the following study procedures:

1. Periodic control visits by an expert dietician
2. Regular consultations with a gastroenterologist/family doctor
3. Structured specific questionnaires
4. Regular control of serum antibody titres for CD
5. Serial endoscopies with duodenal biopsies
6. Determination of derived peptides from gluten, in faeces/urine

The dieticians are the health professionals best placed to assess the degree of compliance with the GFD but it also needed to have regular consultations with a gastroenterologist to measure other clinical aspects. Patients with more complex needs will require a multidisciplinary approach, including various medical specialists, to assess their associated diseases and their compliance with the GFD.

Other important tools during the follow up of CD patients are the structured short questionnaires. These are used as an alternative to consultations with a dietician to obtain a rapid assessment of the adherence to the GFD. It is easy to complete this type of questionnaire in the patient's usual clinic. Most of them have been validated and the responses are highly correlated with antibody levels and the presence of villous atrophy in duodenal biopsies and useful for monitoring [59–64].

The main limitation of these questionnaire that is important to know before choosing one is that each author group has developed and validated these tools in their own countries, considering diverse clinical contexts and ones are specific validated for children [61, 62] and others for adults [63, 64]. So, before use a questionnaire we need to know all the variables that is considering and check if it is applicable to our patients.

Regarding the regular control of serum antibody titres for CD to measure the GFD compliance have an important limitation: antibody titres cannot identify the existence of small dietary transgressions, so its use is limited to indicating an obvious lack of compliance but is of no value for evaluating whether there is strict adherence to the GFD [59].

The utility of endoscopies with duodenal biopsies and the determination of derived peptides from gluten, in faeces and urine will be covered in depth in other chapters of this book.

Finally, one of the concepts that have emerged in the last decade with the rise of the self-efficacy and self-management of chronic diseases though the new technology is the Mobile health (mHealth). mHealth is a way to promote health by applying mobile technologies to improve health outcomes.

This has been also investigated in the impact of the adherence to GFD in CD patients. Nowadays, applications, not specifically defined as medical devices, that are dedicated to CD or gluten-free diets are available in many countries and languages. They are often offered by celiac organizations, mostly to provide information about gluten-free diets, recipes, products, stores, and restaurants. Most of these applications are focused on self-management strategies such as diet tracking, symptom journaling, meal-plan content, education, supplements, and recommended foods. Others include the option to interact among people with CD and share experiences. Recent studies have demonstrated that this type of application, developed by health professionals and celiac organizations are effective to increase the adherence to GFD. But it is extremely important to be sure that the application is reliable [65–68].

For that reason, it is recommendable to encourage CD people use applications endorsed by celiac organization and scientific associations to increase the adherence to GFD. But always remembering that applications are not a substitute for medical advice.

References

1. Codex Alimentarious Commission. Foods for special dietary use for persons intolerant to gluten. Rome, Italy; 2008.
2. Radhika V, Rao VSH. Computational approaches for the classification of seed storage proteins. J Food Sci Technol. 2015;52(7):4246–55.
3. Wieser H, Kieffer R. Correlations of the amount of gluten protein types to the technological properties of wheat flours determined on a micro-scale. J Cereal Sci. 2001;34(1):19–27.
4. Wieser H. Chemistry of gluten proteins. Food Microbiol. 2007;24(2):115–9.
5. Shewry PR, Halford NG. Cereal seed storage proteins: structures, properties and role in grain utilization. J Exp Bot. 2002;53(370):947–58.
6. Lionetti E, Gatti S, Pulvirenti A, Catassi C. Celiac disease from a global perspective. Best Pract Res Clin Gastroenterol. 2015;29(3):365–79.

7. Leffler DA, Edwards-George J, Dennis M, Schuppan D, Cook F, Franko DL, et al. Factors that influence adherence to a gluten-free diet in adults with celiac disease. Dig Dis Sci. 2008;53(6):1573–81.
8. Christoph MJ, Larson N, Hootman KC, Miller JM, Neumark-Sztainer D. Who values gluten-free? Dietary intake, behaviors, and sociodemographic characteristics of young adults who value gluten-free food. J Acad Nutr Diet. 2018;118(8):1389–98.
9. Vici G, Belli L, Biondi M, Polzonetti V. Gluten free diet and nutrient deficiencies: a review. Clin Nutr. 2016;35(6):1236–41.
10. Theethira TG, Dennis M, Leffler DA. Nutritional consequences of celiac disease and the gluten-free diet. Expert Rev Gastroenterol Hepatol. 2014;8(2):123–9.
11. Wierdsma NJ, Van Bokhorst-de van der Schueren MA, Berkenpas M, Mulder CJ, Van Bodegraven AA. Vitamin and mineral deficiencies are highly prevalent in newly diagnosed celiac disease patients. Nutrients. 2013;5(10):3975–92.
12. Melini V, Melini F. Gluten-free diet: gaps and needs for a healthier diet. Nutrients. 2019;11 (1):170.
13. Penagini F, Dilillo D, Meneghin F, Mameli C, Fabiano V, Zuccotti GV. Gluten-free diet in children: an approach to a nutritionally adequate and balanced diet. Nutrients. 2013;5 (11):4553–65.
14. Welstead L. The gluten-free diet in the 3rd millennium: rules, risks and opportunities. Diseases. 2015;3(3):136–49.
15. Grace-Farfaglia P. Bones of contention: Bone mineral density recovery in celiac disease—a systematic review. Nutrients. 2015;7(5):3347–69.
16. Gopalsamy GL, Alpers DH, Binder HJ, Tran CD, Ramakrishna B, Brown I, et al. The relevance of the colon to zinc nutrition. Nutrients. 2015;7(1):572–83.
17. Smulders MJ, van de Wiel CCM, van den Broeck HC, van der Meer IM, Israel-Hoevelaken T, Timmer RD, et al. Oats in healthy gluten-free and regular diets: a perspective. Food Res Int. 2018;110:3–10.
18. Poley JR. The gluten-free diet: can oats and wheat starch be part of it? J Am Coll Nutr. 2017;36(1):1–8.
19. Fritz RD, Chen Y. Oat safety for celiac disease patients: theoretical analysis correlates adverse symptoms in clinical studies to contaminated study oats. Nutr Res. 2018;60:54–67.
20. La Vieille S, Pulido OM, Abbott M, Koerner TB, Godefroy S. Celiac disease and gluten-free oats: a Canadian position based on a literature review. Can J Gastroenterol Hepatol. 2016;2016.
21. Hoffmanová I, Sánchez D, Szczepanková A, Tlaskalová-Hogenová H. The pros and cons of using oat in a gluten-free diet for celiac patients. Nutrients. 2019;11(10):2345.
22. Al-Toma A, Volta U, Auricchio R, Castillejo G, Sanders DS, Cellier C, et al. European Society for the Study of Coeliac Disease (ESsCD) guideline for coeliac disease and other gluten-related disorders. United Eur Gastroenterol J. 2019;7(5):583–613.
23. Hosseini SM, Soltanizadeh N, Mirmoghtadaee P, Banavand P, Mirmoghtadaie L, Shojaee-Aliabadi S. Gluten-free products in celiac disease: nutritional and technological challenges and solutions. J Res Med Sci. 2018;28(23):109.
24. Niro S, D'Agostino A, Fratianni A, Cinquanta L, Panfili G. Gluten-free alternative grains: nutritional evaluation and bioactive compounds. Foods. 2019;8(6):208.
25. Schoenlechner R, Drausinger J, Ottenschlaeger V, Jurackova K, Berghofer E. Functional properties of gluten-free pasta produced from amaranth, quinoa and buckwheat. Plant Foods Hum Nutr. 2010;65(4):339–49.
26. Doğan H, Karwe M. Physicochemical properties of quinoa extrudates. Food Sci Technol Int. 2003;9(2):101–14.
27. Wronkowska M, Haros M, Soral-Śmietana M. Effect of starch substitution by buckwheat flour on gluten-free bread quality. Food Bioprocess Technol. 2013;6(7):1820–7.
28. Martínez-Villaluenga C, Peñas E, Hernández-Ledesma B. Pseudocereal grains: nutritional value, health benefits and current applications for the development of gluten-free foods. Food Chem Toxicol. 2020;137:111178.

29. European Commission. 2021. https://ec.europa.eu/food/safety/labelling_nutrition/special_groups_food/gluten_en. Accessed 26 Apr 2021.
30. Association of European Coeliac Societies. 2021. https://www.aoecs.org/. Accessed 24 Apr 2021.
31. El Khoury D, Balfour-Ducharme S, Joye IJ. A review on the gluten-free diet: technological and nutritional challenges. Nutrients. 2018;10(10):1410.
32. Lee AR, Wolf RL, Lebwohl B, Ciaccio EJ, Green PH. Persistent economic burden of the gluten free diet. Nutrients. 2019;11(2):399.
33. Burden M, Mooney PD, Blanshard RJ, White WL, Cambray-Deakin DR, Sanders DS. Cost and availability of gluten-free food in the UK: in store and online. Postgrad Med J. 2015;91 (1081):622–6.
34. Panagiotou S, Kontogianni M. The economic burden of gluten-free products and gluten-free diet: a cost estimation analysis in Greece. J Hum Nutr Diet. 2017;30(6):746–52.
35. Fry L, Madden A, Fallaize R. An investigation into the nutritional composition and cost of gluten-free versus regular food products in the UK. J Hum Nutr Diet. 2018;31(1):108–20.
36. Miranda J, Lasa A, Bustamante M, Churruca I, Simon E. Nutritional differences between a gluten-free diet and a diet containing equivalent products with gluten. Plant Foods Hum Nutr. 2014;69(2):182–7.
37. Missbach B, Schwingshackl L, Billmann A, Mystek A, Hickelsberger M, Bauer G, et al. Gluten-free food database: the nutritional quality and cost of packaged gluten-free foods. PeerJ. 2015;3:e1337.
38. Calvo-Lerma J, Crespo-Escobar P, Martinez-Barona S, Fornes-Ferrer V, Donat E, Ribes-Koninckx C. Differences in the macronutrient and dietary fibre profile of gluten-free products as compared to their gluten-containing counterparts. Eur J Clin Nutr. 2019;73 (6):930–6.
39. Wu JH, Neal B, Trevena H, Crino M, Stuart-Smith W, Faulkner-Hogg K, et al. Are gluten-free foods healthier than non-gluten-free foods? An evaluation of supermarket products in Australia. Br J Nutr. 2015;114(3):448–54.
40. do Nascimento AB, Fiates GMR, Dos Anjos A, Teixeira E. Analysis of ingredient lists of commercially available gluten-free and gluten-containing food products using the text mining technique. Int J Food Sci Nutr. 2013;64(2):217–22.
41. Kulai T, Rashid M. Assessment of nutritional adequacy of packaged gluten-free food products. Can J Diet Pract Res. 2014;75(4):186–90.
42. Mazzeo T, Cauzzi S, Brighenti F, Pellegrini N. The development of a composition database of gluten-free products. Public Health Nutr. 2015;18(8):1353–7.
43. Estévez V, Ayala J, Vespa C, Araya M. The gluten-free basic food basket: a problem of availability, cost and nutritional composition. Eur J Clin Nutr. 2016;70(10):1215–7.
44. Allen B, Orfila C. The availability and nutritional adequacy of gluten-free bread and pasta. Nutrients. 2018;10(10):1370.
45. Scaramuzza AE, Mantegazza C, Bosetti A, Zuccotti GV. Type 1 diabetes and celiac disease: the effects of gluten free diet on metabolic control. World J Diabetes. 2013;4(4):130–4.
46. Tortora R, Capone P, De Stefano G, Imperatore N, Gerbino N, Donetto S, et al. Metabolic syndrome in patients with coeliac disease on a gluten-free diet. Aliment Pharmacol Ther. 2015;41(4):352–9.
47. Ciccone A, Gabrieli D, Cardinale R, Di Ruscio M, Vernia F, Stefanelli G, et al. Metabolic alterations in celiac disease occurring after following a gluten-free diet. Digestion. 2019;100 (4):262–8.
48. Jnawali P, Kumar V, Tanwar B. Celiac disease: Overview and considerations for development of gluten-free foods. Food Sci Hum Wellness. 2016;5(4):169–76.
49. Catassi C, Fabiani E, Iacono G, D'Agate C, Francavilla R, Biagi F, et al. A prospective, double-blind, placebo-controlled trial to establish a safe gluten threshold for patients with celiac disease. Am J Clin Nutr. 2007;85(1):160–6.
50. West J, Logan RF, Smith CJ, Hubbard RB, Card TR. Malignancy and mortality in people with coeliac disease: population based cohort study. BMJ. 2004;329(7468):716–9.

51. Corrao G, Corazza GR, Bagnardi V, Brusco G, Ciacci C, Cottone M, et al. Mortality in patients with coeliac disease and their relatives: a cohort study. The Lancet. 2001;358 (9279):356–61.
52. Tau C, Mautalen C, De Rosa S, Roca A, Valenzuela X. Bone mineral density in children with celiac disease. Effect of a gluten-free diet. Eur J Clin Nutr. 2006;60(3):358–63.
53. Dewar DH, Ciclitira PJ. Clinical features and diagnosis of celiac disease. Gastroenterology. 2005;128(4):S19–24.
54. Green PH, Fleischauer AT, Bhagat G, Goyal R, Jabri B, Neugut AI. Risk of malignancy in patients with celiac disease. Am J Med. 2003;115(3):191–5.
55. Zarkadas M, Cranney A, Case S, Molloy M, Switzer C, Graham I, et al. The impact of a gluten-free diet on adults with coeliac disease: results of a national survey. J Hum Nutr Diet. 2006;19(1):41–9.
56. Mustalahti K, Lohiniemi S, Collin P, Vuolteenaho N, Laippala P, Maki M. Gluten-free diet and quality of life in patients with screen-detected celiac disease. Eff Clin Pract. 2002;5 (3):105–13.
57. Pynnönen PA, Isometsä ET, Verkasalo MA, Kähkönen SA, Sipilä I, Savilahti E, et al. Gluten-free diet may alleviate depressive and behavioural symptoms in adolescents with coeliac disease: a prospective follow-up case-series study. BMC Psychiatry. 2005;5(1):1–6.
58. Addolorato G. Anxiety but not depression decreases in coeliac patients after one-year gluten-free diet: a longitudinal study. Scand J Gastroenterol. 2001;36(5):502–6.
59. Rodrigo L, Pérez-Martinez I, Lauret-Braña E, Suárez-González A. Descriptive study of the different tools used to evaluate the adherence to a gluten-free diet in celiac disease patients. Nutrients. 2018;10(11):1777.
60. Biagi F, Bianchi PI, Marchese A, Trotta L, Vattiato C, Balduzzi D, et al. A score that verifies adherence to a gluten-free diet: a cross-sectional, multicentre validation in real clinical life. Br J Nutr. 2012;108(10):1884–8.
61. Wessels MM, Te Lintelo M, Vriezinga SL, Putter H, Hopman EG, Mearin ML. Assessment of dietary compliance in celiac children using a standardized dietary interview. Clin Nutr. 2018;37(3):1000–4.
62. Pedoto D, Troncone R, Massitti M, Greco L, Auricchio R. Adherence to gluten-free diet in coeliac paediatric patients assessed through a questionnaire positively influences growth and quality of life. Nutrients. 2020;12(12):3802.
63. Leffler DA, Dennis M, George JBE, Jamma S, Magge S, Cook EF, et al. A simple validated gluten-free diet adherence survey for adults with celiac disease. Clin Gastroenterol Hepatol. 2009;7(5):530–6.e2.
64. Fueyo-Díaz R, Magallón-Botaya R, Gascón-Santos S, Asensio-Martínez Á, Palacios-Navarro G, Sebastián-Domingo JJ. Development and validation of a specific self-efficacy scale in adherence to a gluten-free diet. Front Psychol. 2018;9:342.
65. Meyer S, Naveh G. Mobile application for promoting gluten-free diet self-management in adolescents with celiac disease: proof-of-concept study. Nutrients. 2021;13(5):1401.
66. Rohde JA, Barker JO, Noar SM. Impact of eHealth technologies on patient outcomes: a meta-analysis of chronic gastrointestinal illness interventions. Transl Behav Med. 2019;11: 1–10.
67. Lee JA, Choi M, Lee SA, Jiang N. Effective behavioral intervention strategies using mobile health applications for chronic disease management: a systematic review. BMC Med Inform Decis Mak. 2018;18:1–18.
68. Pérez YIV, Medlow S, Ho J, Steinbeck K. Mobile and web-based apps that support self-management and transition in young people with chronic illness: systematic review. J Med Internet Res. 2019;21:e13579.

Follow-Up of Paediatric Patients with Celiac Disease

Ana S. C. Fernandes and Ana Isabel Lopes

1 Introduction

Celiac disease (CD) is a chronic immune-mediated systemic disorder triggered and maintained by the ingestion of gluten and related prolamins (present in cereals like wheat, barley, and rye), that occur in genetically predisposed individuals who have the human leukocyte antigen DQ2 and/or DQ8 haplotypes. It is characterized by an inflammatory enteropathy with variable degrees of severity, as well as a diversity of extra-intestinal symptoms and the presence of celiac-specific autoantibodies. Currently, the only treatment for CD is a strict lifelong gluten-free diet (GFD). Poor dietary compliance, with persistence of symptoms and villous atrophy has been associated with adverse long-term health outcomes, increased morbidity (such as fertility problems and osteopenia), increased risk of malignancy (e.g., lymphoma) and increased mortality [1, 2]. Thus, the main aims of follow-up are: to ensure adherence to a gluten-free and nutritionally adequate diet, that allows symptoms and enteropathy resolution, maintenance of normal growth and development, preventing disease complications and improving quality of life [3]. Follow-up is challenging, variable and highly influenced by the expertise of each clinical centre, since there are no published standardized evidence-based recommendations. This chapter aims to review and summarize the available evidence regarding the follow-up of children with CD.

A. S. C. Fernandes · A. I. Lopes (✉)
Paediatrics Department, Gastroenterology Unit, Hospital de Santa Maria, Centro Hospitalar Universitário Lisboa Norte, Lisbon, Portugal

A. I. Lopes
Faculty of Medicine, Paediatrics University Clinic, University of Lisbon, Lisbon, Portugal

© The Author(s), under exclusive license to Springer Nature Switzerland AG 2022   137
J. Amil-Dias and I. Polanco (eds.), *Advances in Celiac Disease*,
https://doi.org/10.1007/978-3-030-82401-3_10

2   Follow-up of Children with CD

2.1   Clinical Evaluation

Follow-up visits are important to evaluate symptoms remission, both gastroin-testinal (appetite, diarrhoea, constipation, abdominal distension and pain, nausea) and extra-gastrointestinal (mouth ulcers, fatigue, headache, school performance, skin, joint inflammation). It is also important to assess mental health status (disease acceptance, copping strategies, anxiety, depression).

Impaired growth is a frequent manifestation of CD in children. After the initi-ation of a GFD, catch-up growth usually follows, most prominently in the first 6 months [3]. Physical examination is essential for the assessment of growth [weight, height, body mass index (BMI)], pubertal stage and signs of nutritional deficiency. A satisfactory increase in height and weight is an indicator of a suc-cessful GFD. However, excessive weight gain due to a more efficient absorption, as well as dietary changes with new food choices, may follow and cardiovascular risk prevention should also be a concern.

In the first 1–2 years after CD diagnosis, follow-up visits should take place every 3–6 months, to assess for resolution of symptoms, dietary compliance, and nutri-tional adequacy. After stabilization, and serological normalization, an annual evaluation is recommended [2].

2.2   Diet Adherence and Nutritional Adequacy

Adherence to a GFD is very challenging. It is positively influenced by knowledge about CD and GFD, support by health providers, family, friends, CD support groups and national societies. However, GFD products are usually more expensive than the equivalent gluten-containing foodstuffs. Younger age and eating in a more controlled environment are associated with higher adherence, whereas eating out-side of home, fear of stigmatization and a change in patient's attitude towards more rebellious or risk-taking behaviours, predominantly during adolescence, lead to dietary transgressions [2].

At present, there is no consensus on the best method to evaluate adherence to a GFD in children with CD. Methods used to assess adherence include interview, self-report, diet records or recall, and bioassay methods such as CD serologies. Consequently, reported rates of nonadherence are very variable, ranging from 5–70% [4, 5].

Registered dieticians with expertise in CD play an essential role in patient and family education, evaluation of adherence to a GFD, identification of inadvertent gluten consumption, and intervention, being an important instrument in maintaining adherence. Furthermore, dietary counselling is vital in assessing the nutritional ade-quacy of the diet. Gluten rich cereals are an important source of fibre, complex B vitamins and iron, whereas gluten-free products are not routinely fortified [6]. Several studies have raised the concern that the GFD may be associated with nutritional imbalances, namely excessive fat, and insufficient fibre, iron, folate, magnesium, zinc, and selenium consumption [7]. A recent case–control study based on the analysis of 3-day food diaries showed that, even though the nutritional status and BMI of CD children did not differ from healthy controls, the GFD of evaluated children was nutritionally less balanced, with a higher intake of fat (total and saturated), a lower intake of fibre, and a higher consumption of processed meats and salty snacks [8].

Compared to a comprehensive dietician assessment, methods based on self-report may overestimate adherence due to reporting biases, lack of objectivity and providence of insufficient detail [4, 5].

Serological testing has been widely used as an indicator of dietary compliance and mucosal recovery, particularly IgA tissue Transglutaminase (tTG IgA) decline over 6–24 months on a GFD [1]. However, this method seems unable to detect minor or infrequent gluten consumption and to not adequately correlate with histological healing, possibly due to antibodies' long half-life and the fact that they reflect immune response rather than mucosal damage [9]. A recent meta-analysis of eleven studies including a total of 1088 paediatric and adult patients (31% with villous atrophy), calculated the diagnostic accuracy of tTG IgA and endomysial antibodies (EMA) IgA for predicting persistent villous atrophy on a GFD: sensitivity was 0.5 and 0.45, and specificity was 0.83 and 0.91 for tTG IgA and EMA IgA, respectively. These findings indicate that persistently positive serologies strongly suggest mucosal damage, indicating ongoing gluten exposure; however, a significant portion of patients with persistent enteropathy, have normal CD serologies. New monitoring tools are needed, that are sensitive to gluten ingestion and highly predictive of mucosal status [10].

Recently, there has been an increasing interest on assays based on the detection of gluten immunogenic peptides (GIPs), such as 33-mer, which result from the incomplete breakdown of gluten in the gastrointestinal tract and are detectable in faeces and urine. Comino et al. described a method to detect GIP in stool samples 6–48 h after ingestion of gluten, using a monoclonal antibody [11]. Currently, tests for detecting GIPs can detect ingestion of gluten in the past 1–7 days, in amounts as little as 50 mg (equivalent to a penne noodle), a clinically significant amount that, ingested daily, has been proven to induce mucosal damage in patients with CD. These assays have been proposed as an effective, non-invasive, objective, and quantitative assessment of short-term gluten exposure, more sensitive than dietary reports [12]. They have been suggested as a possible simple home-based method for self-assessment of dietary indiscretions and validation of the correctness of GFD in the initial period after diagnosis, when patients are still getting familiar with the diet, confirmation of inadvertent exposure to gluten in the event of acute symptoms, and assistance in the management of non-responsive and refractory CD [13]. However, there are no recommendations on the number and frequency of GIPs testing, and since these assays only reflect recent exposure, studies have suggested that with increased testing, more patients with positive GIPs will be identified, possibly leading to unnecessary and costly investigations, but not necessarily reflecting mucosal damage [14].

In the absence of non-invasive biomarkers, follow-up duodenal biopsy remains the gold standard to assess mucosal healing. Its role in the evaluation of patients with unexplained persistent or newly developed symptoms is fairly consensual. However, its invasiveness and cost make it an impractical method for routine monitoring of disease activity. Several non-invasive or minimally invasive biomarkers of mucosal status have emerged in recent investigation. Among them are several cytokines and chemokines associated with disease activity (such as IL-4, IL-10, IL-1a, IL-1b, IL-8 and IL-21) and increased serum levels seem to correlate with villous atrophy. Fatty acid binding protein 2 (I-FABP), a small cytosolic protein present in mature enterocytes and released to circulation from damaged cells, has also been shown to be present in higher levels in CD patients' plasma and to correlate with the degree of villous atrophy. Similarly, plasma citrulline, a recognized marker of functional enterocyte mass, has also been suggested as a candidate biomarker of villous atrophy [15]. Permeability tests, such as lactulose/mannitol or lactulose/rhamnose, have also been revisited in this setting as markers of mucosal damage [16]. However, all these biomarkers lack specificity, being unable to discriminate between uncontrolled CD or another gastrointestinal disorder.

2.3 Prevention of Complications and Early Diagnosis of Comorbidities

2.3.1 Delayed Growth and Puberty

Small intestinal damage, leading to malabsorption of essential nutrients can result in impaired growth and delayed puberty. Initiation of a GFD is usually followed by catch-up growth, which is usually maximal in the first 6 months. Within 1–2 years on a GFD, children generally return to their normal growth curve, depending on the extent of the disease, the age of diagnosis, and the extent of growth impairment. Earlier diagnosis is associated with greater growth recovery, while delayed diagnosis, after puberty or in adulthood, may impair the reach of their target height.

Several studies have reported a dysfunction of the endocrine axis in children with CD, suggesting that there is more to the pathogenesis of short stature in this group than nutritional deficits. Despite having normal basal levels of growth hormone (GH), CD children seem to have reduced secretion of GH in response to pharmacological stimulus, low levels of insulin-like growth factor I (IGF-I) and IGF-binding protein (IGFBP)-3, increased levels of IGFBP-2 and IGFBP-1, and a partial insensitivity to GH, since exogenous administration of GH may not restore IGF-1 levels during active CD. Additionally, there have been several reports of CD children on a GFD without catch-up growth, who tested positive for anti-pituitary or anti-hypothalamus autoantibodies, which could reflect an autoimmune hypophysitis involving somatotropic cells [17].

Absence of catch-up growth in CD children after one year on a GFD should prompt the reassessment of treatment adherence and nutritional deficiencies. If the previous show no abnormalities, and seronegativity for CD specific antibodies has been reached, an endocrinological investigation is thus required [3, 17].

2.3.2 Nutrient Deficiency

The duration of CD at diagnosis, the extent of intestinal damage and subsequent malabsorption, as well as the energy and nutritional requirements at the age of disease onset, all influence the nutritional status of the newly diagnosed child. Deficiency of nutrients absorbed in the proximal small bowel, such as iron, folate and calcium, is common. Nevertheless, delayed diagnosis and disease progression along the intestine, can lead to malabsorption of carbohydrates, fat and fat-soluble vitamins and other micronutrients. On a recent critical review of the available evidence, a group of experts has suggested that multivitamin supplementation should routinely be offered to all children with celiac disease at the time of diagnosis (quality of data: D, grade of evidence: moderate, strength of recommendation: weak) [18].

During follow-up it has been common practice to check for micronutrient deficiency, such as iron (full blood count, ferritin), calcium, folate, vitamin D and vitamin B12. As mentioned before, gluten-containing cereals are important sources of dietary calcium, folate and vitamin B12, and commercially available gluten-free products are not routinely fortified or usually have a lower content of these micronutrients than the wheat-based products that they intend to replace. However, there is limited evidence on the incidence of nutritional deficiencies in children with treated CD and guidelines from many respectable societies make no recommendation on this matter. A recent study found only mild deficiencies on a minority of the evaluated children (5–10%) and questioned the importance of routine screening for nutritional deficiencies during follow-up, once the patient is on a GFD, unless driven by specific clinical signs or symptoms, such as fatigue or growth abnormalities [19].

2.3.3 Excessive Weight Gain and Cardiovascular Risk Prevention

The nutritional profile of gluten-free food products has been increasingly questioned, with several studies performed on paediatric and adult populations reporting GFD as an unbalanced diet. A lower protein and a higher fat and sodium content of gluten-free formulations are some of the main nutritional concerns. Nonetheless, divergence among studies exists, which may be due to differences in dietary habits in populations from diverse countries, the season during which the study takes place and the variability of gluten-free products' content among brands.

The association between excessive fat, sugar and sodium ingestion and cardio-vascular disease and metabolic syndrome is well known. Several studies have reported an increase in the BMI of CD patients adhering to a GFD, sometimes leading to overweight and obesity. Several explanations have been proposed, namely: a more positive caloric balance due to the recovery of absorption capacity (compensatory hypothesis), a higher fat content and a higher glycaemic index of gluten-free products compared to their gluten containing equivalents, and the fact that the families' perception that gluten-free products are safe may lead to a higher intake of these food products [20]. In addition to this, a recent study reported a change in dietary habits in children with CD and their families after GFD initiation, towards obesogenic habits, such as an increase in junk food intake (e.g., snacks and candies), eating from the cooking pot and eating while doing other activities (e.g., eating in front of the tv, while standing or in the bedroom) [21].

2.3.4 Bone Health

Childhood is a critical period for bone health, with the maximum skeleton growth and bone mass being reached at the end of puberty. Factors that influence bone mineral accrual and bone resorption determine the peak of bone mass formation. CD can affect the bone health of children, not only due to the nutritional defi-ciencies secondary to the malabsorption syndrome (e.g., calcium, vitamin D), but also due to the chronic inflammatory status that can alter bone metabolism and lead to bone mineral loss. In untreated CD, increased production of proinflammatory cytokines (such as IL-6) and reduced levels of inhibitory cytokines (such as IL-12 and IL-18), may directly affect osteoclastogenesis and osteoblast activity, leading to bone mineral loss [22]. This can explain the findings of Blazina et al. [23], who compared the BMD of paediatric CD patients on a strict GFD with nonadherent patients (adherence being defined by negative EMA serology), concluding that, besides both groups had similar deficient calcium intake (with normal calcium serum levels) and vitamin D levels, noncompliant patients had significantly lower BMD Z-scores.

Signs and symptoms of bone health compromise in CD children include bone pain, rickets, tetany, fractures with minimal trauma, osteomalacia, osteopenia and osteoporosis, which have become progressively scarcer, due to the tendency to the earlier diagnosis of CD [18].

Several studies have reported a low bone mass density (BMD) in paediatric patients with newly diagnosed CD compared to healthy controls. However, contrary to adult patients, in children, BMD can be restored, and a normal peak BMD can be reached, especially in the youngest, by strict adherence to a GFD and age-appropriate intake of calcium and vitamin D [22].

When CD diagnosis occurs at a young age, with a short duration of symptoms, and in the absence of clinical signs of severe malabsorption, growth retardation and bone compromise, an extensive laboratorial evaluation (calcium, phosphate, alkaline phosphatase, and parathyroid hormone) and bone density imaging studies may be unnecessary and not cost-effective both at diagnosis and during follow-up. On the other hand, vitamin D levels have frequently been reported to be suboptimal, even in healthy children, and for this reason it may be prudent to check its levels [18]. After diagnosis, adherence to GFD and growth should be monitored. BMD re-evaluation is recommended for children with osteopenia/osteoporosis documented at diagnosis after one year on a GFD (and every 1–2 years until normalization), as well as those without adequate catch-up growth, or noncompliance with GFD and persistently positive specific antibodies for CD [18, 22]. In paediatric patients, BMD Z-scores, regarding healthy age and sex-matched populations should be used.

All children with CD should receive nutritional counselling regarding age-appropriate intake of calcium and vitamin D and be informed about the importance of counter-resistance exercises to promote bone health [18]. Apart from these recommendations, evidence on the adequate management of CD patients with low BMD is limited. Despite frequently prescribed, the role of supplementation with calcium and vitamin D is controversial, and the role of bone resorption inhibitors is even more unclear [22].

2.3.5 Autoimmune Comorbidities

CD can often coexist with other autoimmune disorders, owing to the shared immunopathological mechanisms linked to particular HLA haplotypes. The most frequent of these are autoimmune thyroiditis (specially Hashimoto's hypothyroidism) and type 1 diabetes mellitus (T1DM), each of them affecting 4–5% of patients with CD [3]. Conversely, CD is diagnosed in 3–12% of patients with T1DM, mostly in the first five years after T1DM diagnosis [24, 25], and up to 7% of patients with autoimmune thyroiditis, which has led NASPGHAN and ESPGHAN to recommend the routine screening for CD in patients with these autoimmune endocrine disorders [26]. Still, due to limited available data, the reverse recommendation, i.e. screening for T1DM or autoimmune thyroiditis in patients with CD, has not been made [18]. Even though formal recommendations and cost-effectiveness studies are lacking, experts argue that, in view of the possibility of subclinical thyroid abnormalities and considering the existence of effective treatment, it is reasonable to include thyroid function tests in the follow-up evaluations of CD patients [2, 18]. Since the presence of anti-thyroid antibodies in children with CD has been reported to have a low predictive value for the development of hypothyroidism during a 3-years follow-up, experts have recommended periodic monitoring of fT4 and TSH, and subsequent autoantibodies evaluation if abnormalities in the former are found [2, 27]. The periodicity of this screening during follow-up, and whether strict adherence to GFD can influence the development of thyroid dysfunction or other autoimmune comorbidities is still a matter of debate. On the other hand, since there are no known effective preventive strategies that can be applied in a prediabetic state, screening for T1DM or prediabetes, such as testing for anti-insulin antibodies, anti-glutamic acid decarboxylase antibodies or anti-zinc transporter 8 antibodies, is not recommended. Counselling for signs and symptoms of diabetes can be considered in this context [2, 27].

2.3.6 Immunization Status

Several studies have reported a diminished immunological response to vaccines in CD patients. The immunogenicity of hepatitis B vaccines has been the focus of most investigation. Several retrospective and prospective paediatric studies have confirmed a significantly higher failure rate of HBV vaccination in children with CD compared to healthy controls, which does not seem to be influenced by age. The mechanisms behind this immune response impairment are not clear, with several theories being proposed including an HLA related genetic susceptibility to impaired immune response, and competition between gliadin peptides and HBsAg for binding to HLADQ2 molecules impairing effective antibody production [28].

Studies on the immunological response to hepatitis A vaccines in patients with CD compared to sex and age-matched controls, have found controversial results, with some identifying lower immunological response to HAV vaccine in CD children [29], while others found no differences [30]. The immune response to vaccines against poliomyelitis, diphtheria, tetanus, measles, mumps, rubella, pertussis, Hib and influenza A, has also been investigated, but no differences between children with CD and sex and age-matched controls' responses have been found so far [28].

The available evidence indicating a higher failure rate of HBV vaccination in children with CD, as well as a predisposition of these patients to lose the immune response to this vaccine, has led some experts to suggest the routine screening for HBV immunization at the time of diagnosis of CD and the administration of booster doses of vaccine. The number of booster doses, reimmunization schedule and the route of administration of the vaccines is not consensual [2, 28].

Finally, the response to pneumococcal vaccine in CD patients is a matter of growing scientific interest. Pneumococcal vaccination is currently included in the routine immunization programmes for children under 2 years of age in most countries. There have been several reports of pneumococcal infection and fatal septicaemia in CD patients, especially in the presence of functional hyposplenism [28, 31], and an increased risk of bacterial pneumonia in children and young patients with CD, especially close to the time of diagnosis [32]. Some expert societies such as the British Society of Paediatric Gastroenterology, Hepatology and Nutrition; Celiac UK; World Gastroenterology Organization, and the European Society for the Study of Celiac Disease, currently recommend immunization of all patients with CD with pneumococcal vaccine [33].

2.3.7 Non-responsive and Refractory Celiac Disease

The persistence of symptoms and mucosal damage, despite apparent compliance with GFD is referred to as Non-responsive Celiac Disease (NRCD). The most common reason for NRCD is the persistent stimulation by gluten, whether intended or inadvertent, through cross-contamination or unexpected presence of gluten in food-products, medicines, or supplements.

Persistence of positive serologies for CD may confirm significant and continued lapses in dietary adherence, while GIPs detection tests can confirm recent gluten intake and thus help identify possible sources.

Dietary assessment can also be crucial in identifying sources of gluten that the patient is unaware of, lapses in cooking practices that can lead to cross-contamination, as well as for continuous patient and family education and reinforcement of compliance.

If strict compliance with GFD is confirmed, and gluten exposure can safely be discarded, other concomitant gastrointestinal conditions should be investigated, such as small intestinal bacterial overgrowth, parasite infections, irritable bowel syndrome, pancreatic insufficiency, lactose intolerance or food allergies.

Refractory celiac disease is a rare complication of CD, considered when enteropathy persists after one year of strict compliance with GFD. It is estimated to affect approximately 7–8% of adult patients, with only sporadic cases reported in paediatric patients [34]. This should be managed at a specialized referral centre, for further investigation (immunohistochemistry, PCR, flow cytometry), treatment (corticosteroids or other immunosuppressants) and follow-up (repeated biopsies and additional studies to detect complications) [6, 35].

2.4 Quality of Life

CD may have potential adverse physical and psychosocial implications for affected children/ adolescents, as any other chronic condition. Life-long adherence to GFD, which represents the cornerstone in CD management, requires significant compliance from young patients and may be difficult to follow, given the major changes in eating habits and lifestyle, taking into account the life-long duration [36].

The need to follow a GFD, associated to the chronic trait of illness, could also cause social stigma (and negative impact on peer-relationships) and a consequent lower quality of life (QoL), particularly in adolescents [37]. Additionally, certain approaches to managing a strict GFD for CD may be associated with maladaptive eating behaviours similar to known risk factors for feeding and eating disorders, experience impaired psychosocial well-being and diminished QOL, as reported by Cadenhead W et al., who recommend ongoing follow-up with gastroenterologists, dieticians and psychosocial support referrals, as needed [38]. Assessing the impact of GFD on the QoL is, therefore, a priority to be addressed with validated tools [39]. Quality of life is separately addressed in another section of the book, therefore not expanded here.

2.5 Transition into Adulthood

Despite its importance, data on the transition and care in adolescents/young adults with CD are scarce (there are no randomized trials on transition in CD) [35]. Preparation for transition to adulthood is an essential part of follow up plans of the paediatrician taking care of adolescents.

In the insightful paper from Mozer-Glassberg Y, including a retrospective electronic chart review of 387 Israeli children diagnosed with CD between January 1999 and December 2008, only 42.7% of the patients had regular out-patient gastroenterologist visits; 22% were followed by their primary care physician and over 35% were completely lost to follow-up. Negative serology on follow-up was present in 91% of the CD patients (150/165) followed at the tertiary centre in comparison to 70% (60/86) in those followed up by their primary physician ($p = 0.0002$).

It is recognized that the rate of adherence to GFD is higher in children compared to adults, but data on long-term follow-up after transition to adult care are missing. In the study of Kori et al. [40], including 441 CD patients, young age at diagnosis, regular follow-up visits in childhood, resolution of symptoms and normalization of serology before age 18, were identified as predictors of negative serology after the age of 18. Recently, Schiepatti et al. [41], have assessed determinants and trends of GFD in a cohort of 248 adult patients on a long-term follow-up from childhood. Adherence improved more frequently than worsened ($p < 0.01$), and classical symptoms at diagnosis of CD predicted stricter long-term GFD adherence. At follow-up, initial GFD adherence ($p = 0.04$) was the major determinant of long-term GFD adherence.

The implementation of a systematic transition policy in CD has been limited by a lack of clinical guidelines based on outcome-related research and clear and consistent definitions. It is not yet clear if a standardized protocol-based transition process is more efficient than a process nationally and institutionally based (particularly concerning long-term adherence to a GFD), accordingly to the different healthcare systems. Also, it has not been established if the young adult should be followed by a gastroenterologist or by a primary care physician. In many countries, depending on availability of skilled personnel (including a dietician with expertise), resources, local care delivery and practices, young adults are cared for by a general practitioner rather than by an adult gastroenterologist, both during and after transition.

The Prague Consensus Report [42], based on the best current evidence concerning transition from childhood to adulthood in CD, inference from data in other chronic diseases and pooled clinical experience, has emerged as a baseline document aiming to provide recommendations on optimal care and transition into CD adult healthcare. Transition issues will be separately addressed in another section of the book and therefore not detailed here.

2.6 Proposed Follow-Up Algorithm

Based on current evidence, Figure 1 is intended to schematically summarize a proposal for the follow-up of children with CD, the parameters to be addressed and their follow-up frequency.

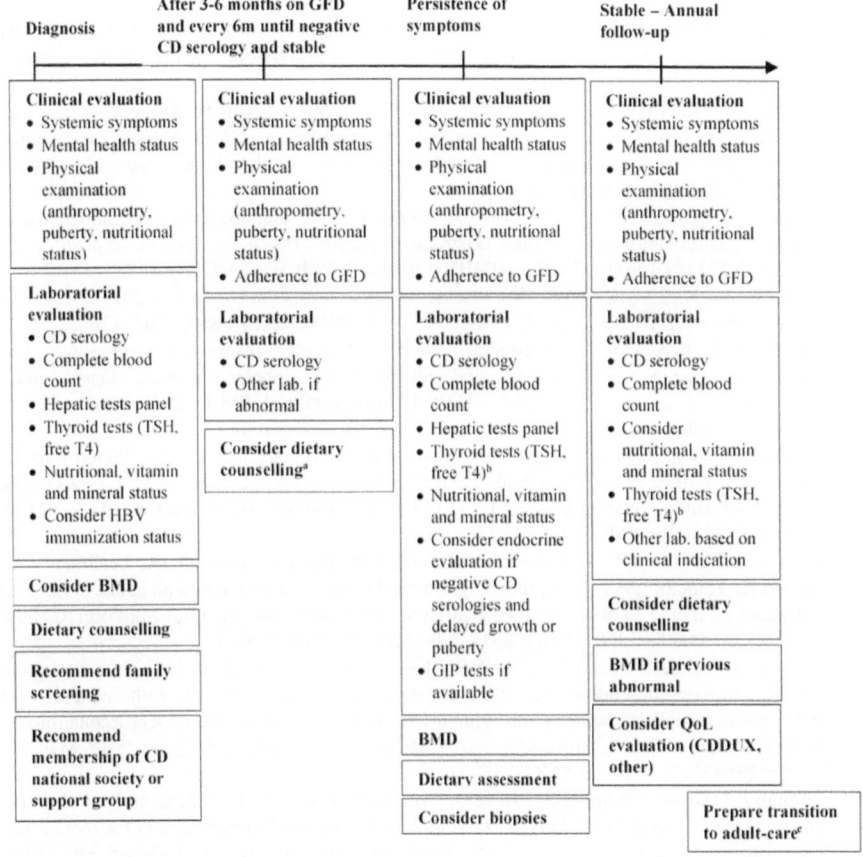

Fig. 1 Proposed follow-up algorithm (by the authors), based on current evidence. [a]Dietary counselling recommended at diagnosis and at least at the second follow-up visit, and thereafter according to need. [b]If previous abnormal thyroid function tests, include anti-thyroid antibodies. [c]Prepare transition accordingly to child and family characteristics and local health resources and practice

References

1. Hill ID, Fasano A, Guandalini S, Hoffenberg E, Levy J, Reilly N, Verma R. NASPGHAN clinical report on the diagnosis and treatment of gluten-related disorders. J Pediatr Gastroenterol Nutr. 2016;63(1):156–65. https://doi.org/10.1097/MPG.0000000000001216. PMID: 27035374.
2. Al-Toma A, Volta U, Auricchio R, Castillejo G, Sanders DS, Cellier C, Mulder CJ, Lundin KEA. European society for the study of coeliac disease (ESsCD) guideline for coeliac disease and other gluten-related disorders. United European Gastroenterol J. 2019;7(5):583–613. https://doi.org/10.1177/2050640619844125. Epub 2019 Apr 13. PMID: 31210940; PMCID: PMC6545713.
3. Valitutti F, Trovato CM, Montuori M, Cucchiara S. Pediatric celiac disease: follow-up in the spotlight. Adv Nutr. 2017;8(2):356–61. https://doi.org/10.3945/an.116.013292. PMID: 28298278; PMCID: PMC5347098.
4. Dowhaniuk JK, Mileski H, Saab J, Tutelman P, Thabane L, Brill H. The gluten free diet: assessing adherence in a pediatric celiac disease population. J Can Assoc Gastroenterol. 2020;3(2):67–73. https://doi.org/10.1093/jcag/gwy067. Epub 2018 Dec 12. PMID: 32328545; PMCID: PMC7165265.
5. Hommel KA, Mackner LM, Denson LA, Crandall WV. Treatment regimen adherence in pediatric gastroenterology. J Pediatr Gastroenterol Nutr. 2008;47(5):526–43. https://doi.org/10.1097/MPG.0b013e318175dda1. PMID: 18955858; PMCID: PMC2605852.
6. Husby S, Bai JC. Follow-up of celiac disease. Gastroenterol Clin North Am. 2019;48(1):127–36. https://doi.org/10.1016/j.gtc.2018.09.009. Epub 2018 Dec 14 PMID: 30711205.
7. Sue A, Dehlsen K, Ooi CY. Paediatric patients with coeliac disease on a gluten-free diet: nutritional adequacy and macro- and micronutrient imbalances. Curr Gastroenterol Rep. 2018;20(1):2. https://doi.org/10.1007/s11894-018-0606-0. PMID: 29356956.
8. Lionetti E, Antonucci N, Marinelli M, Bartolomei B, Franceschini E, Gatti S, Catassi GN, Verma AK, Monachesi C, Catassi C. Nutritional status, dietary intake, and adherence to the mediterranean diet of children with celiac disease on a gluten-free diet: a case-control prospective study. Nutrients. 2020;12(1):143. https://doi.org/10.3390/nu12010143. PMID: 31947949; PMCID: PMC7019969.
9. Leonard MM, Weir DC, DeGroote M, Mitchell PD, Singh P, Silvester JA, Leichtner AM, Fasano A. Value of IgA tTG in predicting mucosal recovery in children with celiac disease on a gluten-free diet. J Pediatr Gastroenterol Nutr. 2017;64(2):286–91. https://doi.org/10.1097/MPG.0000000000001460. PMID: 28112686; PMCID: PMC5457911.
10. Silvester JA, Kurada S, Szwajcer A, Kelly CP, Leffler DA, Duerksen DR. Tests for serum transglutaminase and endomysial antibodies do not detect most patients with celiac disease and persistent villous atrophy on gluten-free diets: a meta-analysis. Gastroenterology. 2017;153(3):689–701.e1. https://doi.org/10.1053/j.gastro.2017.05.015. Epub 2017 May 22. PMID: 28545781; PMCID: PMC5738024.
11. Comino I, Fernández-Bañares F, EsteveM, Ortigosa L, Castillejo G, Fambuena B, et al. Fecal gluten peptides reveal limitations of serological tests and food questionnaires for monitoring gluten-free diet in celiac disease patients. Am J Gastroenterol. 2016;111:1456–65. https://doi.org/10.1038/ajg.2016.439
12. Costa AF, Sugai E, Temprano MP, Niveloni SI, Vázquez H, Moreno ML, et al. Gluten immunogenic peptide excretion detects dietary transgressions in treated celiac disease patients. World J Gastroenterol. 2019;25:1409–20. https://doi.org/10.3748/wjg.v25.i11.1409.
13. Moreno ML, Rodríguez-Herrera A, Sousa C, Comino I. Biomarkers to monitor gluten-free diet compliance in celiac patients. Nutrients. 2017;9(1):46. https://doi.org/10.3390/nu9010046. PMID: 28067823; PMCID: PMC5295090.

14. Laserna-Mendieta EJ, Casanova MJ, Arias Á, Arias-González L, Majano P, Mate LA, Gordillo-Vélez CH, Jiménez M, Angueira T, Tébar-Romero E, Carrillo-Ramos MJ, Tejero-Bustos MÁ, Gisbert JP, Santander C, Lucendo AJ. Poor sensitivity of fecal gluten immunogenic peptides and serum antibodies to detect duodenal mucosal damage in celiac disease monitoring. Nutrients. 2020;13(1):98. https://doi.org/10.3390/nu13010098. PMID: 33396719; PMCID: PMC7824460.

15. Ramírez-Sánchez AD, Tan IL, Gonera-de Jong BC, Visschedijk MC, Jonkers I, Withoff S. Molecular biomarkers for celiac disease: past, present and future. Int J Mol Sci. 2020;21 (22):8528. https://doi.org/10.3390/ijms21228528. PMID: 33198309; PMCID: PMC7697360.

16. Khan MR, Faubion WA, Dyer R, Singh R, Larson JJ, Absah I. Role of lactulose rhamnose permeability test in assessing small bowel mucosal damage in children with celiac disease. Glob Pediatr Health. 2020;7:2333794X20969278. https://doi.org/10.1177/ 2333794X20969278. PMID: 33241082; PMCID: PMC7672748.

17. Meazza C, Pagani S, Gertosio C, Bozzola E, Bozzola M. Celiac disease and short stature in children. Expert Rev Endocrinol Metab. 2014;9(5):535–42. https://doi.org/10.1586/ 17446651.2014.932248. Epub 2014 Jun 23 PMID: 30736215.

18. Snyder J, Butzner JD, DeFelice AR, Fasano A, Guandalini S, Liu E, Newton KP. Evidence-informed expert recommendations for the management of celiac disease in children. Pediatrics. 2016;138(3):e20153147. https://doi.org/10.1542/peds.2015-3147. Epub 2016 Aug 26. PMID: 27565547.

19. Wessels MM, van Veen II, Vriezinga SL, Putter H, Rings EH, Mearin ML. Complementary serologic investigations in children with celiac disease is unnecessary during follow-up. J Pediatr. 2016;169:55–60. https://doi.org/10.1016/j.jpeds.2015.09.078. Epub 2015 Nov 5 PMID: 26547400.

20. Melini V, Melini F. Gluten-free diet: gaps and needs for a healthier diet. Nutrients. 2019;11 (1):170. https://doi.org/10.3390/nu11010170. PMID: 30650530; PMCID: PMC6357014.

21. Levran N, Wilschanski M, Livovsky J, Shachar E, Moskovitz M, Assaf-Jabrin L, Shteyer E. Obesogenic habits among children and their families in response to initiation of gluten-free diet. Eur J Pediatr. 2018;177(6):859–66. https://doi.org/10.1007/s00431-018-3128-8. Epub 2018 Mar 29 PMID: 29594339.

22. Fouda MA, Khan AA, Sultan MS, Rios LP, McAssey K, Armstrong D. Evaluation and management of skeletal health in celiac disease: position statement. Can J Gastroenterol. 2012;26(11):819–29. https://doi.org/10.1155/2012/823648. Erratum. In: Can J Gastroenterol Hepatol. 2017;2017:1323607. PMID: 23166906; PMCID: PMC3495700.

23. Blazina S, Bratanic N, Campa AS, Blagus R, Orel R. Bone mineral density and importance of strict gluten-free diet in children and adolescents with celiac disease. Bone. 2010;47(3):598–603. https://doi.org/10.1016/j.bone.2010.06.008. Epub 2010 Jun 19 PMID: 20601293.

24. Unal E, Demiral M, Baysal B, Ağın M, Devecioğlu EG, Demirbilek H, Özbek MN. Frequency of celiac disease and spontaneous normalization rate of celiac serology in children and adolescent patients with type 1 diabetes. J Clin Res Pediatr Endocrinol. 2021;13(1):72–79. https://doi.org/10.4274/jcrpe.galenos.2020.2020.0108. Epub 2020 Aug 21. PMID: 32820875; PMCID: PMC7947719.

25. Odeh R, Alassaf A, Gharaibeh L, Ibrahim S, Khdair Ahmad F, Ajlouni K. Prevalence of celiac disease and celiac-related antibody status in pediatric patients with type 1 diabetes in Jordan. Endocr Connect. 2019;8(6):780–7. https://doi.org/10.1530/EC-19-0146. PMID: 31085767; PMCID: PMC6590199.

26. Husby S, Koletzko S, Korponay-Szabó I, Kurppa K, Mearin ML, Ribes-Koninckx C, Shamir R, Troncone R, Auricchio R, Castillejo G, Christensen R, Dolinsek J, Gillett P, Hróbjartsson A, Koltai T, Maki M, Nielsen SM, Popp A, Størdal K, Werkstetter K, Wessels M. European society paediatric gastroenterology, hepatology and nutrition guidelines for diagnosing coeliac disease 2020. J Pediatr Gastroenterol Nutr. 2020;70(1):141–56. https:// doi.org/10.1097/MPG.0000000000002497. PMID: 31568151.

27. Bozzola M, Meazza C, Villani A. Auxo-endocrinological approach to celiac children. Diseases. 2015;3(2):111–21. https://doi.org/10.3390/diseases3020111. PMID: 28943613; PMCID: PMC5548236.
28. Passanisi S, Dipasquale V, Romano C. Vaccinations and immune response in celiac disease. Vaccines (Basel). 2020;8(2):278. https://doi.org/10.3390/vaccines8020278. PMID: 32517026; PMCID: PMC7349995.
29. Urganci N, Kalyoncu D. Response to hepatitis A and B vaccination in pediatric patients with celiac disease. J Pediatr Gastroenterol Nutr. 2013;56(4):408–11. https://doi.org/10.1097/MPG.0b013e31827af200. PMID: 23132166.
30. Sari S, Dalgic B, Basturk B, Gonen S, Soylemezoglu O. Immunogenicity of hepatitis A vaccine in children with celiac disease. J Pediatr Gastroenterol Nutr. 2011;53(5):532–5. https://doi.org/10.1097/MPG.0b013e318223b3ed. PMID: 21587080.
31. Röckert Tjernberg A, Bonnedahl J, Inghammar M, Egesten A, Kahlmeter G, Nauclér P, Henriques-Normark B, Ludvigsson JF. Coeliac disease and invasive pneumococcal disease: a population-based cohort study. Epidemiol Infect. 2017;145(6):1203–9. https://doi.org/10.1017/S0950268816003204. Epub 2017 Jan 23 PMID: 28112068.
32. Canova C, Ludvigsson J, Baldo V, Barbiellini Amidei C, Zanier L, Zingone F. Risk of bacterial pneumonia and pneumococcal infection in youths with celiac disease: a population-based study. Dig Liver Dis. 2019;51(8):1101–5. https://doi.org/10.1016/j.dld.2019.02.010. Epub 2019 Feb 28 PMID: 30926284.
33. Murch S, Jenkins H, Auth M, Bremner R, Butt A, France S, Furman M, Gillett P, Kiparissi F, Lawson M, McLain B, Morris MA, Sleet S, Thorpe M; BSPGHAN. Joint BSPGHAN and Coeliac UK guidelines for the diagnosis and management of coeliac disease in children. Arch Dis Child. 2013;98(10):806–11. https://doi.org/10.1136/archdischild-2013-303996. Epub 2013 Aug 28. PMID: 23986560.
34. Mubarak A, Oudshoorn JH, Kneepkens CM, Butler JC, Schreurs MW, Mulder CJ, Houwen RH. A child with refractory coeliac disease. J Pediatr Gastroenterol Nutr. 2011;53(2):216–8. https://doi.org/10.1097/MPG.0b013e318214553a. PMID: 21788766.
35. Pinto-Sanchez MI, Bai JC. Toward new paradigms in the follow up of adult patients with celiac disease on a gluten-free diet. Front Nutr. 2019;1(6):153. https://doi.org/10.3389/fnut.2019.00153. PMID: 31632977; PMCID: PMC6781794.
36. Crocco M, Calvi A, Gandullia P, Malerba F, Mariani A, Di Profio S, Tappino B, Bonassi S. Assessing health-related quality of life in children with coeliac disease: the Italian version of CDDUX. Nutrients. 2021;13(2):485. https://doi.org/10.3390/nu13020485. PMID: 33540585; PMCID: PMC7912899.
37. Olsson C, Lyon P, Hörnell A, Ivarsson A, Sydner YM. Food that makes you different: the stigma experienced by adolescents with celiac disease. Qual Health Res. 2009;19(7):976–84. https://doi.org/10.1177/1049732309338722. PMID: 19556403.
38. Cadenhead JW, Wolf RL, Lebwohl B, Lee AR, Zybert P, Reilly NR, Schebendach J, Satherley R, Green PHR. Diminished quality of life among adolescents with coeliac disease using maladaptive eating behaviours to manage a gluten-free diet: a cross-sectional, mixed-methods study. J Hum Nutr Diet. 2019;32(3):311–20. https://doi.org/10.1111/jhn.12638. Epub 2019 Mar 5. PMID: 30834587; PMCID: PMC6467807
39. Zingone F, Swift GL, Card TR, Sanders DS, Ludvigsson JF, Bai JC. Psychological morbidity of celiac disease: a review of the literature. United European Gastroenterol J. 2015;3(2):136–45. https://doi.org/10.1177/2050640614560786. PMID: 25922673; PMCID: PMC4406898.
40. Kori M, Goldstein S, Hofi L, Topf-Olivestone C. Adherence to gluten-free diet and follow-up of pediatric celiac disease patients, during childhood and after transition to adult care. Eur J Pediatr. 2021. https://doi.org/10.1007/s00431-021-03939-x. Epub ahead of print. PMID: 33515069.

41. Schiepatti A, Maimaris S, Nicolardi ML, Alimenti E, Vernero M, Costetti M, Costa S, Biagi F. Determinants and trends of adherence to a gluten-free diet in adult celiac patients on a long-term follow-up (2000–2020). Clin Gastroenterol Hepatol. 2020:S1542–3565(20)31672–4. https://doi.org/10.1016/j.cgh.2020.12.015. Epub ahead of print. PMID: 33338656
42. Ludvigsson JF, Agreus L, Ciacci C, Crowe SE, Geller MG, Green PH, Hill I, Hungin AP, Koletzko S, Koltai T, Lundin KE, Mearin ML, Murray JA, Reilly N, Walker MM, Sanders DS, Shamir R, Troncone R, Husby S. Transition from childhood to adulthood in coeliac disease: the Prague consensus report. Gut. 2016;65(8):1242–51. https://doi.org/10.1136/gutjnl-2016-311574. Epub 2016 Apr 18. PMID: 27196596; PMCID: PMC4975833.

Celiac Disease Prevention

M. Luisa Mearin

The incidence and prevalence of celiac disease (CD) have risen over time, the clinical presentation has changed dramatically in the last decades and the disease remains frequently unrecognized or undiagnosed [1, 2]. There are good biomarkers for CD and evidence based guidelines for its diagnosis [3, 4], but patients often report a delay in diagnosis that may last for years [5, 6]. In addition, CD remains frequently unrecognized and, therefore, untreated. Untreated disease is associated with long-term complications, such as chronic anaemia, delayed puberty, neuropsychiatric disturbances, infertility, small-for-date-births, osteoporosis, and, rarely, malignancy and it can reduce the quality of life [7–9]. Treatment with a gluten-free diet (GFD) reduces the burden of morbidity and mortality associated with untreated CD. Thus, prevention would be beneficial [10].

Prevention is defined as any activity that reduces the burden of mortality or morbidity from disease, taking place at the primary (avoiding disease development), secondary (early detection and treatment) or tertiary level (avoiding complications by improved treatment) [11].

The purpose of this chapter is to review the knowledge on the primary prevention of CD.

A summary of the effectivity of some primary preventive strategies is presented in Table 1.

M. L. Mearin (✉)
Department of Pediatrics, Leiden University Medical Center, Willem Alexander Children's
Hospital, Leiden, Netherlands
e-mail: l.mearin@iumc.ni

Table 1 Summary of effectivity of some primary preventive strategies for celiac disease

Strategy	Effectiveness
Breastfeeding	No
Breastfeeding at gluten introduction into the diet	No
Age at gluten introduction into the diet	No
Quantity of gluten intake early in life	Probably
Type of diet early in life	Perhaps
(Intestinal) infections	Probably
Type of delivery	No
Antibiotics	No
Microbiota	Unknown

1 Early Feeding

Data from prospective studies of large cohorts such as PREVENTCD, CELIPREV [12, 13], Generation Rotterdam [14], the Norwegian Mother and Child Cohort Study (MoBa) [15] and the Environmental Determinants of Diabetes in the Young (TEDDY) have shown that *breastfeeding and/or gluten introduction during the period of breast feeding*, do not protect against the development of CD [16].

Two randomized trials on the *age at gluten introduction into the diet of young children* did not show a relationship between early (4 months of age) or late (6 or 12 months of age) age at gluten introduction and CD development at 5 years of age [12, 13]. Interest in the quantity of gluten consumed by young children as a possible preventive risk for CD development has been present since the results of a retrospective observational study in Sweden indicated that large amounts of gluten (>16 g/day) at the time of first introduction increased the risk of CD [17]. The same group of investigators found a lower risk of CD in a big population of children born in 1997, who ingested till the age of 2 years significantly less gluten-containing cereal (24 g/day), compared to another matched population born in 1993 with a higher gluten intake (38 g/day) [18]. Also, the results from the observational TEDDY cohort, in which gluten intake was assessed by dietary questionnaires, found that high intake (>5.0 g/day) of gluten during the first 2 years of life was associated with an increased risk of CD [19]. They also found that the risk for CD increased for every 1-g/d increase in gluten consumption (HR, 1.50 [95% CI, 1.35–1.66] with an absolute risk by age of 3 years if the reference amount of gluten was consumed of 20.7% and of 27.9% if gluten intake was 1-g/d higher than the reference amount [20]. Analysis of the data from the PREVENTCD cohort showed that the amount of gluten consumed at 11–36 months of age did not influence the risk for CD development [20, 21], but further analyses of the data in this cohort are ongoing.

Recently, a secondary analysis of the Enquiring About Tolerance (EAT) trial suggested that high gluten intake from age 4 months reduced later CD development

[22]. However, the small sample size and methodological limitations of the study do not permit drawing conclusions on advisable gluten intake in infancy to prevent CD [23].

A new field of interest is the type of overall diet of young children after the weaning period and its relationship with CD development. In the prospective study of dietary patterns of young children in the Generation R project in the Netherlands, it was found that a diet characterized by high consumption of vegetables and grains and low consumption of refined cereals and sweet beverages, was associated with lower odds of CD autoimmunity [24].

Thus, *modulating the diet early on life* represents a possible preventive strategy for CD development and prospective, randomized trials, especially using different quantities of gluten in well characterized cohorts are mandatory.

2 Infections

(Intestinal) infections might change gut permeability and lead to the passage of immunogenic gluten peptides through the epithelial barrier, activating an autoimmune reaction against gluten peptides in genetic predisposed children. In such a case, prevention of infections may offer opportunities for primary prevention of CD.

Data from the PREVENTCD cohort showed no correlation between the risk for CD development and the parental-reported gastrointestinal infections in the first 18 months of life [12]. However, the TEDDY study found that parental-reported early gastrointestinal infections increased the risk of CD autoimmunity within the following 3 months (HR 1.33; 95% CI 1.11–1.59). This effect was observed particularly in those children with non-HLA-DQ2 genotypes who had been breastfed for <4 months, as well as in children born in winter and introduced to gluten before the age of 6 months [25].

Viral infections, especially Reovirus and Enterovirus have been reported as a trigger for CD development [26, 27]. In vitro, Reovirus infection induced a disruption of intestinal immune homeostasis and initiated loss of oral tolerance and T-helper inflammatory immunity to dietary antigens. In CD patients anti-Reovirus antibodies were significantly overrepresented in comparison to health controls [26]. Recently, metagenomics of the faecal virome of the TEDDY cohort showed that there is an interaction between cumulative enteroviral exposures between 1 and 2 years of age with cumulative gluten intake by 2 years of age in relation to the risk of CD and that the effect of Enteroviruses on the risk for CD autoimmunity is higher when greater amounts of gluten are consumed [28].

Seroreactivity to microbial antigens has been found in patients with freshly diagnosed CD, indicating that microbial infection might have a role in the early development of the disease [29]. Recently, crystal structures of T cell receptors in complex with HLA-DQ2 bound to bacterial peptides, demonstrate that molecular mimicry underpins cross-reactivity towards the gliadin epitopes suggesting microbial exposure as a potential environmental factor in CD [30].

3 Type of Delivery

It has been hypothesized that the mode of delivery (vaginal or caesarean section) may influence the risk for CD development, since infants born vaginally and during emergency caesarean section are colonized by faecal and vaginal bacteria of the mother, have a more diversified microbiota and this might influence the development of the mucosal immune system [31]. However, prospective studies have found no association between the type of delivery and the risk of developing CD [32–34].

4 Antibiotics

Analysis of prospective cohorts have shown that there is no evidence between the exposure to antibiotics during pregnancy or during the first years of life and CD development [35].

5 Microbiota

CD development has been linked to the composition of the gut microbiome involved in the development of early oral tolerance [36]. An association between the HLA-DQ genotype associated to CD (HLA-DQ2 and/or DQ8) and the intestinal microbiota composition has been reported in a prospective cohort of high-risk children [37]. A sub analysis of 10 CD cases and 10 matched controls, suggested altered early proportions of *Firmicutes* and members of the *Actinobacteria* phylum (*B. Longum*) in children who later progressed to CD [38]. Also, analysis of the breastmilk of the mothers of children in the PREVENTCD cohort that later developed CD showed more abundance of certain microbial species that the milk samples from mothers whose children remained healthy [39]. A recent Scottish study found a distinct microbiota profile in children with CD representing a specific biomarker of active CD [40]. However, at this moment, it is not clear whether the microbes identified in CD contribute to the pathogenesis of the disease or are the result of it. Results of prospective studies such as the Celiac Disease Genomic, Environmental, Microbiome, and Metabolomic (CDGEMM) will possibly provide answers to these open questions [41].

References

1. Ludvigsson JF, Rubio-Tapia A, Van Dyke CT, Melton LJ III, Zinsmeister AR, Lahr BD, et al. Increasing incidence of celiac disease in a North American population. Am J Gastroenterol. 2013;108:818–24. https://doi.org/10.1038/ajg.2013.60.
2. Meijer C, Shamir R, Szajewska H, Mearin L. Celiac disease prevention. Front Pediatr. 2018;6:368. https://doi.org/10.3389/fped.2018.00368. eCollection 2018. Front Pediatr. 2018. PMID: 30555808.
3. Husby S, Koletzko S, Korponay-Szabó I, Kurppa K, Mearin ML, Ribes-Koninckx C, Shamir R, Troncone R, Auricchio R, Castillejo G, Christensen R, Dolinsek J, Gillett P, Hróbjartsson A, Koltai T, Maki M, Nielsen SM, Popp A, Størdal K, Werkstetter K, Wessels M. European society paediatric gastroenterology, hepatology and nutrition guidelines for diagnosing coeliac disease 2020. J Pediatr Gastroenterol Nutr. 2020;70(1):141–56. https://doi.org/10.1097/MPG.0000000000002497.
4. Al-Toma A, Volta U, Auricchio R, Castillejo G, Sanders DS, Cellier C, Mulder CJ, Lundin KEA. European society for the study of coeliac disease (ESsCD) guideline for coeliac disease and other gluten-related disorders. United European Gastroenterol J. 2019;7(5):583–613. https://doi.org/10.1177/2050640619844125.
5. Vavricka SR, Vadasz N, Stotz M, Lehmann R, Studerus D, Greuter T, et al. Celiac disease diagnosis still significantly delayed—doctor's but not patients' delay responsive for the increased total delay in women. Dig Liver Dis. 2016;48:1148–54. https://doi.org/10.1016/j.dld.2016.06.016.
6. Riznik P, De Leo L, Dolinsek J, Gyimesi J, Klemenak M, Koletzko B, Koletzko S, Koltai T, Korponay-Szabó IR, Krencnik T, Milinovic M, Not T, Palcevski G, Sblattero D, Werkstetter KJ, Dolinsek J. The knowledge about celiac disease among healthcare professionals and patients in Central Europe. J Pediatr Gastroenterol Nutr. 2021;72(4):552–7. https://doi.org/10.1097/MPG.0000000000003019.
7. Husby S, Koletzko S, Korponay-Szabó IR, Mearin ML, Phillips A, Shamir R, et al. European society for pediatric gastroenterology, hepatology, and nutrition guidelines for the diagnosis of coeliac disease. J Pediatr Gastroenterol Nutr. 2012;54:136–60. https://doi.org/10.1097/MPG.0b013e31821a23d0.
8. Kiefte-de Jong JC, Jaddoe VW, Uitterlinden AG, Steegers EA, Willemsen SP, Hofman A, et al. Levels of antibodies against tissue transglutaminase during pregnancy are associated with reduced fetal weight and birth weight. Gastroenterology. 2013;144:726–35. https://doi.org/10.1053/j.gastro.2013.01.003.
9. Lindfors K, Ciacci C, Kurppa K, Lundin KEA, Makharia GK, Mearin ML, Murray JA, Verdu EF, Kaukinen K. Coeliac disease. Nat Rev Dis Primers. 2019;5(1):3. https://doi.org/10.1038/s41572-018-0054-z.Nat. Rev Dis Primers. 2019. PMID: 30631077.
10. Mearin ML. The prevention of coeliac disease. Best Pract Res Clin Gastroenterol. 2015;29:493–501. https://doi.org/10.1016/j.bpg.2015.04.003.
11. Maars van der PJ, Mackenbach JP. Volksgezondheid en Gezondheidszorg. Elsevier; Bunge (1999). Tweede druk [Dutch].
12. Vriezinga SL, Auricchio R, Bravi E, Castillejo G, Chmielewska A, Crespo Escobar P, et al. Randomized feeding intervention in infants at high risk for celiac disease. N Engl J Med. 2014;371:1304–15. https://doi.org/10.1056/NEJMoa1404172.
13. Lionetti E, Castellaneta S, Francavilla R, Pulvirenti A, Tonutti E, Amarri S, et al. Introduction of gluten, HLA status, and the risk of celiac disease in children. N Engl J Med. 2014;371:1295–303. https://doi.org/10.1056/NEJMoa1400697.
14. Jansen MA, Tromp II, Kiefte-de Jong JC, Jaddoe VW, Hofman A, Escher JC, et al. Infant feeding and anti-tissue transglutaminase antibody concentrations in the generation R study. Am J Clin Nutr. 2014;100:1095–101. https://doi.org/10.3945/ajcn.114.090316.
15. Størdal K, White RA, Eggesbo M. Early feeding and risk of celiac disease in a prospective birth cohort. Pediatric. 2013;132:1202–9. https://doi.org/10.1542/peds.2013-1752.

16. Andrén Aronsson CA, Lee HS, Liu E, Uusitalo U, Hummel S, Yang J, et al. Age at gluten introduction and risk of celiac disease. Pediatrics. 2015;135:239–45. https://doi.org/10.1542/peds.2014-1787.

17. Ivarsson A, Hernell O, Stenlund H, Persson LA. Breast-feeding protects against celiac disease. Am J Clin Nutr. 2002;75:914–21. https://doi.org/10.1093/ajcn/75.5.914.

18. Ivarsson A, Myléus A, Norström F, van der Pals M, Rosén A, Högberg L, et al. Prevalence of childhood celiac disease and changes in infant feeding. Pediatrics. 2013;131:e687–94. https://doi.org/10.1542/peds.2012-1015.

19. Andrén Aronsson C, Lee HS, Koletzko S, Uusitalo U, Yang J, Virtanen SM, et al. TEDDY Study Group. Effects of gluten intake on risk of celiac disease: a case-control study on a Swedish Birth Cohort. Clin Gastroenterol Hepatol. 2016;14:403–9. https://doi.org/10.1016/j.cgh.2015.09.030.

20. Andrén Aronsson C, Lee HS, Hård Af Segerstad EM, Uusitalo U, Yang J, Koletzko S, Liu E, Kurppa K, Bingley PJ, Toppari J, Ziegler AG, She JX, Hagopian WA, Rewers M, Akolkar B, Krischer JP, Virtanen SM, Norris JM, Agardh D; TEDDY Study Group. Association of gluten intake during the first 5 years of life with incidence of celiac disease autoimmunity and celiac disease among children at increased risk. JAMA. 2019;322(6):514–23. https://doi.org/10.1001/jama.2019.10329.

21. Crespo-Escobar P, MearinML, Hervás D, Auricchio R, Castillejo G, Gyimesi J, et al. The role of gluten consumption at an early age in celiac disease development: a further analysis of the prospective PreventCD cohort study. Am J Clin Nutr. 2017;105:890–6. https://doi.org/10.3945/ajcn.116.144352.

22. Logan K, Perkin MR, Marrs T, et al. Early gluten introduction and celiac disease in the EAT study: a prespecified analysis of the EAT randomized clinical trial. JAMAPediatr. 2020;174(11):1041–7. https://doi.org/10.1001/jamapediatrics.2020.2893.

23. Koletzko S, Mearin ML. Early high-dose gluten intake to prevent celiac disease: data do not allow conclusions. JAMA Pediatr. 2021. https://doi.org/10.1001/jamapediatrics.2020.6516. Online ahead of print.

24. Barroso M, Beth SA, Voortman T, Jaddoe VWV, van Zelm MC, Moll HA, Kiefte-de Jong JC. Dietary patterns after the weaning and lactation period are associated with celiac disease autoimmunity in children. Gastroenterology. 2018;154(8):2087–2096.e7. https://doi.org/10.1053/j.gastro.2018.02.024 Epub 2018 Mar 2 PMID: 29481779.

25. Kemppainen KM, Lynch KF, Liu E, Lönnrot M, Simell V, Briese T, et al. TEDDY Study Group. Factors that increase risk of celiac disease autoimmunity after a gastrointestinal infection in early life. Clin Gastroenterol Hepatol. 2017;15:694–702. https://doi.org/10.1016/j.cgh.2016.10.033.

26. Bouziat R, Hinterleitner R, Brown JJ, Stencel-Baerenwald JE, Ikizler M, Mayassi T, et al. Reovirus infection triggers inflammatory responses to dietary antigens and development of celiac disease. Science. 2017;356:44–50. https://doi.org/10.1126/science.aah5298.

27. Kahrs CR, Chuda K, Tapia G, Stene LC, Mårild K, Rasmussen T, Rønningen KS, Lundin KEA, Kramna L, Cinek O, Størdal K. Enterovirus as trigger of coeliac disease: nested case-control study within prospective birth cohort. BMJ. 2019;364:l231. https://doi.org/10.1136/bmj.l231.

28. Lindfors K, Lin J, Lee HS, Hyöty H, Nykter M, Kurppa K, Liu E, Koletzko S, Rewers M, Hagopian W, Toppari J, Ziegler AG, Akolkar B, Krischer JP, Petrosino JF, Lloyd RE, Agardh D, TEDDY Study Group. Metagenomics of the faecal virome indicate a cumulative effect of enterovirus and gluten amount on the risk of coeliac disease autoimmunity in genetically at risk children: the TEDDY study. Gut 2020;69(8):1416–22. https://doi.org/10.1136/gutjnl-2019-319809.

29. Viitasalo L, Niemi L, Ashorn M, Ashorn S, Braun J, Huhtala H, Collin P, Mäki M, Kaukinen K, Kurppa K, Iltanen S. Early microbial markers of celiac disease. J Clin Gastroenterol. 2014;48(7):620–4. https://doi.org/10.1097/MCG.0000000000000089.

30. Petersen J, Ciacchi L, Tran MT, Loh KL, Kooy-Winkelaar Y, Croft NP, Hardy MY, Chen Z, McCluskey J, Anderson RP, Purcell AW, Tye-Din JA, Koning F, Reid HH, Rossjohn J. T cell

receptor cross-reactivity between gliadin and bacterial peptides in celiac disease. Nat Struct Mol Biol. 2020;27(1):49–61. https://doi.org/10.1038/s41594-019-0353-4.

31. Penders J, Thijs C, Vink C, Stelma FF, Snijders B, Kummeling I, et al. Factors influencing the composition of the intestinal microbiota iin early infancy. Pediatrics. 2006;118:511–21. https://doi.org/10.1542/peds.2005-2824.

32. Lionetti E, Castellaneta S, Francavilla R, Pulvirenti A, Catassi C, SIGENP Working Group of Weaning and CD Risk. Mode of delivery and risk of celiac disease: risk of celiac disease and age at gluten introduction cohort study. J Pediatr. 2017;184:81–6. https://doi.org/10.1016/j.jpeds.2017.01.023.

33. Koletzko S, Lee HS, Beyerlein A, Aronsson CA, Hummel M, Liu E, et al. TEDDY Study Group. Cesarean section on the risk of celiac disease in the offspring: the Teddy study. J Pediatr Gastroenterol Nutr. 2018;66:417–24. https://doi.org/10.1097/MPG.0000000000001682.

34. Dydensborg S, Hansen AV, Størdal K, Andersen AN, Husby S. Mode of delivery is not associated with celiac disease. Clin Epidemiol. 2018;10:323–32. https://doi.org/10.2147/CLEP.S152168.

35. Kemppainen KM, Vehik K, Lynch KF, Larsson HE, Canepa RJ, Simell V, Koletzko S, Liu E, Simell OG, Toppari J, Ziegler AG, Rewers MJ, Lernmark Å, Hagopian WA, She JX, Akolkar B, Schatz DA, Atkinson MA, Blaser MJ, Krischer JP, Hyöty H, Agardh D, Triplett EW; Environmental Determinants of Diabetes in the Young (TEDDY) Study Group. Association between early-life antibiotic use and the risk of islet or celiac disease autoimmunity. JAMA Pediatr. 2017;171(12):1217–25.

36. Galipeau HJ, McCarville JL, Huebener S, Litwin O, Meisel M, Jabri B, et al. Intestinal microbiota modulates gluten-induced immunopathology in humanized mice. Am J Pathol. 2015;185:2969–82. https://doi.org/10.1016/j.ajpath.2015.07.018.

37. Olivares M, Neef A, Castillejo G, Palma GD, Varea V, Capilla A, et al. The HLA-DQ2 genotype selects for early intestinal microbiota composition in infants at high risk of developing coeliac disease. Gut. 2015;64:406–17. https://doi.org/10.1136/gutjnl-2014-306931.

38. Olivares M, Walker AW, Capilla A, Benítez-Páez A, Palau F, Parkhill J, et al. Gut microbiota trajectory in early life may predict development of celiac disease. Microbiome. 2018;6:36. https://doi.org/10.1186/s40168-018-0415-6.

39. Benítez-Páez A, Olivares M, Szajewska H, Pieścik-Lech M, Polanco I, Castillejo G, Nuñez M, Ribes-Koninckx C, Korponay-Szabó IR, Koletzko S, Meijer CR, Mearin ML, Sanz Y. Breast-milk microbiota linked to celiac disease development in children: a pilot study from the PreventCD cohort. Front Microbiol. 2020;11:1335. https://doi.org/10.3389/fmicb.2020.01335. eCollection 2020.

40. Zafeiropoulou K, Nichols B, Mackinder M, Biskou O, Rizou E, Karanikolou A, Clark C, Buchanan E, Cardigan T, Duncan H, Wands D, Russell J, Hansen R, Russell RK, McGrogan P, Edwards CA, Ijaz UZ, Gerasimidis K. Alterations in intestinal microbiota of children with celiac disease at the time of diagnosis and on a gluten-free diet. Gastroenterology. 2020;159(6):2039–2051.e20. https://doi.org/10.1053/j.gastro.2020.08.007.

41. Leonard MM, Camhi S, Huedo-Medina TB, Fasano A. Celiac disease genomic, environmental, microbiome, and metabolomic (CDGEMM) study design: approach to the future of personalized prevention of celiac disease. Nutrients. 2015;7:9325–36. https://doi.org/10.3390/nu7115470.

How to Organize the Transition from Paediatric Care to Adult Health Care

Miguel A. Montoro-Huguet and Blanca Belloc-Barbastro

1 Introduction

Celiac disease (CD) is a chronic, multiorgan autoimmune disease that affects the small bowel in genetically predisposed persons precipitated by the ingestion of gluten [1]. CD is one of the most common chronic gastrointestinal diseases [2, 3]. The treatment is primarily a gluten-free diet (GFD), which requires significant patient education, motivation, and follow-up. People with CD should be monitored regularly for residual or new symptoms, adherence to GFD, and assessment for complications. In children, special attention to assure normal growth and development is recommended [4].

The concept of transition consists of a gradual process of empowerment that equips young people with the skills and knowledge necessary to manage their own healthcare in paediatric and adult services. Effective transition has been shown to improve long-term outcomes [5]. The organization of transition is a dynamic process, aiming at ensuring continuity, coordination, flexibility, and sensitivity in a multi-disciplinary context, to meet the adolescent's clinical, psycho-social, and educational needs as well as enhance his/her abilities [6].

For chronic gastrointestinal conditions such as inflammatory bowel disease, CD, and chronic liver diseases with a paediatric onset, patients should undergo a transition process during adolescence. The transition of adolescents from paediatric to adult care is a crucial moment in managing chronic diseases such as CD. The

M. A. Montoro-Huguet (✉)
Departamento de Medicina, Psiquiatría y Dermatología, Facultad de Ciencias de la Salud y del Deporte, University of Zaragoza, 50009 Zaragoza, Spain

M. A. Montoro-Huguet · B. Belloc-Barbastro
Unidad de Gastroenterología, Hepatología y Nutrición, Hospital Universitario San Jorge de Huesca, 22004 Huesca, Spain

M. A. Montoro-Huguet · B. Belloc-Barbastro
Aragonese Institute of Health Sciences (IACS), Aragon, Spain

© The Author(s), under exclusive license to Springer Nature Switzerland AG 2022 161
J. Amil-Dias and I. Polanco (eds.), *Advances in Celiac Disease*,
https://doi.org/10.1007/978-3-030-82401-3_12

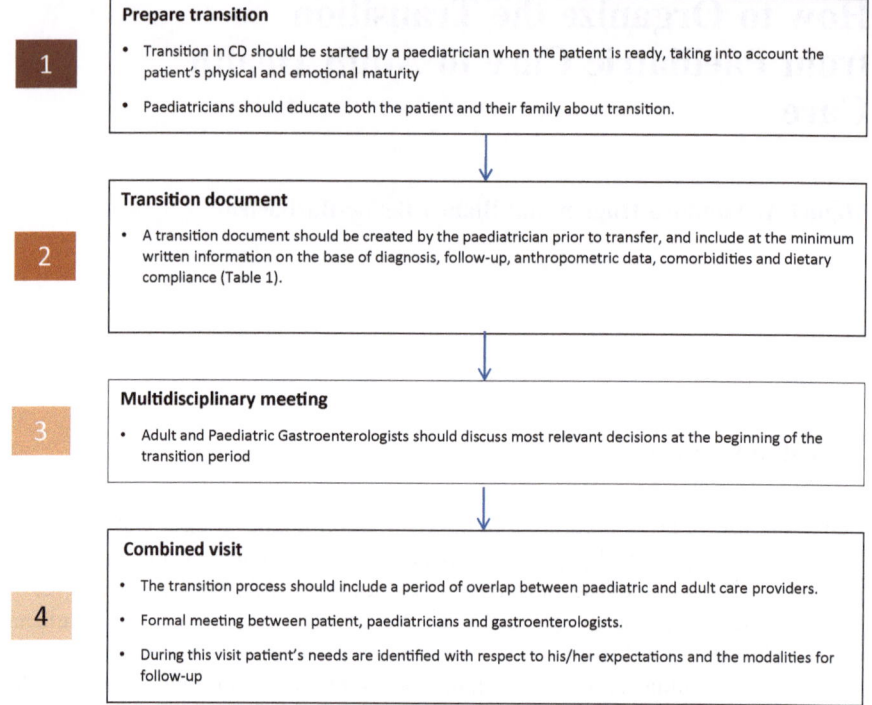

Fig. 1 Transition steps from paediatric to adult care

transition phase for young CD patients is pivotal in maintaining optimal quality of life and a long-term prognosis comparable to the general population (Fig. 1).

2 Specific Aims for Transition

The Prague consensus report [7] proposes recommendations for the management of CD in adolescents and young adults, and how to facilitate the transition to adult healthcare for patients with CD. The transition process should gradually parallel the evolution of child to adult and include an incremental transfer of responsibility for self-care to the adolescent patient with CD. Transition is a complex process, and specific aims in adolescents and young adults are:

– Encourage maturation of communication and decision-making skills.
– Allow patients to take responsibility for medical self-management.
– Education and counselling of the adolescent/young adult to manage a gluten-free diet and consequences of non-adherence.

- Recognition and treatment of psychological problems: discouragement, feeling overwhelmed, anxiety about the future and complications such as depression and eating disorders.
- Increase disease knowledge and its potential complications.
- Help the patient develop good health habits and self-care skills that encourage autonomy and establish good health habits.
- Address the family's anxieties or questions.

3   The Actual Transfer of Care

The process of transition from childhood to adulthood is characterized by physical, mental, and psychosocial development. Data on the transition and transfer of care in adolescents/young adults with CD are scarce.

Generally, paediatric transition to adult care should involve patients, their parents or caregivers, the physician, and the dietician. Although difficult to establish, a position statement [6] by the Italian Societies of Gastroenterology suggests that the ideal age for transition is between 16 and 20 years, depending on physical and emotional maturity, disease activity, adherence to treatment, and autonomy disease management. Thus, paediatricians should decide when their young patients are ready to start the transition programme. In a joint statement [8], three physician organizations suggest that the physician starts a discussion about transition when the adolescent is 12–13 years old and develops a transition plan at 14–15, with the actual transfer taking place at ≥ 18 years of age. Cultural and social differences, as well as individual patient preferences, mean variations may occur.

Ideally, at the beginning of adolescence, the paediatric Gastroenterologist should educate both the patient and their family about transition, make the patient gradually autonomous in managing his/her chronic disease, and prepare him/her for the later transfer to an adult facility. Research should take advantage of new tools to assess transition readiness (as measured by self-management and advocacy skills, rather than chronological age) to determine when a young person may be ready to transfer.

Adolescent patients are often characterized by low adherence to therapy and should be strictly monitored before starting the transition. Poor compliance to GFD among teenagers can negatively affect both quality of life and clinical course. Regarding the psychological aspects involved, children and adolescents with chronic diseases are at greater risk of long-lasting psychological distress than the general patient population, resulting in non-adherence to their treatment and follow-up regimens.

Growth impairment is a known consequence of untreated or undertreated CD [9], though many children with short stature diagnosed with CD in childhood demonstrate good catch-up growth. Untreated CD, or diagnosed after attainment of adult height, usually results in shorter final height than seen in healthy controls.

While the precise pathophysiology may be poorly understood, some adolescents and young adults with CD will experience a delay in pubertal development. When the transition is anticipated, the paediatrician should provide data regarding the patient's history of physical development and should note to the adult physician whether the patient has achieved his/her final adult height. For those patients who have experienced significant pubertal delay the paediatrician may be better suited to provide guidance and coordinate the transition to adult care after puberty, particularly if other paediatric specialists, as endocrinologists try to manage growth failure. A bone age X-ray may inform growth expectations and timing of transition [7].

Implementing a systematic transition policy in CD has been limited by a lack of clinical guidelines based on outcome-related research. In the absence of solid evidence, different models of transition will likely be developed locally. The actual transfer can take many forms. In some settings, the paediatric and adult gastroenterologists see the patient simultaneously; in others, paediatric and adult gastroenterologists meet annually to discuss patients in transition. Optimally, joint transition clinics with paediatric and adult service clinicians can be established for information delivery and generating trust in the new physician. The Prague consensus [7] recommends that the actual transfer from paediatric to adult care should be structured and include the minimum written information based on diagnosis, follow-up, anthropometric data, comorbidities, and dietary compliance.

These patients may have difficulty communicating with health providers for many reasons about communicating with adolescents and young adults. The presence of a parent can be helpful if the adolescent has not been prepared for independent visits. Young patients may have difficulty expressing sensitive concerns in person to a provider. Still, they may do so more readily by different types of electronic communication, including email, videoconferencing, SMS messaging, and online consultations. This has also been tested in paediatric groups with some success [10].

During the transition period, several issues may be discussed [7].

(1) Some adolescents/young adults may question their diagnosis and feel the transition period is a natural point for discussing how the diagnosis was made and whether re-evaluation is appropriate.
(2) In adolescence, patients with CD should gradually assume the exclusive responsibility for their care, although parental support is still important. The responsibility of keeping a GFD must be shared by the patient and his/her parents.
(3) Adolescence is recognized to be a period when adherence is poor, and these patients report lower adherence than younger children. Therefore, dietary adherence, consequences of non-adherence, and complications despite being asymptomatic are key components for discussion in a transition setting.
(4) Dietary non-adherence in adolescents is associated with increased disease burden, poorer quality of life, and increased physical symptoms. Moreover, patients should know that dietary adherence is essential before conception and during pregnancy as women with untreated CD are more likely to suffer an adverse pregnancy outcome.

Another issue that needs to be discussed is medical monitoring with laboratory tests and healthcare visits. Otherwise, allocating time and space to discuss with experts about psychological aspects could be necessary.

4 Factors Affecting the Transition

The transition of adolescents from paediatric to adult care is crucial in managing chronic diseases such as CD. A smooth transition may encounter obstacles linked to the experience of the patients and their families (caregivers) and the paediatric and adult health care providers. The most effective way to achieve a smooth transition has become a subject of considerable debate. A planned and organized transition of care for adolescents with CD is recommended, though little data are available regarding factors associated with successful transition [11].

The crucial issue of switching from a family-centred (paediatric) care model, with parents' direct involvement in the diagnostic and therapeutic decision-making process, to self-managed (adult) care, may cause a young patient to experience a sense of exclusion and fear. Some barriers to a successful transfer include:

- Lack of coordination between adult and paediatric services,
- Lack of planning and resistance of patient and families to the transition of their healthcare.

An inappropriate transition or the incomplete transmission of data from the paediatrician to the adult gastroenterologist can decrease compliance to a young patient's treatment and prognosis. Otherwise, lack of regular follow-up seems to be a particular problem for the phase of transition between paediatric and adult care [7].

A recent study [12] provides new insights in the transition of care of young adults with CD: patients diagnosed younger show poorer transition rates, and those lacking symptoms are less likely to transition to adult care. Moreover, this study suggests that lifelong adherence to a GFD may differ depending upon age of diagnosis. Individuals with CD diagnosed early in childhood have demonstrated better dietary adherence than individuals diagnosed as older children or adults. Those diagnosed in adolescence may be less adherent than their younger and older comparators.

Transitional periods, such as starting school, have been associated with diminished dietary adherence. Nevertheless, living circumstances (living with a parent or relatives living independently) did not impact the likelihood of transition. Similarly, the level of education attained, when controlling for age, did not influence follow-up. However, another study [13] suggests that compliance and quality of life improve with a better knowledge of the disease. Patients diagnosed later in life might better follow the GFD independently of their knowledge. Moreover, this study reports that self-management and knowledge improved as age increased. A direct correlation between age at diagnosis and dietary adherence is in line with a previous study [14]. Younger age at diagnosis and current age were related to dietary non-adherence. Those currently in their teens were likely to be non-adherent.

A planned and organized transition of care for adolescents with CD is recommended, though little data are available regarding factors associated with a successful transition. Table 1 shows the document used in our institution to obtain relevant information about the child who is going to be transferred to an adult unit. Further studies are needed to identify and remove barriers to transition.

5   Use of Biopsy, CD Serology and Genetic Testing in Transition to Adulthood

European Society of Paediatric Gastroenterology, Hepatology and Nutrition (ESPGHAN) 2012 guidelines [15] suggested, for the first time, that the diagnosis of CD can be made without biopsies in a subgroup of paediatric patients.

New ESPGHAN 2020 guidelines [16] support that the no-biopsy approach for CD diagnosis is confirmed to be safe in children with high TGA IgA values ≥ 10 times the upper limit of normal with accurate, appropriate tests and positive endomysial antibodies (EMA IgA) in a second serum sample. The updated review of the 2012 criteria in 2020 provide new evidence on some aspects, such as the role of HLA and the diagnosis of asymptomatic patients. This test would only be indicated for screening of at-risk individuals and in case of uncertain diagnosis. On the other hand, this guideline gives a conditional recommendation that, taking available evidence into account, CD can be diagnosed without duodenal biopsies in asymptomatic children, using the same criteria as in patients with symptoms.

Therefore, the are some differences in the use of histology for diagnostic purposes in children and adults:

- One is the no-biopsy approach in children, in selected cases; while in adults, guidelines [4, 17] emphasize the combined use of biopsy and serological analyses for diagnosis.
- Other difference is that normal architecture with increased intraephitelial lymphocytes (IELs) is considered non-specific in paediatric guidelines whereas IELs $\geq 25/100$ enterocytes have been validated as a cut-off point in adults.
- In children, Marsh 1 is not considered sufficient to diagnose CD, but some observations suggest that potential CD cases with Marsh 1 small bowel lesions have a higher chance to evolve to villous atrophy in comparison to Marsh 0. Patients with no/mild histological changes (Marsh 0/I) but confirmed autoimmunity (TGA IgA/EMA–IgA+) should be followed closely.

The no biopsy policy adopted by ESPGHAN guidelines may present a topic for discussion in paediatric to adult care. If the patient was diagnosed according to the ESPGHAN criteria is necessary to review the symptoms, results of serology, HLA status and response to GFD.

If the existing diagnostic guidelines have not been met, and the diagnosis needs re-evaluation, a new diagnostic approach should be instituted. Serology and

Table 1 Items to be included in transition document that should accompany the adolescent/young adult to adult healthcare in CD

Name
Date of birth
Diagnosis of celiac disease, year, and name of the Institution
Has the patient (or their relatives) been registered with an official patient association?
Weight and height at the time of transition (BMI)
Presentation pattern (e.g., anaemia, growth retardation, malabsorptive diarrhoea)
History of fractures (YES/NO) If so, specify
Age at the time of menarche
Serology at diagnosis (please indicate the value with range of normality)
Histology at diagnosis (please indicate grade of lesions)
HLA status if available
Associated diseases (thyroid diabetes, other)
Clinical response to gluten-free diet • Symptomatic response (YES/NO) • Histological response (if available) (None, partial, total)
Is there an associated intestinal condition as a cause of *"unresponsive celiac disease"*? – Sugar intolerance – Intestinal bacterial overgrowth – Pancreatic exocrine insufficiency – Microscopic colitis – Irritable bowel syndrome – Crohn's disease – Giardiasis – Others • Should any relevant psychosocial factor be named? (YES/NO) If so, Specify • Is there any identifiable psychiatric comorbidity? (YES/NO) If so, Specify

CD: celiac disease; BMI: Body mass index; HLA: human leukocyte antigen

histology may be part of this approach. In adolescents and young adults, biopsy to reconfirm a childhood diagnosis of CD may be considered when the tenfold positive TGA-IgA result has not been confirmed by positive EMA in a second serology at the time of diagnosis or when the ESPGHAN diagnostic criteria have not been met in a child without duodenal biopsies. Biopsies may also be relevant when the adolescent has ceased a GFD because he or she doubts the diagnosis, the patient or the physician requires documentation of healing, and the presence of symptoms suggests active CD. A gluten challenge is indicated before the biopsy. Moreover, HLA testing can be used to rule out CD in unclear cases. As the adult patient depends on his/her own judgement to follow dietary instructions it is strongly recommended that a definite diagnostic decision, based on the above-mentioned criteria is established before transition. If diagnosis is in doubt or there was inconsistent protocol, compliance may be questioned.

It is also important that the paediatric and adult physicians agree on the same criteria to avoid confusing the patient or questioning the real need for the GFD.

6   Follow up

Follow-up of patients with CD is recommended to ensure dietary adherence, prevent, or detect complications or associated conditions, including autoimmune thyroid disease, and promote optimal health. Data suggest continued follow-up improves dietary adherence. Based on expert opinion, all paediatric patients should be seen at 3–6 months intervals for the first year after diagnosis. Once symptoms have resolved and serological tests for CD have normalized, an annual follow-up visit is recommended.

CD is associated with fracture risk [18], predominantly before treatment or in the setting of non-adherence to GFD. Bone mineral density is frequently depressed in both children and adults with CD at the time of diagnosis, and deficits have been shown to correlate with the degree of histological severity. Most children recover from bone mineral density abnormalities following appropriate therapy. Thus, dual-energy X-ray absorptiometry should only be considered for young adults at high risk.

7   Primary Care Involvement

In many countries, adolescents leaving paediatric care are often cared for by a general practitioner rather than by an adult gastroenterologist. Primary care physicians (PCPs) are then also responsible for the healthcare during and after transition. In adults, PCPs may take a major role in care. Some adolescent/young adult patients are also referred to primary care when they are considered healthy after diagnostic workup information and initial follow-up in secondary care (either with a paediatrician or an adult gastroenterologist). Primary care may be a suitable care provider if adequate personnel skills and laboratory facilities are sufficient for long-term follow-up, and this may depend on local practice. In the authors' opinion, a joint follow-up by both bodies (PCPs and gastroenterologists) will be necessary in many cases.

8 Conclusions

The transition between paediatric and adult care for young people is now recognized as a key component of care, across the spectrum of physical and mental illness and disability, though there has been little high-quality evaluation published. Transferring care in an organized manner has been associated with improved outcomes, such as a greater feeling of preparedness in young patients with chronic illness and improved adherence with medical care. The transition team has the delicate task of assisting young adults and their families in understanding and appreciating the cultural and practical differences between paediatric and adult medicine. An effective transition can avoid gaps in medical care and ensure physical and mental well-being during this difficult time.

In the absence of solid evidence, different models of transition could vary both nationally and internally. Models of transition will eventually need to be evaluated in randomized controlled trials with clear patient outcome measures. Socio-economic effectiveness and outcomes of care of the different models should also be carefully evaluated.

References

1. European Society for the Study of Celiac Disease (ESsCD) guideline for celiac disease and other gluten-related disorders. United European Gastroenterol J. 2019;7(5):583–613.
2. Mustalahti K, Catassi C, Reunanen A, Fabiani E, Heier M, McMillan S. The prevalence of celiac disease in Europe: results of a centralized, international mass screening project. Ann Med. 2010;42:587–95.
3. Rubio-Tapia A, Ludvigsson JF, Brantner TL, Murray JA, Everhart JE. The prevalence of celiac disease in the United States. Am J Gastroenterol. 2012;107:1538–44.
4. Rubio-Tapia A, Hill ID, Kelly CP, Calderwood AH, Murray JA, for the American College of Gastroenterology. ACG clinical guidelines: diagnosis and management of celiac disease. Am J Gastroenterol. 2013;108:656–76.
5. Nagra A, McGinnity PM, Davis N, Salmon AP. Implementing transition: ready steady go. Arch Dis Child Educ Pract Ed. 2015;100(6):313–20.
6. Elli, Elli E, Maieron R, Martelossi S, Guariso G, Buscarini E, Conte D, et al. Transition of gastroenterological patients from paediatric to adult care: a position statement by the Italian societies of gastroenterology. Dig Liver. 2015;47(9):734–40.
7. Transition from childhood to adulthood in coeliac disease: the Prague consensus report. Gut. 2016;65(8):1242–51.

8. Cooley WC, Sagerman PJ. American Academy of Pediatrics; American Academy of Family Physicians; American College of Physicians; Transitions Clinical Report Authoring Group. Supporting the health care transition from adolescence to adulthood in the medical home. Pediatrics. 2011;128:182–200.
9. Meazza C, Pagani S, Laarej K, Cantoni F, Civallero P, Boncimino A. Short stature in children with coeliac disease. Pediatr Endocrinol Rev. 2009;6:457–63.
10. Gentles SJ, Lokker C, McKibbon KA. Health information technology to facilitate communication involving health care providers, caregivers, and pediatric patients: a scoping review. J Med Internet Res. 2010;12:e22.
11. Improving the transition between paediatric and adult healthcare: a systematic review. Arch Dis Child. 2011;96(6):548–53.
12. Reilly NR, Hammer ML, Ludvigsson JF, Green PH. Frequency and predictors of successful transition of care for young adults with childhood celiac disease. J Pediatr Gastroenterol Nutr. 2020;70(2):190–4.
13. Zingone F, Massa S, Malamisura B, Pisano P, Ciacci C. Coeliac disease: factors affecting the transition and a practical tool for the transition to adult healthcare. United European Gastroenterol J. 2018;6(9):1356–62.
14. Kurppa K, Lauronen O, Collin P, Ukkola A, Laurila K, Huhtala H, et al. Factors associated with dietary adherence in celiac disease: a nationwide study. Digestion. 2012;86:309–14.
15. Husby S, Koletzko S, Korponay-Szabó IR, Mearin ML, Phillips A, Shamir R, et al. European society for pediatric gastroenterology, hepatology, and nutrition guidelines for the diagnosis of celiac disease. J Pediatr Gastroenterol Nutr. 2012;54:136–60.
16. Husby S, Koletzko S, Korponay-Szabó I, Kurppa K, Mearin ML, Ribes-Koninckx C, et al. European society paediatric gastroenterology, hepatology and nutrition guidelines for diagnosing coeliac disease 2020. J Pediatr Gastroenterol Nutr. 2020;70(1):141–56.
17. Ludvigsson JF, Bai JC, Biagi F, Card TR, Ciacci C, Ciclitira PJ, et al. Diagnosis and management of adult celiac disease: guidelines from the British society of gastroenterology. Gut. 2014;63:1210–28.
18. Heikkilä K, Pearce J, Mäki M, kaukinen K. Celiac disease and bone fractures: a systematic review and meta-analysis. J Clin Endocrinol Metab. 2015;100:25–34.

New Therapeutic Strategies in Celiac Disease

Carmen Gianfrani, Serena Vitale, and Riccardo Troncone

1 Introduction

(a) *Celiac disease: epidemiology and pathogenesis*

Celiac disease (CD) in a chronic intestinal disorder with autoimmune treats that affects approximately 1 in 100 individuals worldwide in gluten consuming countries and is caused by a dysregulated immune response to gluten proteins of wheat and related proteins of barley and rye [1]. CD is characterized by a large spectrum of clinical presentations, with either gastrointestinal and extra-intestinal manifestations, though symptomless cases are not uncommon [2]. Both symptomatic and asymptomatic CD are characterized by the presence of HLA-risk genes, namely the DQ2- and DQ8-heterodimers encoding alleles, and serum positivity of anti-tissue transglutaminase (tTG2) antibodies. Gluten ingestion by CD patients causes a chronic inflammatory process that may lead to profound morphological changes characterized by villous atrophy and crypt hyperplasia and marked functional dysregulation [3]. However, there is a spectrum of histological alterations with at the mildest end only infiltration of the epithelium (potential CD). The diverse grade of enteropathy does not exactly match with clinical manifestations, as some patients with villous atrophy might be asymptomatic, whilst those with potential-CD may complain of severe symptoms [4].

A complex interaction between genetic, environmental and inflammatory pathways contributes to CD aetiology, with many of these factors not fully understood. As documented by several studies, CD is defined as a cell-mediated immune dis-

C. Gianfrani · S. Vitale
Institute of Biochemistry and Cell Biology, National Research Council of Italy, Naples, Italy

C. Gianfrani · R. Troncone (✉)
Department of Translational Medicine & European Laboratory for the Investigation of Food-Induced Diseases, University Federico II, Naples, Italy
e-mail: troncone@unina.it

© The Author(s), under exclusive license to Springer Nature Switzerland AG 2022   171
J. Amil-Dias and I. Polanco (eds.), *Advances in Celiac Disease*,
https://doi.org/10.1007/978-3-030-82401-3_13

ease, in which CD4+ T cells reactive to gluten peptides and restricted by HLA DQ2/DQ8 molecules have a central role, as reviewed in [5]. A step-forward in the understanding of CD pathogenesis came from the elucidation of the gluten peptide deamidation by tTG2, the CD-associated autoantigen [6]. Due to high contents of glutamine and proline, gluten proteins basically lack negatively charged amino acids, a mandatory condition to bind the HLA DQ2/DQ8 molecules. The tTG2-catalyzed deamidation of specific glutamine residues strongly increases the capability of digested gluten peptides to stimulate intestinal CD4+ T cells to produce inflammatory cytokines, mainly Interferon(IFN)-γ and Interleukin(IL)-21 [7]. To date, more than 50 different immunogenic peptides have been identified in all three gliadin families and in glutenins [8]. These gluten T cell epitopes are active both in children and adult celiac patients with no substantial differences in their immunogenicity [9, 10]. However, further investigations carried out on gut mucosal explants of either acute patients and murine models of CD-enteropathy demonstrated no direct involvement of gluten-specific CD4+ T lymphocytes in the epithelial destruction and mucosa histological changes [11]. Further evidences pinpointed the involvement of cytotoxic natural killer T cells (NKT) in the villous atrophy, through an inflammatory mechanism triggered by gluten but independent from a T cell receptor (TCR) mediated activation [11]. Notably, IL-15 cytokine, massively released by enterocytes of acute CD patients upon gluten exposure, is the main activator of NKT-mediated lysis of epithelium [11]. It was elegantly demonstrated that in the gut mucosa of acute CD patients, the adaptive anti-gluten immunity drives the massive expansion of intraepithelial NKT cells (IE-CTL), whilst IL-15 and gluten-stressed enterocytes synergistically drive their activation, by licencing to kill enterocytes [12]. Based on these relevant findings, HLA-DQ8 and IL-15 double transgenic mice have been recently developed, as model of CD-enteropathy [13]. The authors demonstrated that the overexpression of IL-15 in the epithelium and lamina propria, as well as IFN-γ released by anti-gluten CD4+ T cells resident in the lamina propria are required for the development of villous atrophy. By contrast, patients with potential CD, characterized by an anti-tTG positivity but a morphologically normal mucosa, have IE-CTL not fully activated or armed to destroy epithelial cells [12]. The same group demonstrated that viral infection, in particular a reovirus strain, is a key environmental factor contributing to the inflammatory processes that allow the loss of immune tolerance to dietary gluten proteins [14].

All together these studies have dissected the complex inflammatory mechanism induced by gluten proteins in CD patients and have provided essential knowledge for the development of new therapeutic strategies alternative to the gluten free diet (GFD).

(b) *Biochemical and immunological properties of gluten proteins*

Gluten, a heterogeneous mix of water-insoluble proteins, can be considered as the "dough treasure" due to its unique visco-elasticity properties [15]. Gluten contains hundreds of high homologous proteins that are grouped in two large families, the gliadins and glutenins families, based on monomeric or polymeric structures, respectively [15]. All gliadins and glutenins proteins are characterized by high contents of glutamine and proline (more than 50% of aminoacidic content, thus named prolamins). This peculiar amino acid composition constitutes a limitation for degradation by gastrointestinal (GI) proteases, indeed, the marked resistance to digestion results in the release into the gut lumen of large gluten fragments with a T-cell immunogenic potential [16].

(c) *Current therapy and unmet needs*

From a nutritional standpoint, gluten proteins are poor nutrients for humans, but necessary to give elastic properties to dough and high palatability to large food stuff products. However, a diet that excludes wheat, barley or rye cereals has accounted for an increased risk of nutrient deficiency and metabolic syndrome [17]. Furthermore, some patients encounter difficulty to maintain the cure over time, especially during travelling and social events, with compliance consequences. Moreover, a consistent number of gluten-free products have a high glycaemic index and are highly caloric, with a not negligible risk to develop obesity and cardio-vascular disease over the time [17]. Last but not least, there are patients suffering of refractory CD characterized by unresponsiveness to the gluten exclusion diet with serious complications including high mortality risk [18]. Based on all these constrains and risk of nutritionally unbalanced diet based on gluten exclusion, there is an unmet need to develop valid and safe drug strategies to treat celiac disease.

(d) *Aims of new strategies: replacing GFD or treating non-responsive CD as addition to GFD?*

Gluten containing cereals are largely used in the diet worldwide, furthermore gluten is an ingredient very common as additive in many foodstuffs, creating substantial risk of "gluten free" food contamination that CD community daily meet. In order to solve these issues, numerous strategies are currently under investigation that are devoted to either in vitro detoxify gluten proteins, or to provide specific drugs that supplied as an oral pill, or systemically injected, may in vivo counteract the gluten immune toxicity.

As the exclusion diet, when strictly followed, guarantees the complete recovery of small intestinal damage, with resolution of all types of symptoms (both intestinal and extraintestinal), and disappearance of CD-antibodies, the main feature of a new therapy is to be as safe and effective as the GFD. However, it still under debate if alternative therapies, whenever efficacious and safe, have to be used to counteract occasional and inadvertent exposure to gluten, or to fully replace GFD.

2 Enzymatic Approaches Reducing Gluten Load by Wheat Flour Pre-treatment

A number of procedures based on enzymatic reaction are currently under investigation to obtain wheat (or barley and rye) flour with detoxified gluten proteins to prepare pasta and baked food with no immune toxicity for CD patients [19]. These novel approaches are designed to achieve the gluten proteins detoxification before the ingestion and include the pre-digestion of flour with proteases from a mixture of acidic microorganisms and a transamidation reaction with microbial tTG2 and methyl-lysine [20–24].

(a) *Bacterial and fungal proteases hydrolysis of wheat flour*

Gobbetti and co-workers exploited the use of selective sour dough *Lactobacilli* combined with fungal proteases to completely degrade proline rich proteins [20]. They found that gluten proteins were almost completely hydrolyzed after 24 hours of fermentation and lost the immunostimulatory activity, as assessed on celiac intestinal T cells. A clinical study reported that 60 days consumption of fully hydrolyzed baked goods was highly tolerated, as no immune activation or mucosal lesion were induced [21]. This approach, although it guarantees a total degradation of gluten immunogenic sequences, alters the dough viscoelasticity and, consequently, requires the flour being integrated with structuring agents, as hydrocolloids or gelatinized supplements.

(b) *Transamidation of wheat flour*

tTG2 is a calcium dependent enzyme with a central physiological role in repairing tissue damages by catalyzing protein cross-linking through a transamidation reaction and formation of lysine-glutamine bonds. By using acyl-acceptor molecules, lysine or lysine methyl ester, as a substrate of tTG2 activity, two different research groups demonstrated that lysine-transamidated gliadin peptides lose binding affinity to HLA-DQ molecules, and consequently the capability to stimulate cognate T cells [22, 23]. A further study demonstrated that it is possible to detoxify whole wheat flour with a food-grade microbial transglutaminase (mTG), largely used by industry in order to ameliorate the texture of foods [24]. In a randomized single blinded study CD patients consumed for 90 days, 3.7 g/day of gluten in transamidated-flour bread slices. Compared to the control group eating untreated flour, a reduction of clinical, serological and gut mucosa histological relapses was observed in the experimental group. Furthermore, no volunteer completed the study in control group, by contrast, 14 out 35 completed the 90-days of transamidated gluten dietary treatment [24]. In a next phase 2 randomized double blinded study (NCT02472119), CD patients in remission were enrolled to consume for 90-days transamidated or unmodified bread and underwent endoscopy. Only a minority of volunteers (14.3%) ingesting modified bread and 57.1% ingesting regular bread presented villous atrophy and positive serology (anti-tTG2 and EMA) [25]. These clinical findings, combined with good baking properties and palatability of treated flour, make this detoxification approach very attractive for CD dietary treatment [24, 25], Table 1.

3 Approaches Reducing Gluten Contact with the Immune System

(a) *Gluten sequestering*

In the recent years, an increasing attention has been paid to biochemical strategies that aim to sequester gluten proteins in the gut lumen, thus avoiding their interaction with the gut immune system. **Chitosan**, a biocompatible aminopolysaccharide was used to produce a supramolecular compound by in vitro assembling gluten proteins. The chitosan-gluten complex displayed a marked reduction of T cell and humoral immunogenicity in preclinical analysis [26].

Similarly, **AGY Gluten Sequestering** is a strategy based on wheat gliadin protein complexation by polyclonal antibody. A single phase 1 clinical study has been completed on biopsy-diagnosed CD patients on GFD for at least 6 months who were orally administered with two capsules of 500 mg each for 4 weeks (for a total of 1 g AGY per day) before meal. The trial showed that AGY was safe and induced an improvement in celiac-associated symptoms, measured by the health-related quality of life (HRQoL) questionnaire and did not increased intestinal permeability measure by lactulose/mannitol excretion ratio (LMER). The authors claimed that AGY-based therapy is designed to neutralize 5 g of gliadin [27], Table 1.

(b) *Glutenases, endopeptidases highly efficient in degrading gluten*

Because of the pronounced resistance of gluten proteins to the intragastric degradation due to the high content of prolamin and glutamine residues, large gluten fragments remain intact in the gut lumen, being potential stimulators for inflammatory T cells. In the last two decades, great efforts were devoted to design proteases highly efficient to degrade gluten at low pH and resistant to pepsin digestion, two conditions occurring in the gastric milieu. Since the pioneristic study by Khosla and co-workers in 2002, that reported bacterial prolyl endopeptidases (PEPs) highly efficient in degrading the most immunogenic gliadin peptides, the a-gliadin 33-mer [28], several "glutenases" have been described from a variety of sources including bacteria and fungi (*Aspergillus niger, Flavobacterium meningosepticum, Sphingomonas capsulate, Actinoallomurus*), plant (*barley*), or in vitro engineered recombinant proteins with a diverse bioactivity [29, 30].

Latiglutenase–ALV003. A combination strategy has been proposed by Khosla, based on a glutamine-specific endoprotease derived from germinating barley (**EP-B2**), active under gastric condition, and on a prolyl endopeptidase from *Sphingomonas* (**SC-PEP**), that synergises with pancreatic proteases in duodenum

Table 1 Ended or current therapies targeting gluten load

Therapeutic strategy	Mechanism of action	Status	Clinical Trial Identifier	Publications	Outcome measures	Clinical trials results
mTG-transamidation	Gluten peptides transamidation	Phase 2	NCT02472119	25	Histopathological changes, serum antibodies, symptoms	A minor percentage of patients ingesting transamidated flour rusks showed positive serology (29% vs. 57%), histological damage (14% vs. 57%), and symptoms than patients taking unmodified flour
AGY	Gluten sequestring (egg yolk-derived anti-gliadin antibody)	Phase 1	NCT01765647	27	Safety and tolerability	AGY is safe and induces an improvement in celiac-associated symptoms, does not increase intestinal permeability (by LMER)
		Phase 2	NCT03707730	Not provided	Safety, symptoms, gut permeability	Not available
Latiglutenase (formerly ALV003)	Gluten proteolysis	Phase 1	NCT00626184 NCT00669825	32	Safety and tolerability	Latiglutenase is well tolerated by patients with celiac disease and healthy individuals
		Phase 2	NCT03585478	Not Provided	Villous height/crypth depth (Vh:Cd), symptom severity	Latiglutenase induces a histological damage protection
		Phase 2 (recruiting)	NCT04839575	Not Provided	Symptom scores	Reduction of symptoms severity
		Phase 2a	NCT00959114	33	Intestinal mucosal morphology, intestinal intraepithelial lymphocyte density/phenotype, serology	Latiglutenase attenuates gluten-induced mucosal injury in celiac disease patients
		Phase 2a	NCT01255696			
		Phase 2b	NCT01917630	34	Safety, intestinal mucosal morphology (Vh:Cd) ratio, intestinal intraepithelial lymphocyte density, serology	Latiglutenase does not improve histology and symptom scores in symptomatic patients
				35		Symptom improvement in seropositive and symptomatic patients

(continued)

Table 1 (continued)

Therapeutic strategy	Mechanism of action	Status	Clinical Trial Identifier	Publications	Outcome measures	Clinical trials results
AN-PEP	Gluten proteolysis	Phase 2	NCT00810654	38	Histopathological changes, serum antibodies, gluten-reactive T cells, symptoms	AN-PEP is well tolerated
		NA (Not Applicable)	NCT01335503	37	Gluten monitoring by ELISA, western blot, HPLC	AN-PEP enhances gluten digestion in the stomach of healthy volunteers
		Phase 4 (recruiting)	NCT04788797	Not Provided	Detection of gliadin immunogenic peptides (GIP) in stool and urine	Not available
TAK-062 (formely PvP001/PvP002)	Gluten proteolysis	Phase 1	NCT03594331	Not provided	Adverse events, efficacy	Not available
		Phase 1	NCT03701555	40	Safety, tolerability, pharmacokinetics, gluten degradation activity	TAK-062 is well tolerated and efficient in degrading gluten
Endopeptidase E40	Gluten proteolysis	Preclinical	NA	41	Detection of glutenasic activity by western blot, HPLC, ELISA, celiac intestinal T cells	E40 degrades gluten peptides in gastric condition and in the absence of pepsin
Larazotide (AT-1001)	Blocking epithelial permeability	Phase 3 (recruiting)	NCT03569007	Not provided	Safety, symptom severity	Not available
		Phase 1b	NCT00386165	45	Safety, pharmacokinetic, intestinal permeability	AT-1001 is well tolerated, reduces intestinal pro-inflammatory cytokine production, gastrointestinal symptoms in celiacs after gluten exposure
		Phase 2b	NCT00492960	46	Intestinal permeability changes (by LAMA), clinical symptoms and serology	AT-1001 is well tolerated, reduces gastrointestinal symptoms in celiac patients undergoing gluten challenge, no effect on intestinal permeability
		Phase 2	NCT00362856	47		
		Phase 2B	NCT01396213	48	Safety and tolerability, Symptom (CeD GSRS)	AT-1001 reduces signs and symptoms in celiac patients on a GFD

milieu [31]. This enzyme mixture was able to proteolyse complex gluten proteins in bread, as indicated by in vitro and in vivo (rat) experimental systems that simulated human gastric digestion [31]. After phase 1 studies proving that the new enzyme combination was well tolerated by CD patients and healthy individuals [32], a first phase 2a study performed in adult CD patients demonstrated that Latiglutenase-IMGX003 (formerly **ALV003**) may prevent the mucosa lesion induced by a 6-weeks daily gluten consumption (2 g/day), assessed by detection of intraepithelial CD8+ lymphocytes infiltration and villous height/crypt depth (VH: CrD) ratio, compared to placebo group. However, no differences in gluten-triggered gastrointestinal and extra-intestinal symptoms were observed between the two treatment groups [33]. A second phase 2a study done in a larger cohort reported that Latiglutenase significantly attenuates clinical manifestation severity in anti-tTG2 seropositive CD patients on GFD [34, 35]. Currently, two additional phase 2a/b studies are ongoing. The first (a single centre, randomized, double-blind, placebo controlled/gluten challenge, NCT03585478) has completed the recruitment phase whilst the second one (multicentre, prospective, randomized, double-blind, placebo-controlled/crossover study, NCT04243551) aims to assess the efficacy and safety of latiglutenase treatment in symptomatic CD patients on GDF occasionally exposed to gluten (Table 1).

AN-PEP. PEPs derived from *Aspergillus Niger* (**AN-PEP**) were shown to efficiently degrade gluten to small non-immunogenic peptides in low and high caloric meals. The glutenasic activity was assessed by either a dynamic in vitro system that closely mimics the human gastrointestinal tract [36], and in vivo in healthy volunteers through a catheter used for meal infusion and for aspiration of gastric-duodenal juices [37]. In a randomized double-blind placebo-controlled pilot study in which CD patients consumed toast containing approximately 7 g of gluten daily for 2 weeks, it was demonstrated that AN-PEP was well tolerated. By contrast, this study failed to assess drug efficacy in preventing gluten immune toxicity as, no sign of duodenal immune activation, duodenal morphological changes, as well as, serum antibody positivity (primary study endpoints) were reported in placebo control group [38], Table 1.

Kuma030–TAK-062. Since the above described pioneristic studies, other proteolytic enzymes are currently under investigations for their ability to degrade gluten under gastric conditions and be suitable as oral enzyme supplementation for treatment of gluten intolerance. A third-generation enzyme, computationally designed and molecular engineered as recombinant protein, the **Kuma030** (now **TAK-062**) was reported to be efficient in catalyzing in vitro the digestion of gluten proteins in the stomach acidic condition, by degrading more than 90% of gluten protein load [39]. In a phase 1 single-centre study (NCT03701555), TAK-062 efficacy in gluten degradation was evaluated either in vitro by a dynamic gastric model (DGM) and in vivo in healthy and CD patients. The DGM simulated gastric digestion of two different meals containing 3–9 g gluten, and TAK-062 (100 and 300 mg) in the presence of pepsin (0.6 g/ml) showed a marked gluten degradation up to 99% within 10 minutes. In vivo experiments showed that in volunteers orally administered with homogenized meals added with gluten proteins (1, 3, 6, and 9 g), and a single scaling dose of TAK-062 (100, 300, 900 mg), the intragastric residual gliadin immunogenic peptides, quantified by competitive R5 and G12 ELISA, were almost totally degraded within 20–65 min. Furthermore, TAK062 was well tolerated, as assessed by pharmacokinetics and symptoms evaluation [40], Table 1.

E40. A recent study reported a novel protease of microbial origin, the endoproptease-40 (**E40**) recombinantly produced in *Streptomyces lividans*. Similarly to the previous proteases, E40 showed a high resistance to pepsin digestion at gastric pH and marked proteolytic activity, demonstrated by the extensive degradation of whole gliadin proteins and of the known immunogenic peptides of alfa-gliadin of 33-aminoacid length [41] as tested by SDS gel electrophoresis, mass spectroscopy, and functional assay with celiac intestinal T cells. Of note, E40 degraded gluten within 30 min of incubation at low gliadin:enzyme weight ratio (20:1) and in the absence of pepsin. Altogether, these studies showed that glutenases are highly promising approaches, Table 1. However, further studies are needed to evaluate: (i) the maximal amount of complex food containing gluten proteins within lipid and starch matrices that may be digested, (ii) if the glutenase-based pills are able to protect patients with CD from an occasional inadvertent gluten exposure or from a long-term consumption of a regular gluten containing meal.

4 Approaches Affecting Intestinal Permeability: Zonulin Receptor Agonist

Larazotide acetate–AT1001. Very little is known about the mechanisms trough gluten peptides resulting from GI digestion cross the epithelial layer before encountering immune cells. In non-inflamed condition, the intestinal epithelium is almost impermeable, due to the intercellular tight junctions. An increased permeability has been reported in gut mucosa of CD patients that could partially be mediated by the gluten-triggered upregulation of zonulin [42]. A zonulin receptor peptide agonist that blocks zonulin binding to specific receptor on enterocytes has been largely investigated as potential drug for CD treatment [43, 44]. Phase 1 and 2 studies have shown that Larazotide-AT1001 successfully passed the safety analysis and was well tolerated. With the exception of the pilot clinical studies [45–47] in which an oral supplementation of Larazotide acetate was given for three times daily to patients challenged with 2.5 g gluten for 2 weeks, the last published trial was designed to assess its beneficial effect to ameliorate health condition in symptomatic patients on gluten free diet [48], Table 1.

Other clinical studies are necessary to investigate the Larazotide acetate efficiency to prevent gluten immune toxicity in CD patients, although recently the nature of AT1001 has been questioned [49].

5   Approaches Based on Immunomodulation

(a) *Gluten tolerization by nanoparticles oral supplementation*

For the treatment of CD, different therapeutic approaches based on immunomodulation could be used targeting T cells reactive to gluten proteins and restoring the
immune tolerance. One of these potential therapeutic approaches involves the use of
"*Tolerogenic Immune Modifying Nanoparticles*" encapsulating gliadin protein
(**TIMP-GLIA**, now **TAK-101**), small particles formed with a food grade polymer
(poly-lactide-co-glycolide-PLGA) containing inside wheat gliadin [50]. It was
shown that TIMP-GLIA nanoparticles are able to induce tolerance to gliadin in
mouse model of CD, after systemic administration, with a reduction in enteropathy
and inflammatory cytokine production [50]. To date, after a phase 1 study
(NCT03486990) that assessed the safety and tolerability of TAK-101 in CD
patients without any serious side effects, a phase 2 randomized, double-blind,
placebo-controlled study (NCT03738475) has been completed. In this phase 2 trial,
a cohort of 34 CD patients on GFD received TIMP-GLIA or placebo treatment and,
after 7 days, consumed gluten for 14 days, as following: 12 g for the first 3 days
followed by 6 g for the next 11 days [51]. The group of CD subjects treated with
TIMP-GLIA, at day 6 of gluten challenge, showed a significant decrease of
gliadin-specific T cells response (change from baseline in circulating
gliadin-specific, IFNγ–producing cells) and reduced number of circulating α4β7+
CD4+ T cells, TCRγδ+ T cells and memory effector Th cells, compared to placebo
group [51]. Another phase 2 study (NCT04530123) is ongoing, although it has not
yet recruited patients, to evaluate the optimal dosing of TIMP-GLIA for the
administration to CD patients during gluten challenge, Table 2.

(b) *Gluten peptide-based immunomodulatory strategies*

Nexvax2, is a potential desensitizing vaccine for CD treatment that consists of a
combo-peptide that includes three gluten peptides of 15–16 amino acid length
responsible for the immune reaction elicited by gluten ingestion in the great
majority of CD patients [9, 10]. Multicentre phase 1 clinical studies
(NCT00879749, NCT02528799, NCT03543540) assessed the safety, tolerability
and bioactivity of Nexvax2 [52–54].

   Moreover, to further investigate the effect of Nexvax2, a phase 2 clinical trial
(NCT03644069) has been started on adult HLA-DQ2.5 CD patients. Despite the
encouraging results on safety and tolerability of Nexvax2 obtained in the phases I,
as indicated by a press release of the Company ImmusanT, the study has been
discontinued since the compound did not provide protection against symptoms
induced by gluten exposure compared with placebo.

(c) *tTG2 inhibitors*

Inhibitors of tissue transglutaminase type 2 (tTG2) have been designed to prevent
the deamidation of gliadin peptides, a key post-translational step conferring high
immunogenicity/immunotoxicity to gluten proteins [5, 6, 8]. Several gluten-

mimetic peptides were developed to block the activity of tTG2, and among these, three gluten peptides analogues: **ZED1098, ZED1219, and ZED1227** that covalently bind with the cysteine in the active site and irreversibly block the enzyme. In particular, following the preclinical data on mouse models showing the reduced intestinal inflammation by the inhibitor ZED1227 [55], a phase 1 clinical trial proved its safety. A phase 2a, double-blind, randomised, placebo-controlled, dose-finding study was carried out (EUDRA CT 2017-002241-30) to test the efficacy and tolerability of a 6-week treatment with ZED1227 capsules vs. placebo, Table 2, in subjects with well-controlled CD during gluten challenge; to date, the results have not yet been published.

(d) *Anti-IL15 therapy*

The widely demonstrated over-expression of IL15 in the intestinal mucosa of CD patients makes of great scientific interest the development of drugs targeting the inhibition of IL15 production and/or its signalling pathways. **AMG714** is the first human monoclonal antibody anti-IL15, developed by Amgen, a biotechnology company, tested for the treatment of CD, being able to block all forms of IL15 stopping its activities. This drug was investigated in two phase IIa randomized, double-blind, placebo-controlled studies. One study (NCT02637141 and EUDRA CT 2015-003647-19) was conducted on 64 CD patients on GFD for at least 12 months, that received 150 mg or 300 mg AMG174 or placebo for a total of six doses, with subcutaneous injections every 2 weeks for 10 weeks, during a gluten challenge (2–4 g daily, for 2–12 weeks). AMG 714 did not prevent mucosal damage due to gluten challenge, though the density of intraepithelial T lymphocytes (IELs) was less increased in the group treated with the highest dose. Moreover, no serious adverse effects were observed in the study [56]. In the other phase 2a trial (NCT02633020), AMG174 was investigated in patients with refractory CD type 2 of which 19 received seven intravenous doses over 10 weeks (8 mg/kg) and 9 received placebo. Ameliorative effects by AMG174 were observed on the symptomatology, whereas no difference between drug and placebo groups was found in the reduction of aberrant IELs from baseline [57], Table 2.

(e) *Blocking cell gut migration (anti-CCR9/α4β7 integrin)*

Aimed to specifically block gut migration of gluten reactive T cells in CD patients, other therapeutic approaches were investigated, as CCR9 receptor antagonist CCX282-B, and α4β7 integrin antagonist PTG-100. A randomized, double-blind, placebo-controlled, phase 2 study (NCT00540657) tested the CCX282-B (250 mg capsule, twice daily for 13 weeks) in mitigating the effects of gluten ingestion in 90 patients, on GFD for at least 24 months, in terms of mucosal damage, serology and symptoms, Table 2. The study has already been completed, although the publication of the results is still pending.

To test PT-100, a phase 1b randomized, double-bind, placebo-controlled study is, to date, in recruitment status (NCT04524221). The clinical trial will evaluate the safety and efficacy of PTG-100 in preventing gluten-induced inflammatory injury to the small intestine in 30 CD patients, to whom will be administrated either

Table 2 Ended or current therapies targeting immuno stimulatory properties

Therapeutic strategy	Mechanism of action	Status	Clinical Trial Identifier	Publications	Outcome measures	Clinical trials results
TAK-101 (TIMP-GLIA)	Gluten tolerization	Phase 1	NCT03486990	51	Safety, tolerability, pharmacokinetics in celiac patients; circulating gliadin-specific T cells after 3- and 14-days gluten oral challenge (by ELISA and ELISPOT), intestinal mucosal morphology (Vh:Cd, and IELs), symptoms	TAK-101 was well tolerated and prevents gluten-induced immune activation
		Phase 2	NCT03738475			
		Phase 2 (Not yet recruiting)	NCT04530123	Not provided	Circulating gliadin-specific T cells after 3- and 14-days gluten oral challenge (by ELISA and ELISPOT)	Not available
ZED1227	tTG2 inhibitor	Phase 2a	EUDRA CT 2017–002,241-30	Not provided	Safety, efficacy and tolerability of a 6-week treatment in subjects with well-controlled CD during gluten challenge. Small-bowel mucosal morphology, inflammation, serology, adverse events	Not available
AMG 714 (PRV-015)	Human monoclonal antibody against interleukin 15	Phase 2b (Recruiting)	NCT04424927	Not provided	Symptoms, IELs density, PRV-015 and anti-PRV-015 antibodies serum concentrations	Not available
		Phase 2a	NCT02637141	56	Levels of serum antibodies; % IELs by flow cytometry and IHC; VH:CD Ratio and Marsh score	AMG 714 does not improve VH:CD ratio, affects the intraepithelial

(continued)

Table 2 (continued)

Therapeutic strategy	Mechanism of action	Status	Clinical Trial Identifier	Publications	Outcome measures	Clinical trials results
		Phase 2a	NCT02633020	57	Levels of serum antibodies; % IELs by flow cytometry and IHC; VH:CD Ratio and Marsh score	lymphocyte density and symptoms / AMG 714 in patients with RCDII does not reduce the number of aberrant intraepithelial lymphocytes, effects on symptoms
CCX282-B	CCR9 antagonist	Phase 2	NCT00540657	Not provided	Mucosal damage (Vh:Cd) and inflammation, serology and symptoms after gluten ingestion in treated celiacs	Not available
PT-100	Anti-integrin α4β7	Phase 1	NCT04524221	Not provided	Mucosal damage (Vh:Cd), intraepithelial lymphocyte density, serology and symptoms after gluten ingestion in treated celiacs	Not available
Hookworm Necator Americanus	Gluten tolerization	Phase 2a	NCT00671138	58	Duodenal histology, quantification of α-gliadin peptide (QE65)-specific systemic interferon-γ-producing cells (by ELISPOT) pre- and post-wheat challenge	Hookworm infection does not induce an improvement in histological damage
		Phase1/2	NCT01661933	60	Safety and efficacy; Duodenal histology (VH:CD Ratio and Marsh score), % IELs, levels of serum antibodies post low-dose of gluten challenge	Hookworm infections promote tolerance and stabilize or improve all tested indices of gluten toxicity in CeD subjects

(continued)

Table 2 (continued)

Therapeutic strategy	Mechanism of action	Status	Clinical Trial Identifier	Publications	Outcome measures	Clinical trials results
VIVOMIXX®	Probiotic supplementation	Phase 4	NCT04160767	Not provided	Vitamin B6, B12, 25' hydroxy vitamin D, folic acid and omocystein serum levels and inflammatory markers, metabolomics on stool samples in the CD patients on GFD	Not available

placebo or PTG-100 (600 mg taken twice daily in capsule form), for 42 days. They will also receive gluten challenge, a cookie or equivalent, twice daily. A small bowel mucosa biopsy will be performed at the start and the end of the treatment period to evaluate villous height-to crypt ratio and IELs density. Blood samples will be routinely taken to evaluate tTG2 antibody levels while the symptoms will be recorded using the celiac symptoms index (CSI) survey, Table 2.

(f) *Immunomodulation with Necator-hookworm*

Several clinical trials are evaluating the suppression of mucosal inflammation in CD by experimental infection with hookworm *Necator americanus*. A phase 2a randomized, double blinded, placebo controlled study tested the effect of *Necator americanus* larvae inoculation (at week 0 and 12) on the suppression of the immune response induced by gluten (16 g of gluten daily for 5 days at week 20-group) in 20 GFD-treated CD patients (NCT00671138). No clear protective effects were reported of this trial by Daveson et al. [58], although the basal production of IFNγ and IL17A from duodenal biopsy culture was suppressed in hookworm-infected compared to uninfected patients [59].

In another clinical trial, a 52-week phase1/2 study (NCT01661933), 12 CD adult patients were inoculated with 20 *Necator americanus* larvae and received escalating gluten challenges as pasta. The results shown the combination of infection and gluten micro-challenge promoted tolerance and stabilized or improved all evaluated indices of gluten toxicity (mucosal damage, symptoms and the percentage of inflammatory and regulatory T cells), [60]. A phase 1b multicentre clinical trial (NCT02754609) was completed in 2019, not followed by the publication of the data yet.

(g) *Evaluation of probiotic supplementation*

Dysbiosis could play a key role in the pathogenesis of CD influencing the intestinal permeability and the regulation of the immune system. The administration of probiotics might potentially represent a novel strategy to treat CD. The effects of several probiotic formulations have been investigated in mouse models, demonstrating a modulatory activity on innate and adaptive immune responses activated in CD [61, 62].

Several clinical trials assessed the safety and efficacy of different probiotic strains in the treatment of CD. A mixture of probiotics containing 8 different strains of bacteria (*Streptococcus thermophilus DSM 24731*, bifidobacteria *B. breve DSM 24732*, *B. longum DSM 24736*, *B. infantis DSM 24737*, lactobacilli *L. acidophilus DSM 24735*, *L. plantarum DSM 24730*, *L. paracasei DSM 24733*, *L. delbrueckii subsp. bulgaricus DSM 24734*) attenuated the inflammation and symptomatology of colitis in induced colitic mice models, [63, 64]. This preparation is currently under investigation in a phase 4 study (NCT04160767). The study started in 2019 involves enrolling 90 CD patients on GFD and will test the effect of this probiotic mixture on vitamin B6, B12, 25'OH D, folic acid and omocystein levels, metabolic and inflammatory status, and gut microbiota metabolomics, in the CD patient group that received the drug compared to placebo CD group.

6 Conclusion

GFD is associated with high economic and societal burden, decreased quality of life and in some cases not satisfactory response. For these reasons the search for therapies alternative to GFD has become a priority. In fact, an increased comprehension of pathogenetic mechanisms has revealed new therapeutic targets and also new biomarkers useful to assess the efficacy of new treatments. Although we still lack an animal model recapitulating all the features of CD, progress in this area have contributed to test new strategies particularly those based on immunomodulation [50, 51, 59, 60].

One of the unsolved problems is how to evaluate the response to the new drugs and which outcomes to privilege: gluten-dependent symptoms, gluten-specific T cell response, CD-specific autoantibodies and gut histology after prolonged gluten challenge have been considered. Ideally, the protection of the intestinal mucosa from gluten-induced damage should represent the gold standard, but so far, only in one study investigating the effect of an enzymatic preparation [33] such a goal has been reached. Most approaches have been tested for their ability to attenuate symptoms in patients non responsive to GFD. In fact, the relationship between symptoms and objective endpoints such as gluten-specific T cell response, autoantibodies and gut histology after gluten challenge has not been fully understood. One other important limitation is the lack hitherto of paediatric studies, a part those based on the use of probiotics; the suitability of such approaches in children remains to be assessed.

In conclusion, at moment no drug has been licenced for the therapy of CD and GFD remains the cornerstone of the treatment. On the other hand, the most promising candidates have entered phase 3 trial and we may expect that advances in the comprehension of CD pathogenesis will help to identify new targets and new strategies (Fig. 1). In such respect, CD remains a model for other autoimmune diseases, such as type 1 diabetes, and progress in this area will certainly impact on their management.

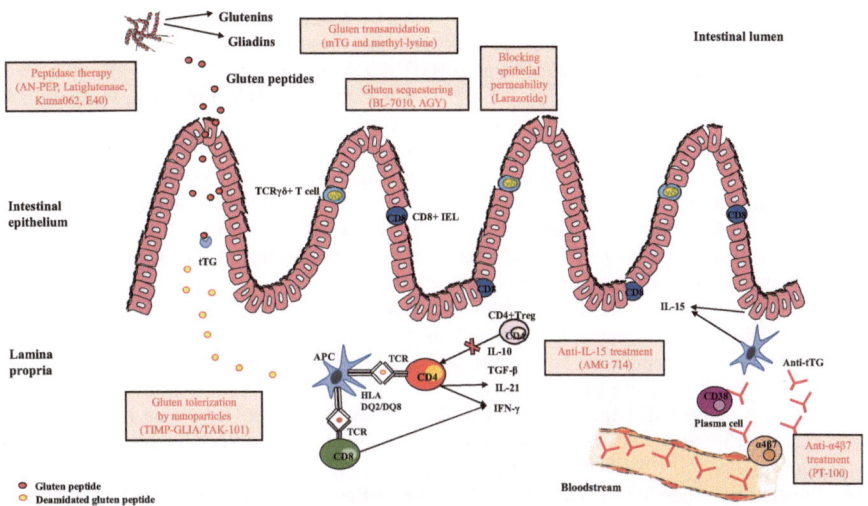

Fig. 1 Celiac Disease pathogenesis and main pathways targeted by drugs under clinical investigation. Several therapies currently in clinical trials are designed to reduce the load of gluten immunotoxic sequences thus inhibiting the contact with the gut-associated immune cells. Among these drugs, glutenases are efficient enzymes that degrade gluten proteins in non-immunogenic short peptides. Transamidation reaction by microbial transglutaminase operates by masking the immunogenic sequences to the recognition of cognate CD4+ T cells resident in lamina propria. Biocompatible polymers and polyclonal antibodies are designed to sequester gluten proteins hampering transport through the epithelial layers. Other experimental molecules act to suppress the proinflammatory CD4+ T cells reactive to gluten, inducing an anergic status

References

1. Scherf KA, Catassi C, Chirdo F, Ciclitira PJ, Feighery C, Gianfrani C, Koning F, Lundin KEA, et al. Recent progress and recommendations on celiac disease from the working group on prolamin analysis and toxity. Front Nutr. 2020. https://doi.org/10.3389/fnut.2020.00029.
2. Rubin ES, Crowe SE. Celiac disease. Ann Intern Med. 2020. https://doi.org/10.7326/AITC202001070.
3. Volta U, Caio G, Ghirardi C, Lungaro L, Mansueto P, Carroccio A, De Giorgio R. Minimal lesions of the small intestinal mucosa: more than morphology. Dig Dis Sci. 2020. https://doi.org/10.1007/s10620-020-06571-1.
4. McAllister BP, Williams E, Clarke K. A comprehensive review of celiac disease/gluten-sensitive enteropathies. Clin Rev Allergy Immunol. 2019. https://doi.org/10.1007/s12016-018-8691-2.
5. Jabri B, Sollid LM. T Cells in celiac disease. J Immunol. 2017. https://doi.org/10.4049/jimmunol.1601693.
6. Sollid LM. Molecular basis of coeliac disease. Ann Rev Immunol. 2000. https://doi.org/10.1146/annurev.immunol.18.1.53.
7. Dunne MR, Byrne G, Chirdo FG, Feighery C. Coeliac disease pathogenesis: the uncertainties of a well-known immune mediated disorder. Front Immunol. 2020. https://doi.org/10.3389/fimmu.2020.01374.

8. Sollid LM, Qiao SW, Anderson RP, Gianfrani C, Koning F. Nomenclature and listing of celiac disease relevant gluten T-cell epitopes restricted by HLA-DQ molecules. Immunogenetics. 2012. https://doi.org/10.1007/s00251-012-0599-z.

9. Tye-Din JA, Stewart JA, Dromey JA, Beissbarth T, van Heel DA, Tatham A, Henderson K, Mannering SI, et al. Comprehensive, quantitative mapping of T cell epitopes in gluten in celiac disease. Sci Transl Med. 2010. https://doi.org/10.1126/scitranslmed.3001012.

10. Camarca A, Auricchio R, Picascia S, Fierro O, Maglio M, Miele E, Malamisura B, Greco L, et al. Gliadin-reactive T cells in Italian children from PreventCD cohort at high risk for celiac disease. Pediatr Allergy Immunol. 2017. https://doi.org/10.1111/pai.12720.

11. Jabri B, Sollid LM. Tissue-mediated control of immunopathology in coeliac disease. Nat Rev Immunol. 2009. https://doi.org/10.1038/nri2670.

12. Setty M, Discepolo V, Abadie V, Kamhawi S, Mayassi T, Kent A, Ciszewski C, Maglio M, et al. Distinct and synergistic contributions of epithelial stress and adaptive immunity to functions of intraepithelial killer cells and active celiac disease. Gastroenterology. 2015. https://doi.org/10.1053/j.gastro.2015.05.013.

13. Abadie V, Kim SM, Lejeune T, Palanski BA, Ernest JD, Tastet O, Voisine J, Discepolo V, et al. IL-15, gluten and HLA-DQ8 drive tissue destruction in coeliac disease. Nature. 2020. https://doi.org/10.1038/s41586-020-2003-8.

14. Bouziat R, Hinterleitner R, Brown JJ, Stencel-Baerenwald JE, Ikizler M, Mayassi T, Meisel M, Kim SM, et al. Reovirus infection triggers inflammatory responses to dietary antigens and development of celiac disease. Science. 2017. https://doi.org/10.1126/science.aah5298.

15. Lamacchia C, Camarca A, Picascia S, Di Luccia A, Gianfrani C. Cereal-based gluten-free food: how to reconcile nutritional and technological properties of wheat proteins with safety for celiac disease patients. Nutrients. 2014. https://doi.org/10.3390/nu6020575.

16. Gianfrani C, Camarca A, Mazzarella G, Di Stasio L, Rotondi Aufiero V, Giardullo N, Ferranti P, Picariello G, et al. Extensive in vitro gastrointestinal digestion markedly reduces the immune-toxicity of Triticum monococcum wheat: implication for celiac disease. Mol Nutr Food Res. 2015. https://doi.org/10.1002/mnfr.201500126.

17. Marciniak M, Szymczak-Tomczak A, Mahadea D, Eder P, Dobrowolska A, Krela-Kaźmierczak I. Multidimensional disadvantages of a gluten-free diet in celiac disease: a narrative review. Nutrients. 2021. https://doi.org/10.3390/nu13020643.

18. Hujoel IA, Murray JA. Refractory celiac disease. Curr Gastroenterol Rep. 2020. https://doi.org/10.1007/s11894-020-0756-8.

19. Kivelä L, Caminero A, Leffler DA, Pinto-Sanchez MI, Tye-Din JA, Lindfors K. Current and emerging therapies for coeliac disease. Nat Rev Gastroenterol Hepatol. 2021. https://doi.org/10.1038/s41575-020-00378-1.

20. Rizzello CG, De Angelis M, Di Cagno R, Camarca A, Silano M, Losito I, De Vincenzi M, De Bari MD, et al. Higly efficient gluten degradation by lactobacilli and fungal proteases durino food processing: new perspectives for celiac disease. Appl Environ Microbiol. 2007. https://doi.org/10.1128/AEM.00260-07.

21. Greco L, Gobbetti M, Auricchio R, Di Mase R, Landolfo F, Paparo F, Di Cagno R, De Angelis M, et al. Safety for patients with celiac disease of baked goods made of wheat flour hydrolyzed during food processing. Safety for patients with celiac disease of baked goods made of wheat flour hydrolyzed during food processing. Clin Gastroenterol Hepatol. 2011. https://doi.org/10.1016/j.cgh.2010.09.025.

22. Zhou L, Kooy-Winkelaar YMC, Cordfunke RA, Dragan I, Thompson A, Drijfhout JW, van Veelen PA, Chen H, Frits Koning. Abrogation of immunogenic properties of gliadin peptides through transamidation by microbial transglutaminase is acyl-acceptor dependent. J Agric Food Chem. 2017. https://doi.org/10.1021/acs.jafc.7b0255

23. Gianfrani C, Siciliano R, Facchiano A, Camarca A, Mazzeo M, Costantini S, Salvati V, Maurano F, et al. Transamidation of wheat flour inhibits the response to gliadin of intestinal T cells in celiac disease. Gastroenterology. 2007. https://doi.org/10.1053/j.gastro.2007.06.023.

24. Mazzarella G, Salvati VM, Iaquinto G, Stefanile R, Capobianco F, Luongo D, Bergamo P, Maurano F, et al. Reintroduction of gluten following flour transamidation in adult celiac patients: a randomized, controlled clinical study. Clin Dev Immunol. 2012. https://doi.org/10.1155/2012/329150.

25. Marino M, Casale R, Borghini R, Di Nardi S, Donato G, Angeloni A, Moscaritolo S, Grasso L, et al. The effects of modified versus unmodified wheat gluten administration in patients with celiac disease. Int Immunopharmacol. 2017. https://doi.org/10.1016/j.intimp.2017.03.012.

26. Ribeiro M, Picascia S, Rhazi L, Gianfrani C, Carrillo JM, Rodriguez-Quijano M, Branlard G, Nunes FM. In Situ Gluten-Chitosan interlocked self-assembled supramolecular architecture reduces T-cell-mediated immune response to Gluten in celiac disease. Mol Nutr Food Res. 2018. https://doi.org/10.1002/mnfr.201800646.

27. Sample DA, Sunwoo HH, Huynh HQ, Rylance HL, Robert CL, Xu BW, Kang SH, Gujral N, et al. AGY, a novel egg yolk-derived anti-gliadin antibody, is safe for patients with celiac disease. Dig Dis Sci. 2017. https://doi.org/10.1007/s10620-016-4426-5.

28. Shan L, Molberg Ø, Parrot I, Hausch F, Filiz F, Gray GM, Sollid LM, Khosla C. Structural basis for gluten intolerance in celiac sprue. Science. 2002. https://doi.org/10.1126/science.107412.

29. Wei G, Helmerhorst EJ, Darwish G, Blumenkranz G, Schuppan D. Gluten degrading enzymes for treatment of celiac disease. Nutrients. 2020. https://doi.org/10.3390/nu12072095.

30. Bethune MT, Khosla C. Oral enzyme therapy for celiac sprue. Methods Enzymol. 2012. https://doi.org/10.1016/B978-0-12-416039-2.00013-6.

31. Gass J, Bethune MT, Siegel M, Spencer A, Khosla C. Combination enzyme therapy for gastric digestion of dietary gluten in patients with celiac sprue. Gastroenterology. 2007. https://doi.org/10.1053/j.gastro.2007.05.028.

32. Siegel M, Garber ME, Spencer AG, Botwick W, Kumar P, Williams RN, Kozuka K, Shreeniwas R, et al. Safety, tolerability, and activity of ALV003: results from two phase 1 single, escalating-dose clinical trials. Dig Dis Sci. 2012. https://doi.org/10.1007/s10620-011-1906-5.

33. Lähdeaho ML, Kaukien K, Laurila K, Vuotikka P, Koivurova OP, Karja-Lahdensuu T, Marcantonio A, Adelman DC, et al. Glutenase ALV003 attenuates gluten-induced mucosal injury in patients with celiac disease. Gastroenterology. 2014. https://doi.org/10.1053/j.gastro.2014.02.031.

34. Murray J, Kelly CP, Green PHR, Marcantonio A, Wu T-T, Mäki M, Adelman DC, CeliAction Study Group of Investigators. No difference between latiglutenase and placebo in reducing villous atrophy or improving symptoms in patients with symptomatic celiac disease. Gastroenterology. 2017. https://doi.org/10.1053/j.gastro.2016.11.004.

35. Syage JA, Murray JA, Green PHR, Khosla C. Latiglutenase improves symptoms in seropositive celiac disease patients while on a gluten-free diet. Dig Dis Sci. 2017. https://doi.org/10.1007/s10620-017-4687-7.

36. Mitea C, Havenaar R, Drijfhout JW, Edens L, Dekking L, Koning F. Efficient degradation of gluten by a prolyl endopeptidase in a gastrointestinal model: implications for celiac disease. Gut. 2008. https://doi.org/10.1136/gut.2006.111609.

37. Salden BN, Monserrat V, Troost FJ, Bruins MJ, Edens L, Bartholomé R, Haenen GR, Winkens B, Koning F, Masclee AA. Randomised clinical study: aspergillus niger-derived enzyme digests gluten in the stomach of healthy volunteers. Aliment Pharmacol Ther. 2015. https://doi.org/10.1111/apt.13266.

38. Tack GJ, van de Water JM, Bruins MJ, Kooy-Winkelaar EM, van Bergen J, Bonnet P, Vreugdenhil AC, Korponay-Szabo I et al. Consumption of gluten with gluten-degrading enzyme by celiac patients: a pilot-study. World J Gastroenterol. 2013. https://doi.org/10.3748/wjg.v19.i35.5837.

39. Wolf C, Siegel JB, Tinberg C, Camarca A, Gianfrani C, Paski S, Guan R, Montelione G, et al. Engineering of Kuma030: a gliadin peptidase that rapidly degrades immunogenic gliadin peptides in gastric conditions. J Am Chem Soc. 2015. https://doi.org/10.1021/jacs.5b08325.

40. Pultz IS, Hill M, Vitanza JM, Wolf C, Saaby L, Liu T, Winkle P, Leffler DA. Gluten degradation, pharmacokinetics, safety, and tolerability of TAK-062, an engineered enzyme to treat celiac disease. Gastroenterology. 2021. https://doi.org/10.1053/j.gastro.2021.03.019.
41. Cavaletti L, Taravella A, Carrano L, Carenzi G, Sigutrà A, Solinas N, De Caro S, Di Stasio L, et al. E40, a novel microbial protease efficiently detoxifying gluten proteins, for the dietary management of gluten intolerance. Sci Rep. 2019. https://doi.org/10.1038/s41598-019-48299-7.
42. Schulzke JD, Bentzel CJ, Schulzke I, Riecken EO, Fromm M. Epithelial tight junction structure in the jejunum of children with acute and treated celiac sprue. Pediatr Res. 1998. https://doi.org/10.1203/00006450-199804000-00001.
43. Fasano A, Not T, Wang W, Uzzau S, Berti I, Tommasini A, Goldblum SE. Zonulin, a newly discovered modulator of intestinal permeability, and its expression in coeliac disease. Lancet. 2000. https://doi.org/10.1016/S0140-6736(00)02169-3.
44. Lammers KM, Lu R, Brownley J, Lu B, Gerard C, Thomas K, Rallabhandi P, Shea-Donohue T, et al. Gliadin induces an increase in intestinal permeability and zonulin release by binding to the chemokine receptor CXCR3. Gastroenterology. 2008. https://doi.org/10.1053/j.gastro.2008.03.023.
45. Paterson BM, Lammers KM, Arrieta MC, Fasano A, Meddings JB. The safety, tolerance, pharmacodynamic effects of single doses of AT-1001 in celiac disease subjects: a proof of concept study. Alimen Pharmacol Ther. 2007. https://doi.org/10.1111/j.1365-2036.2007.03413.x.
46. Kelly CP, Green PH, Murray JA, Dimarino A, Colatrella A, Leffler DA, Alexander T, Arsenescu R, et al. Larazotide acetate in patients with coeliac disease undergoing a gluten challenge: a randomised placebo-controlled study. Alimen Pharmacol Ther. 2013. https://doi.org/10.1111/apt.12147.
47. Leffler DA, Kelly CP, Abdallah HZ, Colatrella AM, Harris LA, Leon F, Arterburn LA, Paterson BM, et al. A randomized, double-blind study of larazotide acetate to prevent the activation of celiac disease during gluten challenge. Am J Gastroenterol. 2012. https://doi.org/10.1038/ajg.2012.211.
48. Leffler DA, Kelly CP, Green PH, Fedorak RN, Di Marino A, Perrow W, Rasmussen H, Wang C, et al. Larazotide acetate for persistent symptoms of celiac disease despite a gluten-free diet: a randomized controlled trial. Gastroenterology. 2015. https://doi.org/10.1053/j.gastro.2015.02.008.
49. Sollid LM, Koning F. Lack of relationship of AT1001 to zonulin and prehaptoglobin-2: clinical implications. Gut. 2020. https://doi.org/10.1136/gutjnl-2020-323829.
50. Freitag TL, Podojil JR, Pearson RM, Fokta FJ, Sahl C, Messing M, Andersson LC, Leskinen K, et al. Gliadin nanoparticles induce immune tolerance to gliadin in mouse models of celiac disease. Gastroenterology. 2020. https://doi.org/10.1053/j.gastro.2020.01.045.
51. Kelly CP, Murray JA, Leffler DA, Getts DR, Bledsoe AC, Smithson G, First MR, Morris A, et al. TAK-101 nanoparticles induce gluten-specific tolerance in celiac disease: a randomized, double-blind. Placebo-Controlled Study Gastroenterology. 2021. https://doi.org/10.1053/j.gastro.2021.03.014.
52. Goel G, King T, Daveson AJ, Andrews JM, Krishnarajah J, Krause R, Brown GJE, Fogel R, et al. Epitope-specific immunotherapy targeting CD4-positive T cells in coeliac disease: two randomised, double-blind, placebo-controlled phase 1 studies. Lancet Gastroenterol Hepatol. 2017. https://doi.org/10.1016/S2468-1253(17)30110-3.
53. Daveson AJM, Ee HC, Andrews JM, King T, Goldstein KE, Dzuris JL, MacDougall JA, Williams LJ, et al. Epitope-specific immunotherapy targeting CD4-positive T cells in celiac disease: safety, pharmacokinetics, and effects on intestinal histology and plasma cytokines with escalating dose regimens of Nexvax2 in a randomized, double-blind, placebo-controlled phase 1 study. EBioMedicine. 2017. https://doi.org/10.1016/j.ebiom.2017.11.018.
54. Truitt KE, Daveson AJM, Ee HC, Goel G, MacDougall J, Neff K, Anderson RP. Randomised clinical trial: a placebo-controlled study of subcutaneous or intradermal NEXVAX2, an

investigational immunomodulatory peptide therapy for coeliac disease. Aliment Pharmacol Ther. 2019. https://doi.org/10.1111/apt.15435.

55. Ventura MA, Sajko K, Hils M, Pasternack R, Greinwald R, Tewes B, Schuppan D. Su1161 - The oral transglutaminase 2 (TG2) inhibitor Zed1227 blocks TG2 activity in a mouse model of intestinal Inflammation. Gastroenterology. 2018. https://doi.org/10.1016/S0016-5085(18) 31861-4.

56. Lähdeaho ML, Scheinin M, Vuotikka P, Taavela J, Popp A, Laukkarinen J, Koffert J, Koivurova OP, et al. Safety and efficacy of AMG 714 in adults with coeliac disease exposed to gluten challenge: a phase 2a, randomised, double-blind, placebo-controlled study. Lancet Gastroenterol Hepatol. 2019. https://doi.org/10.1016/S2468-1253(19)30264-X.

57. Cellier C, Bouma G, van Gils T, Khater S, Malamut G, Crespo L, Collin P, Green PHR, et al. Safety and efficacy of AMG 714 in patients with type 2 refractory coeliac disease: a phase 2a, randomised, double-blind, placebocontrolled, parallel-group study. Lancet Gastroenterol Hepatol. 2019. https://doi.org/10.1016/S2468-1253(19)30265-1.

58. Daveson AJ, Jones DM, Gaze S, McSorley H, Clouston A, Pascoe A, Cooke S, Speare R, et al. Effect of hookworm infection on wheat challenge in celiac disease–a randomised double-blinded placebo controlled trial. PLoS ONE. 2011. https://doi.org/10.1371/journal. pone.0017366.

59. McSorley HJ, Gaze S, Daveson J, Jones D, Anderson RP, Clouston A, Ruyssers NE, Speare R, et al. Suppression of inflammatory immune responses in celiac disease by experimental hookworm infection. PLoS ONE. 2011. https://doi.org/10.1371/journal.pone. 0024092.

60. Croese J, Giacomin P, Navarro S, Clouston A, McCann L, Dougall A, Ferreira I, Susianto A, et al. Experimental hookworm infection and gluten microchallenge promote tolerance in celiac disease. J Allergy Clin Immunol. 2015. https://doi.org/10.1016/j.jaci.2014.07.022.

61. McCarville JL, Dong J, Caminero A, Bermudez-Brito M, Jury J, Murray JA, Duboux S, Steinmann M, et al. A commensal Bifidobacterium longum strain prevents gluten-related immunopathology in mice through expression of a serine protease inhibitor. Appl Environ Microbiol. 2017. https://doi.org/10.1128/AEM.01323-17.

62. Ferrari E, Monzani R, Saverio V, Gagliardi M, Pańczyszyn E, Raia V, Villella VR, Bona G, et al. Probiotics supplements reduce ER stress and gut inflammation associated with gliadin intake in a mouse model of gluten sensitivity. Nutrients. 2021. https://doi.org/10.3390/ nu13041221.

63. Biagioli M, Carino A, Di Giorgio C, Marchianò S, Bordoni M, Roselli R, Distrutti E, Fiorucci S, et al. Discovery of a novel multi-strains probiotic formulation with improved efficacy toward intestinal inflammation. Nutrients. 2020. https://doi.org/10.3390/nu12071945.

64. Carino A, Di Giorgio C, Marchianò S, Bordoni M, Roselli R, Distrutti E, Fiorucci S. Nutrients. 2020. https://doi.org/10.3390/nu12071945.

Quality of Life in Celiac Disease

Josefa Barrio⑩ and María Luz Cilleruelo⑩

Abbreviations

CD	Celiac disease
GFD	Gluten-free diet
HRQoL	Health Related Quality of Life
QoL	Quality of Life
WHO	World Health Organization

Questionnaires

SF-36	Short Form 36-Item Health Survey
EuroQoL-5D (EQ_5D)	European Quality of Life-5 Dimensions visual analogic scale
PGWB	Psychological General well-being Index
CDQ	Celiac Disease Questionnaire
CD_QOL	Celiac Quality of Life Survey
PedsQL	Paediatric Quality of Life Inventory
KIDSCREEN-52	Generic KIDSCREEN questionnaire which consists of 52 items
DISABKIDS	Quality Of Life generic questionnaire to evaluate HRQoL in children and adolescents with DISABilities and their families
CDDUX	Celiac Disease DUX
CDPQOL	Celiac Disease Paediatric Quality of life
SF-12	Short-Form 12-Item Health Survey

J. Barrio (✉)
Department of Paediatrics, Hospital Universitario de Fuenlabrada, Madrid, Spain
e-mail: jbarrio.hflr@gmail.com

M. L. Cilleruelo
Department of Paediatrics, Hospital Universitario Puerta Hierro, Madrid, Spain

© The Author(s), under exclusive license to Springer Nature Switzerland AG 2022
J. Amil-Dias and I. Polanco (eds.), *Advances in Celiac Disease*,
https://doi.org/10.1007/978-3-030-82401-3_14

1 Introduction

Celiac disease (CD) is an immune-mediated systemic disorder elicited by gluten and related prolamines which appears in genetically susceptible individuals. It is characterized by the presence of a variable combination of gluten-dependent clinical manifestations, CD-specific antibodies, HLA-DQ2 or HLA-DQ8 haplotypes, and enteropathy. Clinical manifestations are heterogeneous, from gastrointestinal signs and symptoms (e.g., chronic diarrhoea) to extraintestinal manifestations (e.g., anaemia, neuropathy, decreased bone density, increased risk of fractures) [1].

Regarding clinical forms, the classical form of CD presents with signs and symptoms of malabsorption whereas these signs and symptoms are not present in the non-classical form. In addition, there are patients that do not have any signs or symptoms at the time of diagnosis and are diagnosed due to having an increased risk for CD [1].

The only treatment for CD is a strict, life-long compliance to a gluten-free diet (GFD) (i.e., exclusion of wheat, rye, barley, and other gluten-containing food products from the diet). A GFD diet results in the disappearance of symptoms and in nutritional status improvement, at least in symptomatic patients.

Several factors contribute to the negative impact of CD on the Health-Related Quality of Life (HRQoL) of the affected patients. Having CD can adversely affect patient's HRQoL due to its chronic nature, its impact on health, the psychological distress, social and family connotations, and the need to permanently follow its treatment. Initiating treatment, by means of a complete and permanent exclusion of gluten from the diet, achieves relief of symptoms and improves the HRQoL in symptomatic patients. However, such a positive response is not so apparent in patients who are asymptomatic at the time of diagnosis [2, 3]. Furthermore, eating is more than just a physiological procedure. It represents a means for socializing and establishing emotional connections with other individuals and family members. Social aspects of food are a priority for individuals, and are a way to gather and connect with family and friends. If dietary restrictions affect the ability to dine out and travel, they can impact the life of CD patients. The limitations that a GFD imposes on family and social activities, and the psychological distress that all these factors generate, negatively affect the HRQoL of celiac patients.

Another concern for CD patients is the availability of gluten free products, which is different depending on the environment where they live. In 2013 Jordan et al. [4] highlighted the lower awareness of CD and the scarce offering of gluten-free products in the United States in comparison with Europe, where increased marketing of gluten-free options at some European restaurants, stores, and hotels, diminished social constraints. In the opinion of these authors, CD has less of a social impact in Europe, given the greater awareness and health resources offered to patients with CD.

Another problem, which adds to the difficulty of finding gluten-free products in certain environments, is how expensive gluten-free products can be. Consequently, the financial burden of additional food costs creates anxiety for many families, which may negatively affect CD patients' HRQoL. However, in some countries, the economic impact is not as great; for example in Britain, where those with CD receive prescriptions for gluten-free food products or, in Italy, where the National Health Care System covers CD-related expenses, including GFDs and associated medical care. In France or Portugal, patients suffering from CD are reimbursed a certain amount of money each month for gluten-free food, while in others, such as Spain, Holland, or United States, among others, GFDs are not covered by their respective National Health Systems [4].

2 Health-Related Quality of Life

Quality of life (QoL) has become increasingly important in health-care practice and research since 1948 when the World Health Organization (WHO) defined health as "the state of complete physical, mental and social well-being and not merely the absence of disease". While the terms QoL and HRQoL are often used interchangeably, they are generally considered different concepts. The WHO defines QoL as "individuals' perception of their position in life in the context of the culture and value systems in which they live in relation to their goals, expectations, standards and concerns" [5]. Based on this definition, the WHO developed a model of QoL that includes physical health, psychological state, independence, social relationships, and environment. This model includes more objective concepts, such as perceived function, but also addresses the meaning or importance of functional levels.

HRQoL is a complex and multidimensional concept that includes the self-reported assessment of social, emotional and physical functioning or well-being, in relation to the patient's state of health [6]. This well-being can be affected by the individual's disease and/or treatment. Many studies have shown that patients with chronic conditions perceive a lower HRQoL than the general population. Comorbidities such as increased difficulties in physical function, mental health, general health, social function, or home management can further decrease HRQoL. Likewise, the need to undergo medical treatments and visits to the hospital, sometimes throughout life, contributes to worsen the perception of HRQoL [7].

The estimation of the relative impact of chronic diseases on HRQoL is necessary so as to better plan and distribute health care resources in order to improve HRQoL. Nowadays, measuring HRQoL is an important outcome indicator in the evaluation of health-care interventions and treatments, as well as in the understanding of disease-related burdens. It is also essential to identify health inequalities, in the allocation of health resources through epidemiological studies and health surveys. In clinical practice, it has been suggested that HRQoL instruments can be useful for identifying and prioritizing health problems in individual patients, and facilitating communication between patients and healthcare providers. It can also help to identify hidden or unexpected health problems, as an aide for decision-making, and for monitoring any changes in patients' state of health or for detecting responses to treatment [8].

3 HRQoL Questionnaires

The best way to evaluate HRQoL is using validated questionnaires which are composed of different items that are grouped into domains covering various aspects of life. Each item has a 5-point Likert-type scale to assess either the frequency (never, seldom or almost never, sometimes, often, almost always or always) of certain behaviours/ feelings or the intensity of an attitude (not at all, slightly, moderately, very, extremely) over a specified period of time (e.g., last week, or last month).

1. Types of questionnaires

There are two main types of questionnaires, generic and specific. Generic questionnaires provide a measure of daily life aspects and health status of different population groups regardless of their demographic, social or clinical characteristics. Generic questionnaires should be designed to be sensitive to the effects of any treatment or condition affecting health status. To facilitate more precise health evaluation, these questionnaires are composed of several domains that evaluate mental, physical, and social aspects of life and they can be used to evaluate HRQoL in individuals affected by diseases or even for the whole population. On the one hand, these questionnaires allow us to compare the HRQoL of patients with different diseases and, on the other hand, let us compare HRQoL in patients with a certain disease with respect to the general population for which reference values are available. However, generic questionnaires are considered to be less sensitive for detecting small but clinically important differences in treatment effects because they do not focus on specific effects of disease [9].

Disease specific questionnaires are designed to assess HRQoL in patients with a specific diagnosis and treatment. They achieve greater specificity by relating health status to the condition under study and they contain items focused on specific aspects related to the illness and its treatment which may be particularly meaningful to patients with that disease.

To appropriately select and use HRQoL questionnaires, it is important to understand the concepts measured, and the psychometric properties of HRQoL questionnaires. Regarding which kind of questionnaires is more complete and useful, generic instruments better capture information about the effects of a disease on overall health, whereas disease-specific instruments are theoretically more sensitive to detect changes before and after a therapeutic intervention. However, the superiority of one over the other needs to be studied in clinical trials, comparing the results obtained with two or more instruments applied together to the same group of patients. For these reasons, despite the inherent difficulties, some authors recommend using both types of questionnaires to better evaluate HRQoL in a particular disease [9].

2. Methodology used to create HRQoL questionnaires

Another important aspect to consider is the methodology used to create HRQoL questionnaires. Traditionally, systems to measure generic HRQoL in adult patients were developed using a top-down type of methodology, in which contents were created based on pre-existing health surveys, obtained from a review of the literature. A more recent, and contrasting approach is the bottom-up type of methodology, which takes patients' point of view on how their lives are affected by their disease. This approach generally requires the use of qualitative methods. The bottom-up methodology ensures that the measurement tool includes the appropriate language and content validity, as well as that the questionnaire is sensitive enough to detect changes before and after therapeutic intervention. An additional benefit of this approach is that it also enables the allocation of healthcare resources based on patient priorities and concerns [10].

Measurement of HRQoL in adults has experienced rapid advances in recent years and, nowadays, more specific disease questionnaires are developed. These specific questionnaires take into account patients' concerns regarding their disease and treatment, which are identified through focus groups of patients affected by a specific disease. Likewise, interest in developing similar instruments for children and adolescents has also increased. Initially, children's HRQoL studies were performed by taking into account the parents' or indirect informants' opinions, instead of considering the direct opinion of the children themselves. However, several studies have shown that parents have a different perception of their children's HRQoL than their children do [11–14], which makes it interesting to know both points of views.

In addition, the importance of understanding the impact of a disease and its treatment on the HRQoL of children is now more recognized. For this reason, children's perception of the disease and their opinions about treatment have increasingly been solicited and given consideration in clinical practice [6].

Originally, generic questionnaires regarding HRQoL in children were developed by adapting existing adult questionnaires. However, it seems logical that HRQoL questionnaires for adult patients are not appropriate for evaluating HRQoL in children because those instruments were designed taking into account adults' concerns, which are substantially different to those of children.[10].

More recently, other questionnaires, such as the one specifically developed for children with the Celiac Disease, DUX (CDDUX) questionnaire [15], were developed using a bottom-up approach which aims to perceive the situation from the child's perspective, and was created based on information obtained from focus groups from children and their families. These focus groups assessed their experiences, feelings and concerns about living with CD and having to follow a GFD throughout their lives. These types of instruments assess HRQoL as perceived by children more accurately. Considering that intellectual development and the ability to understand concepts, as well as the fact that concerns vary among the different stages of childhood and adolescence, it seems necessary to develop age-group specific questionnaires. Most questionnaires have been developed for children aged

8–18 who have already acquired language and reading abilities. The minimum age limit for self-reported instruments is 5–6 years old, although at those ages it is common for children to require help to complete them. For children under five years, HRQoL questionnaires should be completed by parents [16].

It is also known that the HRQoL of an individual can vary over time, mainly after an intervention or treatment. Therefore, a test–retest strategy over time provides a better idea of the changes caused by treatments, and also allows us to determine if the individual has become better adapted to the disease and the treatment over time.

3. **Translation, cross-cultural adaptation and validation**

Questionnaire development is a difficult process, hence many researchers prefer using previously validated questionnaires when they are available. In order to use questionnaires in a language other than the original language they were developed in, literal translation is not suitable. A transcultural adaptation and validation should be performed which takes into account the idiosyncrasy and peculiarities of the language into which the questionnaire is to be translated. It should be performed using methods that ensure conceptual and semantic equivalence to the original questionnaire according to international guidelines. The methodology proposed by the Mapi Research Trust is the most widely used [17] Transcultural questionnaire adaptions allow for comparison between different countries and cultures.

The adaptation process involves the following three phases: (a) a forward translation (from original language to target language), (b) a backward translation (from the target language back into the original language), and (c) a cognitive interview so as to verify that patients understand each of the items and their meanings. Afterwards, a validation process should be performed which consists of analyzing and reporting the validity, reliability and the equivalence of test scores to ensure the usefulness of the instrument in different populations.

4   Celiac Disease and HRQoL

1. HRQoL questionnaires used to assess HRQoL in celiac patients

   1.1   Adult celiac patients

- Generic questionnaires

HRQoL impairment in adults with CD has mainly been tested using generic questionnaires such as **SF-36 (Short Form 36-Item Health Survey)** [18] **EuroQol-5D (EQ-5D) (European Quality of Life-5 Dimensions visual analogic scale)** [19] or the **PGWB index (Psychological General Well-being Index)**, [20] among others.

- Specific questionnaires

More recently, two specific instruments were designed for adult celiac patients, **the CDQ (Celiac Disease Questionnaire)** [21] **and the CD-QOL (Celiac Quality of Life Survey)** [22].

In 2007 Hauser et al. [21] developed and validated the CDQ in Germany, which contains four domains with 7 items each: emotional and social problems, disease-related concerns, and gastrointestinal symptoms. The questionnaire was transculturally adapted and validated in Italy by Marchese et al. [23] and in France by Pouchot et al. [24].

Subsequently, Dorn et al. [22] developed and validated another specific questionnaire for CD in the USA, the celiac disease quality of life Survey (CD-QOL). This questionnaire consists of 20 items grouped into 4 domains (limitations, dysphoria, health-related problems, and inappropriate treatment) allowing us to obtain both overall and by domain scores. It has been translated into Spanish by Casellas et al. [25] and into Italian by Zingone et al. [26].

Table 1 shows the most used generic and specific questionnaires in adult celiac patients.

1.2 Children and adolescent celiac patients

Over the last few years, the growing interest in evaluating HRQOL in CD children has led to the publication of many studies. Most of them have been performed using generic questionnaires. Some of the most frequently used questionnaires are described below.

Table 1 HRQoL questionnaires used to assess HRQoL in adults with celiac disease

Generic HRQoL questionnaires
• SF-36 (Short-Form 36-Item Health Survey)
• EuroQol-5D (EQ-5D) (EuroQol's visual analogic scale)
• PGWB Index (The Psychological General Wellbeing)
• SF-12 (Short-Form 12-Item Health Survey)
Specific HRQoL questionnaires
• CDQ (The Celiac Disease Questionnaire) *Germany. Hauser W. J Clin Gastroenterol 2007*
• Transcultural adaptation
– Italy. Marchese A. Eur J Intern Med 2013
– France. Pouchot J. PLoS ONE 2014
• CD-QOL (The Celiac Disease Quality of Life Survey) *USA. Dorn SD. Aliment Pharmacol Ther 2010*
• Transcultural adaptation
– Spain: Casellas F. Rev Esp Enferm Dig 2013
– Italy: Zingone F. Dig Liver Dis 2013

- Generic questionnaires

The PedsQL (Paediatric Quality of Life Inventory) [27] is one of the most widely used, well-validated paediatric tools for measuring HRQoL in both healthy children and those with a variety of chronic paediatric GI diseases, such as CD.[3, 28–33] The **PedsQL version 4.0** [34] is a brief 23-item measurement tool which assesses 4 HRQoL domains: physical, emotional, social, and school functioning domains. Questionnaire scores are recorded on a scale from 1 to 100 and responses are transformed into a 5-point Likert-type scale. A score of 100 represents the best possible HRQoL, a score of 0 the worst.

The KIDSCREEN-52 (Generic KIDSCREEN questionnaire) has been used by Myleus et al. [35] in Swedish children with CD and by our group in Spanish celiac children [36]. This questionnaire consists of 52 items covering ten HRQoL domains: social acceptance (bullying), moods and emotions, physical well-being, psychological wellbeing, self-perception, school environment, parent relationship and family life, economic resources, autonomy, and social support and peers. KIDSCREEN-52 scores are computed for each domain and transformed with a mean of 50 for the general population, and a standard deviation (SD) of 10.

The DISABKIDS (Quality of Life generic questionnaire to evaluate HRQoL in children and adolescents with DISABilities and their families) was used in its short version (DCGM-12) by Bystrom et al. [37] in Swedish celiac children. It contains four domains: mental health, with questions about independence and ability to live without restrictions due to the disease; emotions, including anxiety, anger, and worries; social health, with questions concerning social community, including acceptance by and good relations with others, social exclusion, including shame and feelings of exclusion and physical health with questions concerning functional limitations and subjective physical health status.

- Specific questionnaires

  The development of specific questionnaires was more recent. Two paediatric disease-specific questionnaires for CD are currently available.

  – The first disease-specific HRQoL questionnaire (Van Dorn et al.) [15] was developed in Holland for children with CD aged 8–18, together with a parent-proxy version. It was known as **the CDDUX (Celiac Disease DUX),** and it contains 12 items distributed as 3 scales: "having CD" (3 items), which provides information on how the child feels when offered food that contains gluten or when thinking about food containing gluten; "communication" (3 items), which provides information about how the child feels when talking about CD to others or when explaining what the disease is; and "diet" (6 items), which provides information on how the child feels about having to follow a strict lifelong diet or not being able to eat things that other people can. Each item has five response options which appear as drawn faces with different expressions to reflect feelings (very good, good, normal, not too good, bad). The responses marked by the patients or parents are transformed into a 5-point Likert-type scale in which HRQoL is described as very poor (score 1–20), poor (score 21–40), neutral (score 41–60), good (score 61–80) or very good (score 81–100). This method allows for assessment of the disease's consequences and how the children themselves and their parents feel about the disease and their need to adhere to a strict diet. It also provides information on the impact of CD on HRQoL from the point of view of the child, which can be different from the information given by researchers or parents. The CDDUX has been widely used after transcultural adaption and validation in many languages, [39–43] including into Spanish [44], according to international consensus guidelines as we previously cited.

- Subsequently, Jordan et al. [4] developed another age and disease-specific HRQoL questionnaire for CD children in the USA based on the opinion that the CDDUX, created and used in the European population, might not be comparable for use in the USA population, where the stigma of having CD and social concerns may be greater due to the lower CD disease awareness when compared to European countries. They used a focus group methodology to develop and validate the age and disease-specific questionnaire **CDPQOL (Celiac Disease Paediatric Quality of life)**. Through the focus group, the authors detected that the challenges of having CD vary by age and stage of life. For example, the 13- to 18-year-old age group faced issues pertaining to buying their own food and becoming an adult which resulted in 2 additional domains for their age group. Thus, two different questionnaires were developed, the CDPQOL 8–12, with 13 items and the CDPQOL 13–18, with 17 items, taking into account differences in comprehension and experiential distinctions in the questions asked. Each item has five response options (never, almost never, sometimes, often, almost always). Scores are recorded on a scale from 1 to 100, and responses marked by children are transformed into a 5-point Likert-type scale. Moreover, this questionnaire takes into consideration not only aspects related to diet like CDDUX, but also the disease-specific implications of living with CD. Unlike the other questionnaires described, CDPQOL does not have a parent-proxy version.

Table 2 shows the most used generic and specific questionnaires in celiac children.

5   Assessment of HRQoL by Celiac Patients

5.1   Adult Celiac Patients

In order to assess the effect of GFD on HRQoL in celiac patients over 16 years of age, Burger et al. [2] published a systematic review and meta-analysis which included studies published up to 2015. In the meta-analysis, only studies which assessed HRQoL using two validated questionnaires, the specific CD-QOL [22] or

Table 2 HRQoL questionnaires used to assess HRQoL in children with celiac disease

Generic HRQoL questionnaires
• PedsQL (Paediatric Quality of Life Inventory)
• KIDSCREEN-52 (Generic KIDSCREEN questionnaire which consists of 52 items)
• DISABKIDS-QOL (Quality Of Life generic questionnaire to evaluate HRQoL in children and adolescents with DISABilities and their families)
Specific HRQoL questionnaires
• CDDUX (Celiac Disease DUX)
Holland. Van Doorn RK. J Pediatr Gastroenterol Nutr. 2008
• Transcultural adaptations
– Argentine. Pico M. Acta Gastroenterol Latinoam 2012
– Brazil. Lins MT. J Pediatr (Rio J). 2015
– Spain. Barrio J. J Pediatr Gastroenterol Nutr 2016
– Israel. Meyer S. Am J Occup Ther. 2016
– Iran. Taghdir M. J Multidiscip Healthc. 2016
– Chile. Rojas M. Rev Chil Pediatr. 2019
• CDPQOL (Celiac Disease Paediatric Quality of Life)
USA. Jordan NE. J Pediatr Gastroenterol Nutr 2013

the generic SF-36, [18] and the PGWB Index, were included [20]. When patients on a GFD were compared with non-celiac controls, patients (n = 2,728) and controls (n = 1,692) showed similar HRQoL scores for the total PGWB. However, celiac patients scored significantly lower on the SF-36 Physical Component Score and Mental Component Score. After this meta-analysis, other studies were published, such as the one by Deepak et al. [45] a study that used the generic **SF-12 (Short-Form 12-Item Health Survey)** and the specific CD-QOL questionnaires, which observed low Physical Component Score and Mental Component Scores in patients with CD; thereby, implicating a poor general health in these patients in both the physical and mental health domains, in accordance with the results observed in Burger's meta-analysis [2]. In addition, Rodríguez A [46] analyzed the responses provided by 1,230 Spanish adult patients obtaining an overall mean value of 56.3 points for the CD-QOL index, with a range from 81.3 points in dysphoria domain to 36.1 points in the inadequate treatment domain, thereby pointing out the difficulties related to following the diet.

Leinonen et al. [47] observed that approximately half of the adult patients diagnosed with CD in childhood consider that their daily lives are restricted. These patients showed lower vitality scores in the PGWB questionnaire while no differences were observed in other aspects, such as self-perceived HRQoL or gastrointestinal symptoms measured with the Gastrointestinal Symptom Rating Scale questionnaire. These restrictions were associated with increased efforts to maintain a GFD, persistent symptoms, and health concerns. It is difficult to establish at diagnosis which patients will present these problems. In the opinion of these authors, the moment of transition from child to adult care can be an opportunity to identify those patients who could benefit from increased medical and social support.

Table 3 HRQoL scores in adult celiac patients with different generic and specific questionnaires

	Patients	Questionnaires	Median score
Usai (2007)	129	**SF-36:** –MCS –PSC	42.9 ± 10 47.6 ± 10
Urkola (2011)	698	**PGWB total**	102.7 ± 17.8
Barrat (2011)	225	**SF-36:** –MCS –PSC	46.2 ± 10 46.6 ± 10
Paavola (2012)	466	**PGWB total**	102.6 ± 16.3
Paarlahti (2013)	596	**PGWB total**	102.8 ± 17.2
Rodriguez (2016)	1,230	**CD-QOL**	52.3 ± 23.43
Deepak (2018)	60	**SF-12** –MCS –PCS **CD-QOL**	50.22 ± 9.04 50.30 ± 9.88 48 (24–84)

Table 3 shows HRQoL scores obtained in the adult celiac patients studies with high number of patients included.

5.2 Celiac Children

5.2.1 Generic Questionnaires

Nikniaz et al. [38] recently published the results of a meta-analysis and systematic review of the main studies assessing HRQoL in children with CD. These authors compared the assessment of HRQoL between children with CD and healthy controls, as well as the assessment of HRQOL between parent-proxy reports and children's self-reports.

The five studies that used the generic PedsQL questionnaire showed a mean HRQoL' score of 77.29 points which, on a scale from 1 to 100, is interpreted as a good HRQoL [38].

Regarding the generic KIDSCREEN-52 questionnaire, the two studies which used it in celiac children showed some differences in their HRQoL assessment. Spanish celiac children [36] reported a relatively high HRQoL, with functioning and wellbeing in the ten HRQoL domains, ranging from a mean score of 91.92 (SD 11.91) in "social acceptance" to 32.18 (SD 11.82) in "social support and peers." Swedish celiac children [35] scored over 76 in all domains, with mean scores ranging from 75.8 (16.7) for "physical wellbeing" to 92.6 (15.1) for "social acceptance". This translates into a self-reported "very good" HRQoL along with significantly higher scores awarded in most domains, compared to Spanish children. Likewise, Bystrom et al. [37], using the generic DISABKIDS QoL questionnaire for another group of Swedish CD children with the same range of age, obtained a median value for the children's total score of 92 points and in the specific domains the scores were: Mental Health 85 points, Social Health 95 points, and Physical Health 100 points. Although it is difficult to explain the differences observed in perceived HRQoL between Spanish and Swedish children, we hypothesize that the higher scores awarded by the Swedish patients could be related, among other factors, to a higher standard of living, financing of gluten free dietary products for children under 16 years, and the idiosyncrasy of each country.

To summarize, although worse evaluations were obtained in some domains, in general, celiac children from several countries perceive their general HRQoL as good when it is assessed with different generic questionnaires.

5.2.2 Specific Questionnaires

Regarding the assessment of HRQoL using specific questionnaires, Nitziak et al. [38] performed a meta-analysis of the six studies which showed complete data [15, 39, 40, 41, 43, 44]. By using the CDDUX, the mean total HRQoL score was 58.81 points. These results revealed that children with CD and their parents have a neutral HRQoL experience when they consider living with CD (scores 40–60 are interpreted as a neutral HRQoL). The standardized total score ranged from 30.20 in Iranian children to 67.12 in Argentinian children.

In all six studies, the best scores were reported in the "communication" domain, while the worst scores were observed in the "having CD" and "diet" domains. Therefore, it seems that children with CD, regardless of their nationality, worry about similar aspects of their disease. Their main concern is having to keep a GFD and thus feeling different from their peers. Only one group, Rojas et al. [42], observed that the dimension worst scored by Chilean CD children was communication. Although it is difficult to know exactly the reason for these differences, Chilean authors hypothesize that health care actions in their country which provide CD awareness are probably insufficient in the communication area.

Comparisons of the overall and by-domain scores of the seven studies that used the CDDUX questionnaire are shown in Table 4.

By using another CD specific questionnaire for children, the CDPQOL, Jordan et al. [4] observed that children aged 8–12 had a mean score of 83.8 and adolescents aged 13–18 years had a mean score of 82.1, which can be interpreted as a good perception of their HRQoL in both groups.

5.2.3 Assessment of HRQoL Using Both Generic and Specific Questionnaires

Four research groups used the CDDUX specific and the generic DUX-25 questionnaire [15], PedsQL [39, 48] and KIDSCREEN-52 [49] in the same group of children. In addition, Jordan et al. [4] assessed HRQoL in the same group of

Table 4 HRQoL Overall and by domains scores obtained in seven different populations, which filled out the CDDUX questionnaire

Country		Holland		Argentina		Brazil		Spain		Israel		Iran		Chile	
Author		Van Doorn (2008)		Pico (2012)		Lins (2015)		Barrio (2016)		Meyer (2016)		Taghdir (2016)		Rojas 2018	
Year publication		2008		2012		2015		2016		2016		2016		2019	
n		510		193		33		214		34		65		37	
CDDUX scale		Mean	SD	Mean	SD	Mean	SD	Mean	SD			Mean	SD		
TOTAL	Children	44	15	67.12	14.4	57.6	12.3	55.02	12.8	62.6	12.8	30	17	66.1	15.3
	Parents	39	15	56.6	14.1	45.4	10.4	53.9	12.3	57.6	9.0	25	14	67.6	16.1
Having CD	Children	36	21	53.3	17.8	44.0	13.8	46.3	13.3	50.4	17.8			70.3	18.2
	Parents	30	18	51.5	14.4	45.4	14.3	49.3	12.9	44.9	8.7			70.5	16.1
Communication	Children	59	21	76.0	20.9	71.3	18.1	71.2	17.1	79.4	15.2			58	21.1
	Parents	53	20	64.1	18.3	38.4	10.7	68.8	16.3	77.6	13.2			58.2	22.6
Diet	Children	36	16	65.1	17.8	57.6	16.7	51.3	16.9	60.4	15.4			68.1	17.4
	Parents	33	18	55.4	17.1	52.3	14.1	48.7	15.7	54.2	12.6			70.9	18.8

children with the CDPQOL specific questionnaire and the PedsQL generic questionnaire. All of them observed higher scores in the generic questionnaires than in the specific ones. Furthermore, the assessment of the correlation between both types of questionnaires in the same group of children [4, 20, 39, 49] showed poor to moderate correlation. The explanation for these differences could be related to the more specific nature of the issues assessed by the specific questionnaires, which focus more on the main concerns of these patients regarding their disease. Therefore, to obtain more complete information on the HRQoL of these children, it is advisable to use both types of questionnaires.

5.2.4   Comparison Between HRQoL in Celiac Children on a GFD Versus Control Population

Regarding studies which compared HRQoL in celiac patients on GFD versus a control population, Nikniaz's meta-analysis [38] showed that total HRQoL scores and sub-scores were not significantly different between celiac children and healthy controls in the five studies which compared them using the PedsQL questionnaire. Likewise, in other studies which used different questionnaires and were analyzed in the same meta-analysis, the conclusion was that the HRQoL of celiac patients and controls is similar, despite slight differences.

5.2.5   Comparison Between Children and Parent Reports

Nitziak's meta-analysis [38] included studies which compared the HRQoL scores of parent-proxy reporting and child self-reporting and found that, in the studies that used the specific CDDUX [15, 40, 41, 43, 44, 48, 50] parents' HRQoL total, diet and communication scores were significatively lower than their children's HRQoL scores. These results indicate a lower perception of HRQoL from the parents' perspective than from the perspective of the CD children themselves. Another study, which used the generic DISABKIDS, [37] observed similar results, with worse scores for parents than for children. However, the three studies which used the generic PedsQL questionnaire [31, 33, 51] reported no significant differences between parents' reports and those of their children. In the opinion of Sawyer et al. [52], the worst perception of parents regarding their children's HRQOL observed in studies performed on families with chronic diseases could be linked to the parents' feeling of responsibility and concerns about what their children have to endure.

Our group [36] used the generic KIDSCREEN-52 questionnaire and also obtained lower scores in parents than children. However, Spanish children awarded significantly lower scores than their parents in the "social support and peers" domain, suggesting that this is an area of particular concern for children.

6   Related Factors to HRQOL

6.1   Adult Celiac Patients

In their meta-analysis, Burger et al. [2] assessed disease-related factors which negatively influenced HRQoL. Eight studies, which included a total of 998 patients, provided prospective data on the effect of a GFD. They observed that GFD significantly improves HRQoL after one year of treatment according to psychological general well-being total scores and SF-36 Mental and Psychological Component Scores.

When they studied HRQoL regarding adherence to diet, they found that HRQoL was significantly higher in patients with strict dietary adherence compared to patients with non-strict adherence based upon the results of 436 patients' SF-36 Mental Component and Psychological Component Scores.

In this meta-analysis, the authors concluded that, although GFD significantly improves adult patients' HRQoL, it does not completely normalize. This could be related to the fact that, although GFD induces disease remission and resolves the symptoms, there are other social and emotional effects that can negatively influence the HRQoL.

Sainsbury et al. [53] performed another meta-analysis of previous retrospective studies to assess the relationship between GFD adherence and depressive symptoms in adults with CD. A total of eight cross-sectional studies which included 1644 patients were selected and it was concluded that, with the available evidence, there is an association between poorer GFD adherence and self-reported depressive symptoms. However, these results have to be taken with caution because of the retrospective nature of the studies. Regarding depressive symptoms, it has been observed that adult celiac patients who present autonomous or intrinsic motivation show a lower level of anxiety symptoms and depression and better physical functioning and adherence to diet than those whose motivation is controlled and guided by external pressure [54].

Rodriguez et al. [46], in a representative sample of adult celiac patients from the general population, observed that a longer period following a GFD, advanced age at diagnosis and a longer period of time until the diagnosis was accurately made were independent factors that negatively affected HRQoL.

There seems to be a subgroup of adult celiac patients who are extremely vigilant regarding dietary compliance and may suffer negative consequences and score worse on their HRQoL tests. The clinician must be aware of the existence of these patients because, just as strict GFD monitoring is advised, simultaneous promotion of an adequate HRQoL which addresses emotional and social well-being should be considered [55].

Studies performed in adults showed discrepancies regarding disease onset and HRQoL. Some authors [21, 56] reported worse HRQoL scores in adults who were diagnosed in childhood, while others, such as Haüser et al. [21] and Zingone et al. [26], found no relationship between the time elapsed since disease onset and

HRQoL. Although, one might speculate that earlier diagnosis would make it easier to become accustomed to a GFD, this has not been confirmed in the studies performed to date.

6.2   Celiac Children

Contrary to what would be expected, Nikniaz et al. [38] observed a significant correlation between HRQoL and GFD compliance in only 4 out of 11 studies. Our group assessed factors related to HRQoL by using the specific CDDUX and the generic KIDSCREEN-52 in Spanish celiac children. According to the multivariate analysis, we observed that the main factors related to having a worse HRQoL with both questionnaires were: having social and/or economic difficulties related to following the diet and having transgression-related symptoms [49] A negative perception of HRQoL in patients reporting non-adherence to the diet and with social or economic difficulties related to following a diet has been reported by other authors in children [39, 41] as well as in adults [5, 18, 21, 57, 58] with CD. Likewise, a negative association between HRQoL and having transgression-related symptoms has been reported by other authors in adult patients [58].

Van Kooper et al. [48] and Shull et al. [33] observed that children recently diagnosed with celiac disease had a worse HRQoL before starting the gluten-free diet than children from the general population. However, Van Koopen's study showed that after one year of treatment, the HRQoL of CD patients was similar to that of children in the general population.

Few studies have evaluated the influence of race in HRQoL. Mager et al. [51] found higher HRQoL scores in non-Caucasian versus Caucasian youth with CD and their parents.

In our study [49], the analysis of other factors related to HRQoL using the specific CDDUX questionnaire showed that non-complete adherence to diet, diagnosis of the non-classical form of CD and being older than 2 years of age upon diagnosis were associated with having worse HRQoL. However, using the generic KIDSCREEN-52 questionnaire, being female and over 12 years of age when the survey was filled out were associated with a worse perception of HRQoL. Using both questionnaires, the group of children with less time since diagnosis experienced some aspects of their HRQoL as lower.

In Byström's study, [37] children who were diagnosed before the age of five scored better than those who were five years or older at diagnosis. Furthermore, children with greater time since diagnosis experienced their current HRQoL as higher. These results are consistent with the report from Högberg et al. [59].

HRQoL perception by adolescent celiac patients has traditionally been considered poor [60], which is in line with our findings with KIDSCREEN-52 Questionnaire [36].

7 Conclusions

1. Studies in children and adults show that recently diagnosed CD patients have a worse HRQoL before starting the GFD than general population. However after GFD, their HRQoL significatively improves and is similar to that of the general population. Although having a chronic disease such as CD and being subjected to a strict and permanent diet could lead to a worse quality of life, HRQoL is not significantly affected in most studies.
2. The main factors related to having a worse HRQoL are the non-adherence to diet and having social and/or economic difficulties related to following the diet.
3. Most studies showed that parents' perception of their children's HRQoL is worse than the perception of children themselves. Therefore, both perspectives should be considered for assessing HRQoL.
4. Assessment of HRQoL should be objectively addressed using validated HRQoL questionnaires. The use of a generic and a specific questionnaire in the same group can provide more complete information.

References

1. Husby S, Koletzko S, Korponay-Szabó IR, Mearin ML, Phillips A, Shamir R, et al. European Society for pediatric gastroenterology, hepatology, and nutrition guidelines for the diagnosis of coeliac disease. J Pediatr Gastroenterol Nutr. 2012;54:136–60. https://doi.org/10.1097/MPG.0b013e31821a23d0.
2. Burger JPW, de Brouwer B, IntHout J, Wahab PJ, Tummers M, Drenth JPH. Systematic review with meta-analysis: dietary adherence influences normalization of health-related quality of life in coeliac disease. Clin Nutr. 2017;36:399–406. https://doi.org/10.1016/j.clnu.2016.04.021.
3. Biagetti C, Gesuita R, Gatti S, Catassi C. Quality of life in children with celiac disease: a paediatric cross-sectional study. Dig Liver Dis. 2015;47:927–32. https://doi.org/10.1016/j.dld.2015.07.009.
4. Jordan NE, Li Y, Magrini D, Simpson S, Reilly NR, DeFelice AR, et al. Development and validation of a celiac disease quality of life instrument for North American children. J Pediatr Gastroenterol Nutr. 2013;57:477–86. https://doi.org/10.1097/MPG.0b013e31829b68a1.
5. - Division of Mental Health and Prevention of Substance Abuse. WHOQOL user manual. Geneva: World Health Organization;1998.
6. Seid M, Varni JW, Jacobs JR. Pediatric health-related quality-of-life measurement technology: intersections between science, managed care, and clinical care. J Clin Psychol Med Settings. 2000;7:17–27. https://doi.org/10.1023/A:1009541218764.
7. Hand C. Measuring health-related quality of life in adults with chronic conditions in primary care settings: critical review of concepts and 3 tools. Can Fam Physician. 2016;62:e375–83.
8. Higginson IJ. Measuring quality of life: using quality of life measures in the clinical setting. BMJ. 2001;322:1297–300. https://doi.org/10.1136/bmj.322.7297.1297.
9. Briançon S, Gergonne B, Guillemin F, Empereur F, Klein S. Disease-specific versus generic measurement of health-related quality of life in cross-sectional and longitudinal studies: an inpatient investigation of the sf-36 and four disease-specific instruments. In: Mesbah M,

Cole BF, Lee M-LT, editors. Statistical methods for quality of life studies. Boston, MA: Springer US;2002. p. 87–99. https://doi.org/10.1007/978-1-4757-3625-0_8.

10. Stevens K, Palfreyman S. The use of qualitative methods in developing the descriptive systems of preference-based measures of health-related quality of life for use in economic evaluation. Value Health. 2012;15:991–8. https://doi.org/10.1016/j.jval.2012.08.2204.

11. Eiser C, Morse R. Can parents rate their child's health-related quality of life? Results of a systematic review. Qual Life Res. 2001;10:347–57. https://doi.org/10.1023/a:1012253723272.

12. Theunissen NCM, Vogels TGC, Koopman HM, Verrips GHW, Zwinderman KAH, Verloove-Vanhorick SP, et al. The proxy problem: child report versus parent report in health-related quality of life research. Qual Life Res. 1998;7:387–97. https://doi.org/10.1023/a:1008801802877.

13. Vance YH, Morse RC, Jenney ME, Eiser C. Issues in measuring quality of life in childhood cancer: measures, proxies, and parental mental health. J Child Psychol Psychiatry. 2001;42:661–7.

14. Addington-Hall J. Measuring quality of life: Who should measure quality of life? BMJ. 2001;322:1417–20. https://doi.org/10.1136/bmj.322.7299.1417.

15. van Doorn RK, Winkler LM, Zwinderman KH, Mearin ML, Koopman HM. CDDUX: a disease-specific health-related quality-of-life questionnaire for children with celiac disease. J Pediatr Gastroenterol Nutr. 2008;47:147–52. https://doi.org/10.1097/MPG.0b013e31815ef87d.

16. Germain N, Aballéa S, Toumi M. Measuring health-related quality of life in young children: how far have we come? J Mark Access Health Policy. 2019;7:1618661. https://doi.org/10.1080/20016689.2019.1618661.

17. -MAPI Research Institute. PedsQL linguistic validation guidelines. Lyon: Mapi Research Institute;2006.

18. Tontini GE, Rondonotti E, Saladino V, Saibeni S, de Franchis R, Vecchi M. Impact of gluten withdrawal on health-related quality of life in celiac subjects: an observational case-control study. Digestion. 2010;82:221–8. https://doi.org/10.1159/000265549.

19. Casellas Jordá J, López Vivancos J. Fatigue as a determinant of health in patients with celiac disease. J Clin Gastroenterol. 2010;44:423–7.https://doi.org/10.1097/MCG.0b013e3181c41d12.

20. Dupuy HJ. The psychological general well-being (PGWB) index. In: Wenger NK, Mattson ME, Furberg CF, Elinson J, editors. Assessment of quality of life in clinical trials of cardiovascular therapies. New York: Le Jacq Publishing; 1984. p. 170–83.

21. Häuser W, Gold J, Stallmach A, Caspary WF, Stein J. Development and validation of the Celiac Disease Questionnaire (CDQ), a disease-specific health-related quality of life measure for adult patients with celiac disease. J Clin Gastroenterol. 2007;41:157–66. https://doi.org/10.1097/01.mcg.0000225516.05666.4e.

22. Dorn SD, Hernandez L, Minaya MT, Morris CB, Hu Y, Leserman J, et al. The development and validation of a new coeliac disease quality of life survey (CD-QOL). Aliment Pharmacol Ther. 2010;31:666–75. https://doi.org/10.1111/j.1365-2036.2009.04220.x.

23. Marchese A, Klersy C, Biagi F, Balduzzi D, Bianchi PI, Trotta L, et al. Quality of life in coeliac patients: Italian validation of a coeliac questionnaire. Eur J Intern Med. 2013;24:87–91. https://doi.org/10.1016/j.ejim.2012.09.015.

24. Pouchot J, Despujol C, Malamut G, Ecosse E, Coste J, Cellier C. Validation of a French version of the quality of life "Celiac Disease Questionnaire". In: Assassi S, editor. PLoS ONE. 2014;9:e96346. https://doi.org/10.1371/journal.pone.0096346.

25. Casellas F, Rodrigo L, Molina-Infante J, Vivas S, Lucendo AJ, Rosinach M, et al. Transcultural adaptation and validation of the Celiac Disease Quality of Life (CD-QOL) survey, a specific questionnaire to measure quality of life in patients with celiac disease. Rev Esp Enfermedades Dig. 2013;105:585–93. https://doi.org/10.4321/s1130-0108201300 01000003.

26. Zingone F, Iavarone A, Tortora R, Imperatore N, Pellegrini L, Russo T, et al. The Italian translation of the celiac disease-specific quality of life scale in celiac patients on gluten free diet. Dig Liver Dis. 2013;45:115–8. https://doi.org/10.1016/j.dld.2012.10.018.

27. Varni JW, Seid M, Rode CA. The PedsQL: measurement model for the pediatric quality of life inventory. Med Care. 1999;37:126–39. https://doi.org/10.1097/00005650-199902000-00003.

28. Fidan T, Ertekin V, Karabağ K. Depression-anxiety levels and the quality of life among children and adolescents with coeliac disease. Dusunen Adam J Psychiatry Neurol Sci. 2013; 232–8. https://doi.org/10.5350/DAJPN2013260301.

29. Talebi S, Jafari SA, Kianifar H, Moharreri F, Mostafavi N. Quality of life in children with celiac disease: a cross-sectional study. Int J Pediatr. 2017;5:5339–49. https://doi.org/10.22038/ijp.2017.23860.2017.

30. Sevinç E, Çetin FH, Coşkun BD. Psychopathology, quality of life, and related factors in children with celiac disease. J Pediatr (Rio J).2017;93:267–73. https://doi.org/10.1016/j.jped.2016.06.012.

31. Stojanović B, Kočović A, Radlović N, Leković Z, Prokić D, Đonović N, et al. Assessment of quality of life, anxiety and depressive symptoms in serbian children with celiac disease and their parents. Indian J Pediatr. 2019;86:427–32. https://doi.org/10.1007/s12098-018-2836-4.

32. Kara A, Demirci E, Ozmen S. Evaluation of psychopathology and quality of life in children with celiac disease and their parents. Gazi Med J. 2019;30. https://doi.org/10.12996/gmj.2019.11.

33. Shull MH, Ediger TR, Hill ID, Schroedl RL. Health-related quality of life in newly diagnosed pediatric patients with celiac disease. J Pediatr Gastroenterol Nutr. 2019;69:690–5. https://doi.org/10.1097/MPG.0000000000002465.

34. Varni JW, Seid M, Kurtin PS. PedsQLTM 4.0: Reliability and validity of the pediatric quality of life InventoryTM Version 4.0 generic core scales in healthy and patient populations: med care. 2001;39:800–12. https://doi.org/10.1097/00005650-200108000-00006.

35. Myléus A, Petersen S, Carlsson A, Hammarroth S, Högberg L, Ivarsson A. Health-related quality of life is not impaired in children with undetected as well as diagnosed celiac disease: a large population based cross-sectional study. BMC Public Health. 2014;14:425. https://doi.org/10.1186/1471-2458-14-425.

36. Barrio J, Cilleruelo ML, Román E, Fernández C. Health-related quality of life in Spanish coeliac children using the generic KIDSCREEN-52 questionnaire. Eur J Pediatr. 2018;177:1515–22. https://doi.org/10.1007/s00431-018-3204-0.

37. Byström I-M, Hollén E, Fälth-Magnusson K, Johansson A. Health-related quality of life in children and adolescents with celiac disease: from the perspectives of children and parents. Gastroenterol Res Pract. 2012;2012:–6. https://doi.org/10.1155/2012/986475.

38. Nikniaz Z, Abbasalizad Farhangi M, Nikniaz L. Systematic review with meta-analysis of the health-related quality of life in children with celiac disease. J Pediatr Gastroenterol Nutr. 2020;70:468–77.https://doi.org/10.1097/MPG.0000000000002604.

39. Pico M, Spirito MF, Roizen M. [Quality of life in children and adolescents with celiac disease: argentinian version of the specific questionnaire CDDUX. Acta Gastroenterol Latinoam. 2012;42:12–9.

40. Lins MTC, Tassitano RM, Brandt KG, Antunes MM de C, Silva GAP da. Translation, cultural adaptation, and validation of the celiac disease DUX (CDDUX). J Pediatr (Rio J). 2015;91:448–54. https://doi.org/10.1016/j.jped.2014.11.005.

41. Taghdir M, Honar N, Mazloomi SM, Sepandi M, Ashourpour M, Salehi M. Dietary compliance in Iranian children and adolescents with celiac disease. J Multidiscip Healthc. 2016;9:365–70. https://doi.org/10.2147/JMDH.S110605.

42. Rojas M, Oyarzún A, Ayala J, Araya M. Health related quality of life in celiac children and adolescents. Rev Chil Pediatr. 2019;90:632–41. https://doi.org/10.32641/rchped.v90i6.1126.

43. Meyer S, Rosenblum S. Children with celiac disease: health-related quality of life and leisure participation. Am J Occup Ther. 2016;70:7006220010 p1–7006220010p8. https://doi.org/10.5014/ajot.2016.020594.

44. Barrio J, Román E, Cilleruelo M, Márquez M, Mearin M, Fernández C. Health-related quality of life in spanish children with coeliac disease. J Pediatr Gastroenterol Nutr. 2016;62:603–8. https://doi.org/10.1097/MPG.0000000000000963.

45. Deepak C, Berry N, Vaiphei K, Dhaka N, Sinha SK, Kochhar R. Quality of life in celiac disease and the effect of gluten-free diet: Quality of life in celiac disease. JGH Open. 2018;2:124–8. https://doi.org/10.1002/jgh3.12056.

46. Rodríguez Almagro J, Hernández Martínez A, Lucendo AJ, Casellas F, Solano Ruiz MC, Siles González J. Health-related quality of life and determinant factors in celiac disease. A population-based analysis of adult patients in Spain. Rev Esp Enfermedades Dig. 2016;108. https://doi.org/10.17235/reed.2016.4094/2015.

47. Leinonen H, Kivelä L, Lähdeaho M-L, Huhtala H, Kaukinen K, Kurppa K. Daily life restrictions are common and associated with health concerns and dietary challenges in adult celiac disease patients diagnosed in childhood. Nutrients. 2019;11:1718. https://doi.org/10. 3390/nu11081718.

48. van Koppen EJ, Schweizer JJ, Csizmadia CGDS, Krom Y, Hylkema HB, van Geel AM, et al. Long-term health and quality-of-life consequences of mass screening for childhood celiac disease: a 10-year follow-up study. Pediatrics. 2009;123:e582–8. https://doi.org/10.1542/ peds.2008-2221.

49. Barrio J, Cilleruelo ML, Román E, Fernández C. Health-related quality of life using specific and generic questionnaires in Spanish coeliac children. Health Qual Life Outcomes. 2020;18:250. https://doi.org/10.1186/s12955-020-01494-x.

50. Khurana B, Lomash A, Khalil S, Bhattacharya M, Rajeshwari K, Kapoor S. Evaluation of the impact of celiac disease and its dietary manipulation on children and their caregivers. Indian J Gastroenterol. 2015;34:112–6. https://doi.org/10.1007/s12664-015-0563-6.

51. Mager DR, Marcon M, Brill H, Liu A, Radmanovich K, Mileski H, et al. Adherence to the gluten-free diet and health-related quality of life in an ethnically diverse pediatric population with celiac disease. J Pediatr Gastroenterol Nutr. 2018;66:941–8. https://doi.org/10.1097/ MPG.0000000000001873.

52. Sawyer MG, Reynolds KE, Couper JJ, French DJ, Kennedy D, Martin J, et al. A two-year prospective study of the health-related quality of life of children with chronic illness? The parents? Perspective. Qual Life Res. 2005;14:395–405. https://doi.org/10.1007/s11136-004-0786-y.

53. Sainsbury K, Marques MM. The relationship between gluten free diet adherence and depressive symptoms in adults with coeliac disease: a systematic review with meta-analysis. Appetite. 2018;120:578–88. https://doi.org/10.1016/j.appet.2017.10.017.

54. Barberis N, Quattropani MC, Cuzzocrea F. Relationship between motivation, adherence to diet, anxiety symptoms, depression symptoms and quality of life in individuals with celiac disease. J Psychosom Res. 2019;124:109787.https://doi.org/10.1016/j.jpsychores.2019. 109787.

55. Wolf RL, Lebwohl B, Lee AR, Zybert P, Reilly NR, Cadenhead J, et al. Hypervigilance to a gluten-free diet and decreased quality of life in teenagers and adults with celiac disease. Dig Dis Sci. 2018;63:1438–48. https://doi.org/10.1007/s10620-018-4936-4.

56. Ukkola A, Mäki M, Kurppa K, Collin P, Huhtala H, Kekkonen L, et al. Patients' experiences and perceptions of living with coeliac disease—implications for optimizing care. J Gastrointest Liver Dis JGLD. 2012;21:17–22.

57. Casellas F, Rodrigo L, Lucendo AJ, Fernández-Bañares F, Molina-Infante J, Vivas S, et al. Benefit on health-related quality of life of adherence to gluten-free diet in adult patients with celiac disease. Rev Espanola Enfermedades Dig Organo Soc Espanola Patol Dig. 2015;107:196–201.

58. Usai P, Manca R, Cuomo R, Lai MA, Boi MF. Effect of gluten-free diet and co-morbidity of irritable bowel syndrome-type symptoms on health-related quality of life in adult coeliac patients. Dig Liver Dis. 2007;39:824–8. https://doi.org/10.1016/j.dld.2007.05.017.

59. Högberg L, Grodzinsky E, Stenhammar L. Better dietary compliance in patients with coeliac disease diagnosed in early childhood. Scand J Gastroenterol. 2003;38:751–4. https://doi.org/10.1080/00365520310003318.
60. Olsson C, Hrnell A, Ivarsson A, Sydner YM. The everyday life of adolescent coeliacs: issues of importance for compliance with the gluten-free diet. J Hum Nutr Diet. 2008;21:359–67. https://doi.org/10.1111/j.1365-277x.2008.00867.x.

New Fields of Research in Celiac Disease

Anat Guz-Mark and Raanan Shamir

Research in celiac disease (CD) spans from pathogenesis, through diagnosis of CD, and up to treatment options and follow-up of patients, covering every aspect of this multifaceted disease. While understanding the complex mechanism of CD pathogenesis represents the scientific main interest of research, in clinical practice most efforts are focused on new tools in diagnosis and management. From the perspective of patients living with CD, the main interest is in aspects of gluten tolerance and therapeutic options. Many of these issues are incorporated in the different chapters of this book under each specific topic. In this chapter we present an overview of the main current research fields and challenges in understanding and managing CD. We do not, however, discuss drug treatment for CD as this is covered in a specific dedicated chapter.

New fields in CD pathophysiology:

The strong genetic association of CD to specific human leucocyte antigen (HLA) haplotypes is known for decades, with nearly all patients with CD being positive to HLA DQ2/DQ8 [1, 2]. This genetic susceptibility is the cornerstone of CD, perceived as a mandatory prerequisite to develop CD. While DQ2 is the common haplotype identified in patient with CD worldwide, DQ8 is present in less than 10% [3]. Studies exploring the various susceptible haplotypes reported a gene-dose effect with different magnitude of the risk to develop CD, based on the number of copies of specific allele [4, 5]. Homozygosity for DQ2 alleles poses a much higher (fivefold) risk of developing CD compared to a single DQ2 allele or DQ8 haplotype [4]. Moreover, this DQ2 homozygosity is also associated with earlier disease pre-

A. Guz-Mark (✉) · R. Shamir
Institute of Gastroenterology, Nutrition and Liver Diseases, Schneider Children's Medical Center of Israel, Petach Tikva, Israel
e-mail: anatgu@clalit.org.il; anatguz@gmail.com

A. Guz-Mark · R. Shamir
Sackler Faculty of Medicine, Tel-Aviv University, Tel-Aviv, Israel

sentation and more severe phenotype [6]. Overall, research on the various associations of HLA haplotypes, with the risk to develop CD and the
haplotype-phenotype associations are important for the future development of
preventive, diagnostic and therapeutic measures.

An interesting new field is the non-HLA heritability, estimated as responsible for
68% of the heritability of CD in a large recent twins study [7]. The role of genetic
factors other than HLA in CD is much less understood and is being currently
assessed. Over 40 different loci, other than HLA, have found to be associated with
CD so far [8]. The TEDDY international multicentre study has been able to identify
dozens of different single-nucleotide polymorphisms (SNPs) in several non-HLA
regions, to be associated with CD, using genotype analysis [9]. Most of these
involved genes encoded proteins act in activation and regulation of the immune
response, and some of them are common with other autoimmune diseases (such as
rheumatoid arthritis, type 1 diabetes, and multiple sclerosis [10–12]). These different identified *loci* are associated with T cell activation, inflammatory cytokines
and cytokine-receptors regulation, and thymic T-cell selection [13]. Other identified
loci are noncoding regions, regulating gene expression [14].

These new identified SNPs, along with the already known HLA genes, could be
combined to form a more accurate genetic risk score to CD, as shown recently in a
pilot study incorporating non-HLA SNPs with HLA gene testing, providing a more
accurate discriminating tool for genetic risk assessment [15].

The main challenges in pathogenesis research in CD is understanding the
mechanisms linking between the genetic susceptibility and the immune response to
gluten exposure. It has been long known that gluten-specific T cells are present in
intestinal tissue of patients with CD [1], being activated to produce an inflammatory
cascade in response to presentation to deaminated gluten peptides. The specific
gluten epitopes, recognized by pathogenic T cells, have been identified in wheat, as
well as in rye and barely [16]. The reasons for the loss of tolerance to dietary gluten
by a minority of HLA-susceptible individuals are still poorly understood. The role
of regulatory T cells (Tregs) in this process is of major interest in contemporary
studies, given the effect of Tregs in inhibiting the immune response to digested
antigens. Recent comparative studies have found an attenuated inhibitory effect of
gluten-specific Tregs on intestinal and circulating T lymphocytes in patients with
CD, suggesting Treg dysfunction to play an important role in the pathogenesis of
the disease [17, 18]. Gene expression studies in patients with CD further demonstrate upregulation of IFN-γ and reduced expression of the transcription factor
BACH2 genes—which is a dominant regulator of T cell differentiation promoting
development of Tregs [19]. Further studies are needed to better identify the specific
characteristics and behaviour of these Tregs and the role they play in the loss of
gluten tolerance and activation of the pathologic response.

The role of environmental factors in the development of CD is detailed in the
specific chapter on this topic. A major contemporary focus, now being actively
studied, is host-microbiome interaction and the potential role of gut microbiota in
CD pathogenesis. This field is of particular interest, due to its potential in exploring
prevention strategies in at-risk populations. In the last few years there is plethora of

publications regarding the role of gut microbiota in the pathogenesis of immune-regulated disorders [20], and among these CD and the loss of gluten oral tolerance has an increasing presence. Previous studies have already shown unbalanced microbiota population in the gut of patients with CD, with higher incidence of pro-inflammatory bacteria [21, 22], and reduction in *Bifidobacterium* species that are considered modulators of gut immunity [23, 24]. Most of these dysbiosis patterns were found to be constant in patient with CD regardless of inflammation and status of gluten-free diet (GFD), suggesting their primary prominent role in CD pathogenesis [25]. Interestingly, HLA-DQ2 genotype was found to influence early life gut microbiota pattern in infants at high family risk of developing CD [26]. In this study, infants with the high-risk HLA-DQ2 had reduced gut *Bifidobacterium*, compared to subjects with lower-risk genotype. These findings may suggest that the genetic susceptibility to CD may in part influence early life microbiota colonization, which could be linked to immune-regulation mechanisms that should further be investigated. In addition, many studies currently focus on identifying specific composition of gut microbiota in patients with CD and their association with patterns of disease presentation and symptoms [27, 28]. Large prospective longitudinal studies currently being performed, such as the CDGEMM study of infants at risk of CD [29], could shed more light on the relationship between early patterns of gut microbiota and later development of CD.

Several studies have tried to explore the effects of different strains of probiotics in patients diagnosed with CD, and more trials are currently ongoing. These studies focus on supplementing GFD with probiotics in order to influence the dysbiosis present in CD, or to attenuate intestinal inflammatory response derived by gluten [30]. A 3-months RCT in newly diagnosed patients with CD on GFD, showed favourable outcomes with *Bifidobacterium longum* CECT supplementations, demonstrating reductions in activated T lymphocytes and some inflammatory cytokines, as well as reduction in the *Bacteroides fragilis* group and in the content of IgA in stools [31]. Another RCT has shown a reduction in TNF-α in a small group of patients with CD supplemented with *B. breve* strains [32]. An improvement in celiac symptom index was demonstrated in a subset of patients treated with *B. infantis* NLS-SS (together with GFD) [33]. One published RCT investigated the effect of two strains of *Lactobacillus* in children with CD autoimmunity under gluten-containing diet [34], and demonstrated some modulations in the peripheral immune response, with no effect on CD serology. Additional ongoing trials with different strains of probiotics, are currently active and pending results (ClinicalTrials.gov: NCT04160767; NCT03775499; NCT04014660; NCT03562221).

Hopefully, future studies will focus on prevention of CD development through alternation of gut microbiota. The effects of probiotics on patients at risk to develop CD, with the potential of altering the interaction between gut microbiota and the host immune response to gluten, should further be explored.

New aspects in CD diagnosis:

Diagnosis of CD in adults still relays on the combination of CD serology and duodenal biopsies [35, 36], while in children and adolescents CD serology could be sufficient to diagnose selected cases with very high titres of TTG (above 10 times upper limit of normal), combined with positive endomysium antibodies [37]. In practice, duodenal biopsies are mandatory worldwide for many cases with lower TTG titres and in adult population. There is a great interest in the search for non-invasive novel biomarkers for the diagnosis of CD [38], in order to overcome challenges with false positive and false negative serologic results, challenges with patients already on GFD, better management of potential CD, improved diagnosis accuracy with borderline histologic findings, and future expansion of non-invasive diagnosis to adult population.

One unique clinical challenge in modern era is the diagnosis of CD in patients that are already practicing a GFD. Starting GFD before or without diagnosis of CD has several reasons, including symptoms related to gluten ingestion making the patient reluctant from performing gluten challenge; other family members already on GFD with limited exposure to gluten in their household; non-celiac gluten or wheat sensitivity; and cultural or personal dietetic preferences. As gluten intake influences the diagnostic accuracy of both serologic markers and histologic findings, a main target of research in this field is identifying biomarkers that are constant even in the absence of dietary gluten ingestion. Such potential markers are gluten-specific T cells in peripheral circulation, identified by HLA-DQ-gluten tetramers [39]. Several recent studies have shown that HLA-DQ-gluten tetramers can accurately identify patients with CD, whether they are with or without GFD, compared to healthy controls [40–43]. These findings hold promise in clinical practice, as they may assist in CD diagnosis in specific individuals without the need for gluten challenge and intestinal biopsies. Currently, the assays of HLA-DQ-gluten tetramers are not commercially available and require additional research to better define their diagnostic accuracy and feasibility in clinical practice.

In addition, there are several markers for intestinal mucosal damage, that although not specific for CD-induced enteropathy, they could be combined with positive TTG serology in the diagnostic process. A combination of positive CD serology with non-invasive markers of small intestinal damage, are in the focus of interest as a mean to diminish the need for confirmatory biopsies in patients with lower than needed serological titres for non-biopsy diagnosis of CD. The two main markers for intestinal damage being investigated in CD are citrulline [44, 45] and Fatty Acid Binding Protein 2 (FABP2 or I-FABP) [44, 46–48], both been shown to correlate with the degree of villous atrophy in patients with CD.

Finally, there is a unique cytokines profile in CD, reflecting the immune dysregulation and inflammatory process, that differ from healthy individuals and correlates with TTG levels [49]. Some current studies focus on identifying specific cytokines to differentiate between patient with CD and healthy individuals or subjects with non-celiac gluten sensitivity and searching for better correlation with intestinal mucosal damage [50, 51].

Specific circulating microRNA were also found to be indicative of CD in recent studies [52, 53]. Continuous research is needed to better define their role both in pathogenesis and in diagnostic yield as biomarkers.

Research priorities in the management of CD

Although GFD is the cornerstone of CD treatment, strict adherence is challenging in both paediatric and adult patients [54, 55]. A major focus in current research, derived mainly by real-life challenges of patients' community, is identifying the minimal amounts of gluten that can produce immunogenic-inflammatory reactions, while defining the threshold of maximal tolerated gluten in daily life. Based on previous studies, the safe amount of gluten to be tolerated by patients with CD was first considered as under 50 grams of gluten per day (for adults), although some patients showed worsening of intestinal histology after ingestion of only 10 grams per day [56]. Other smaller studies showed different thresholds with wide variation between studies, with the combined conclusion that daily gluten intake below 10 grams is probably safe to patients with CD [57]. A different question relates to the amount of gluten required to elicit a quick pathologic response or in other words what would be the ideal quantity of gluten and the optimal marker to follow when challenging a patient on a GFD. A recent RCT compared various endpoints and biomarkers after 15 days of micro-challenge with 10 and 3 grams of gluten in 14 adult patients [58]. With 3 grams daily gluten consumption, the only significant change from baseline was in patients' self-reported symptoms. Other markers including duodenal histology, video capsule endoscopy findings, and gluten-specific T-cells, showed significant changes at 10 grams gluten only. This trial provides a framework for future studies to incorporate modern biomarkers in the investigation of gluten challenge, in order to better define the accurate threshold of tolerated gluten in CD. Moreover, specific cut-offs for paediatric population are lacking and merit further studies.

In clinical practice, follow-up of patients with CD includes clinical and dietary assessments, as well as monitoring for CD serology as a surrogate marker for treatment adherence. However, CD antibodies have low sensitivity to detect villous atrophy in patients with CD on self-reported GFD, shown to be as low as below 50% in a large recent meta-analysis [59]. For that reason, better markers are needed in order to monitor intentional and unintentional gluten consumption in patients with CD. Gluten immunogenic peptides (GIP) are novel markers for gluten consumption, based on their resistance to gastric and intestinal degradation. GIP could be detected in faeces of patients, as soon as 3 days after minimal gluten ingestion [60]. Several studies have proved detection of faecal or urine GIP in patients with CD who were reporting good adherence to GFD [61–65], highlighting the gap between patients' perception or reported adherence and real-life exposure to gluten. Furthermore, as these tests provide short term information, exposures a few days before the test (stool GIP) and even testing more than 24 hours after exposure (urine GIP) may miss incidental or non-incidental transgressions. There are ongoing studies aiming to explore further utilization of GIP testing, including a point-of-care home test for patients with CD (NCT03462979, clinicaltrials.gov).

Another interesting arm of research is the genetic manipulation of wheat, aiming to develop wheat with reduced gluten toxicity. Several studies so far have failed to identify sufficient changes in wheat genome that will both preserve wheat gastronomic and technical properties, together with prevention of immunogenic-inflammatory response in CD [66, 67].

Most recently, the International Wheat Genome Sequencing Consortium analyzed and published a detailed annotated reference genome sequence of wheat, covering 94% of its genome [68]. This development could assist in future studies in wheat engineering exploring the production of non-immunogenic gluten to be appropriate for ingestion by patients with CD.

Another approach being investigated in the food industry, is modification of pre-digested wheat in order to lower gluten presence in the product or reduce its immunogenicity. For that purpose several processes are being studied, including fermentation of wheat by microorganisms that release proteases able to digest gliadin peptides [69–71], thermal processing by microwave [72, 73], and the use of microbial transglutaminase [74]. The role of these modifications and their long-term safety in patients with CD are yet to be defined.

Finally, returning from the research arena to the clinical practice, many CD patients are eager to consume a gluten containing diet. As stated above, a specific chapter in this book is dedicated to novel treatment strategies beyond a gluten diet, and thus are not covered here. We hope that the various research priorities delineated in our chapter will pave way to better therapeutic options in pathways that are already studied as well as new therapeutic modalities, that will enable patients to enjoy better quality of life in full remission.

References

1. Lundin KE, Scott H, Hansen T, Paulsen G, Halstensen TS, Fausa O, et al. Gliadin-specific, HLA-DQ (alpha 1*0501, beta 1*0201) restricted T cells isolated from the small intestinal mucosa of celiac disease patients. J Exp Med. 1993;178(1):187–96.
2. Sollid LM, Markussen G, Ek J, Gjerde H, Vartdal F, Thorsby E. Evidence for a primary association of celiac disease to a particular HLA-DQ alpha/beta heterodimer. J Exp Med. 1989;169(1):345–50.
3. Lionetti E, Catassi C. Co-localization of gluten consumption and HLA-DQ2 and -DQ8 genotypes, a clue to the history of celiac disease. Dig Liver Dis. 2014;46(12):1057–63.
4. Liu E, Lee H-S, Aronsson CA, Hagopian WA, Koletzko S, Rewers MJ, et al. Risk of pediatric celiac disease according to HLA haplotype and country. N Engl J Med. 2014;371(1):42–9.
5. Lionetti E, Castellaneta S, Francavilla R, Pulvirenti A, Tonutti E, Amarri S, et al. Introduction of gluten, HLA status, and the risk of celiac disease in children. N Engl J Med. 2014;371 (14):1295–303.
6. Karinen H, Kärkkäinen P, Pihlajamäki J, Janatuinen E, Heikkinen M, Julkunen R, et al. Gene dose effect of the DQB1*0201 allele contributes to severity of coeliac disease. Scand J Gastroenterol. 2006;41(2):191–9.
7. Kuja-Halkola R, Lebwohl B, Halfvarson J, Wijmenga C, Magnusson PKE, Ludvigsson JF. Heritability of non-HLA genetics in coeliac disease: a population-based study in 107 000 twins. Gut. 2016;65(11):1793–8.

8. Sallese M, Lopetuso LR, Efthymakis K, Neri M. Beyond the HLA genes in gluten-related disorders. Front Nutr. 2020;7:575844.
9. Sharma A, Liu X, Hadley D, Hagopian W, Liu E, Chen WM, et al. Identification of non-HLA genes associated with celiac disease and country-specific differences in a large, international pediatric cohort. PLoS One. 2016;11(3):e0152476.
10. Smyth DJ, Plagnol V, Walker NM, Cooper JD, Downes K, Yang JH, et al. Shared and distinct genetic variants in type 1 diabetes and celiac disease. N Engl J Med. 2008;359(26):2767–77.
11. Zhernakova A, Stahl EA, Trynka G, Raychaudhuri S, Festen EA, Franke L, et al. Meta-analysis of genome-wide association studies in celiac disease and rheumatoid arthritis identifies fourteen non-HLA shared loci. PLoS Genet. 2011;7(2):e1002004.
12. Östensson M, Montén C, Bacelis J, Gudjonsdottir AH, Adamovic S, Ek J, et al. A possible mechanism behind autoimmune disorders discovered by genome-wide linkage and association analysis in celiac disease. PLoS One. 2013;8(8):e70174.
13. Dubois PC, Trynka G, Franke L, Hunt KA, Romanos J, Curtotti A, et al. Multiple common variants for celiac disease influencing immune gene expression. Nat Genet. 2010;42(4): 295–302.
14. Castellanos-Rubio A, Fernandez-Jimenez N, Kratchmarov R, Luo X, Bhagat G, Green PH, et al. A long noncoding RNA associated with susceptibility to celiac disease. Science. 2016;352(6281):91–5.
15. Sharp SA, Jones SE, Kimmitt RA, Weedon MN, Halpin AM, Wood AR, et al. A single nucleotide polymorphism genetic risk score to aid diagnosis of coeliac disease: a pilot study in clinical care. Aliment Pharmacol Ther. 2020;52(7):1165–73.
16. Tye-Din JA, Stewart JA, Dromey JA, Beissbarth T, van Heel DA, Tatham A, et al. Comprehensive, quantitative mapping of T cell epitopes in gluten in celiac disease. Sci Transl Med. 2010;2(41):41ra51.
17. Hmida NB, Ben Ahmed M, Moussa A, Rejeb MB, Said Y, Kourda N, et al. Impaired control of effector T cells by regulatory T cells: a clue to loss of oral tolerance and autoimmunity in celiac disease? Am J Gastroenterol. 2012;107(4):604–11.
18. Cook L, Munier CML, Seddiki N, van Bockel D, Ontiveros N, Hardy MY, et al. Circulating gluten-specific FOXP3(+)CD39(+) regulatory T cells have impaired suppressive function in patients with celiac disease. J Allergy Clin Immunol. 2017;140(6):1592-603.e8.
19. Quinn EM, Coleman C, Molloy B, Dominguez Castro P, Cormican P, Trimble V, et al. Transcriptome analysis of CD4+ T cells in coeliac disease reveals imprint of BACH2 and IFNγ regulation. PLoS One. 2015;10(10):e0140049.
20. McLean MH, Dieguez D Jr, Miller LM, Young HA. Does the microbiota play a role in the pathogenesis of autoimmune diseases? Gut. 2015;64(2):332–41.
21. Nadal I, Donant E, Ribes-Koninckx C, Calabuig M, Sanz Y. Imbalance in the composition of the duodenal microbiota of children with coeliac disease. J Med Microbiol. 2007;56(Pt 12):1669–74.
22. Schippa S, Iebba V, Barbato M, Di Nardo G, Totino V, Checchi MP, et al. A distinctive "microbial signature" in celiac pediatric patients. BMC Microbiol. 2010;10:175.
23. Collado MC, Donat E, Ribes-Koninckx C, Calabuig M, Sanz Y. Imbalances in faecal and duodenal Bifidobacterium species composition in active and non-active coeliac disease. BMC Microbiol. 2008;8:232.
24. De Palma G, Nadal I, Medina M, Donat E, Ribes-Koninckx C, Calabuig M, et al. Intestinal dysbiosis and reduced immunoglobulin-coated bacteria associated with coeliac disease in children. BMC Microbiol. 2010;10:63.
25. Sanz Y, Palma GD, Laparra M. Unraveling the ties between celiac disease and intestinal microbiota. Int Rev Immunol. 2011;30(4):207–18.
26. Olivares M, Neef A, Castillejo G, Palma GD, Varea V, Capilla A, et al. The HLA-DQ2 genotype selects for early intestinal microbiota composition in infants at high risk of developing coeliac disease. Gut. 2015;64(3):406–17.

27. Di Biase AR, Marasco G, Ravaioli F, Dajti E, Colecchia L, Righi B, et al. Gut microbiota signatures and clinical manifestations in celiac disease children at onset: a pilot study. J Gastroenterol Hepatol. 2021;36(2):446–54.
28. Schiepatti A, Bacchi S, Biagi F, Panelli S, Betti E, Corazza GR, et al. Relationship between duodenal microbiota composition, clinical features at diagnosis, and persistent symptoms in adult Coeliac disease. Dig Liver Dis. 2021.
29. Leonard MM, Camhi S, Huedo-Medina TB, Fasano A. Celiac Disease Genomic, Environmental, Microbiome, and Metabolomic (CDGEMM) study design: approach to the future of personalized prevention of celiac disease. Nutrients. 2015;7(11):9325–36.
30. Marasco G, Cirota GG, Rossini B, Lungaro L, Di Biase AR, Colecchia A, et al. Probiotics, prebiotics and other dietary supplements for gut microbiota modulation in celiac disease patients. Nutrients. 2020;12(9).
31. Olivares M, Castillejo G, Varea V, Sanz Y. Double-blind, randomised, placebo-controlled intervention trial to evaluate the effects of Bifidobacterium longum CECT 7347 in children with newly diagnosed coeliac disease. Br J Nutr. 2014;112(1):30–40.
32. Klemenak M, Dolinšek J, Langerholc T, Di Gioia D, Mičetić-Turk D. Administration of bifidobacterium breve decreases the production of TNF-α in children with celiac disease. Dig Dis Sci. 2015;60(11):3386–92.
33. Smecuol E, Constante M, Temprano MP, Costa AF, Moreno ML, Pinto-Sanchez MI, et al. Effect of Bifidobacterium infantis NLS super strain in symptomatic coeliac disease patients on long-term gluten-free diet—an exploratory study. Benef Microbes. 2020;11(6):527–34.
34. Håkansson Å, Andrén Aronsson C, Brundin C, Oscarsson E, Molin G, Agardh D. Effects of lactobacillus plantarum and lactobacillus paracasei on the peripheral immune response in children with celiac disease autoimmunity: a randomized, double-blind, placebo-controlled clinical trial. Nutrients. 2019;11(8).
35. Al-Toma A, Volta U, Auricchio R, Castillejo G, Sanders DS, Cellier C, et al. European Society for the Study of Coeliac Disease (ESsCD) guideline for coeliac disease and other gluten-related disorders. United Eur Gastroenterol J. 2019;7(5):583–613.
36. Bai JC, Ciacci C. World gastroenterology organisation global guidelines: celiac disease february 2017. J Clin Gastroenterol. 2017;51(9):755–68.
37. Husby S, Koletzko S, Korponay-Szabó I, Kurppa K, Mearin ML, Ribes-Koninckx C, et al. European society paediatric gastroenterology, hepatology and nutrition guidelines for diagnosing coeliac disease 2020. J Pediatr Gastroenterol Nutr. 2020;70(1):141–56.
38. Ramírez-Sánchez AD, Tan IL, Gonera-de Jong BC, Visschedijk MC, Jonkers I, Withoff S. Molecular biomarkers for celiac disease: past, present and future. Int J Mol Sci. 2020;21(22).
39. Christophersen A, Ráki M, Bergseng E, Lundin KE, Jahnsen J, Sollid LM, et al. Tetramer-visualized gluten-specific CD4+ T cells in blood as a potential diagnostic marker for coeliac disease without oral gluten challenge. United Eur Gastroenterol J. 2014;2(4):268–78.
40. Sarna VK, Lundin KEA, Mørkrid L, Qiao SW, Sollid LM, Christophersen A. HLA-DQ-gluten tetramer blood test accurately identifies patients with and without celiac disease in absence of gluten consumption. Gastroenterology. 2018;154(4):886-96.e6.
41. Petersen J, van Bergen J, Loh KL, Kooy-Winkelaar Y, Beringer DX, Thompson A, et al. Determinants of gliadin-specific T cell selection in celiac disease. J Immunol. 2015;194 (12):6112–22.
42. Brottveit M, Ráki M, Bergseng E, Fallang LE, Simonsen B, Løvik A, et al. Assessing possible celiac disease by an HLA-DQ2-gliadin tetramer test. Am J Gastroenterol. 2011;106(7):1318–24.
43. Sarna VK, Skodje GI, Reims HM, Risnes LF, Dahal-Koirala S, Sollid LM, et al. HLA-DQ: gluten tetramer test in blood gives better detection of coeliac patients than biopsy after 14-day gluten challenge. Gut. 2018;67(9):1606–13.
44. Singh A, Verma AK, Das P, Prakash S, Pramanik R, Nayak B, et al. Non-immunological biomarkers for assessment of villous abnormalities in patients with celiac disease. J Gastroenterol Hepatol. 2020;35(3):438–45.

45. Ioannou HP, Fotoulaki M, Pavlitou A, Efstratiou I, Augoustides-Savvopoulou P. Plasma citrulline levels in paediatric patients with celiac disease and the effect of a gluten-free diet. Eur J Gastroenterol Hepatol. 2011;23(3):245–9.
46. Adriaanse MPM, Mubarak A, Riedl RG, Ten Kate FJW, Damoiseaux J, Buurman WA, et al. Progress towards non-invasive diagnosis and follow-up of celiac disease in children; a prospective multicentre study to the usefulness of plasma I-FABP. Sci Rep. 2017;7(1):8671.
47. Adriaanse MP, Leffler DA, Kelly CP, Schuppan D, Najarian RM, Goldsmith JD, et al. Serum I-FABP detects gluten responsiveness in adult celiac disease patients on a short-term gluten challenge. Am J Gastroenterol. 2016;111(7):1014–22.
48. Oldenburger IB, Wolters VM, Kardol-Hoefnagel T, Houwen RHJ, Otten HG. Serum intestinal fatty acid-binding protein in the noninvasive diagnosis of celiac disease. Apmis. 2018;126(3):186–90.
49. Manavalan JS, Hernandez L, Shah JG, Konikkara J, Naiyer AJ, Lee AR, et al. Serum cytokine elevations in celiac disease: association with disease presentation. Hum Immunol. 2010;71 (1):50–7.
50. Masaebi F, Azizmohammad Looha M, Rostami-Nejad M, Pourhoseingholi MA, Mohseni N, Samasca G, et al. The predictive value of serum cytokines for distinguishing celiac disease from non-celiac gluten sensitivity and healthy subjects. Iran Biomed J. 2020;24(6):340–6.
51. Iervasi E, Auricchio R, Strangio A, Greco L, Saverino D. Serum IL-21 levels from celiac disease patients correlates with anti-tTG IgA autoantibodies and mucosal damage. Autoimmunity. 2020;53(4):225–30.
52. Amr KS, Bayoumi FS, Eissa E, Abu-Zekry M. Circulating microRNAs as potential non-invasive biomarkers in pediatric patients with celiac disease. Eur Ann Allergy Clin Immunol. 2019;51(4):159–64.
53. Bascuñán KA, Pérez-Bravo F, Gaudioso G, Vaira V, Roncoroni L, Elli L, et al. A miRNA-based blood and mucosal approach for detecting and monitoring celiac disease. Dig Dis Sci. 2020;65(7):1982–91.
54. Villafuerte-Galvez J, Vanga RR, Dennis M, Hansen J, Leffler DA, Kelly CP, et al. Factors governing long-term adherence to a gluten-free diet in adult patients with coeliac disease. Aliment Pharmacol Ther. 2015;42(6):753–60.
55. Myléus A, Reilly NR, Green PHR. Rate, risk factors, and outcomes of nonadherence in pediatric patients with celiac disease: a systematic review. Clin Gastroenterol Hepatol. 2020;18(3):562–73.
56. Catassi C, Fabiani E, Iacono G, D'Agate C, Francavilla R, Biagi F, et al. A prospective, double-blind, placebo-controlled trial to establish a safe gluten threshold for patients with celiac disease. Am J Clin Nutr. 2007;85(1):160–6.
57. Akobeng AK, Thomas AG. Systematic review: tolerable amount of gluten for people with coeliac disease. Aliment Pharmacol Ther. 2008;27(11):1044–52.
58. Leonard MM, Silvester JA, Leffler D, Fasano A, Kelly CP, Lewis SK, et al. Evaluating responses to gluten challenge: a randomized, double-blind, 2-dose gluten challenge trial. Gastroenterology. 2021;160(3):720-33.e8.
59. Silvester JA, Kurada S, Szwajcer A, Kelly CP, Leffler DA, Duerksen DR. Tests for serum transglutaminase and endomysial antibodies do not detect most patients with celiac disease and persistent villous atrophy on gluten-free diets: a meta-analysis. Gastroenterology. 2017;153(3):689-701.e1.
60. Comino I, Real A, Vivas S, Síglez M, Caminero A, Nistal E, et al. Monitoring of gluten-free diet compliance in celiac patients by assessment of gliadin 33-mer equivalent epitopes in feces. Am J Clin Nutr. 2012;95(3):670–7.
61. Comino I, Fernández-Bañares F, Esteve M, Ortigosa L, Castillejo G, Fambuena B, et al. Fecal gluten peptides reveal limitations of serological tests and food questionnaires for monitoring gluten-free diet in celiac disease patients. Am J Gastroenterol. 2016;111(10):1456–65.
62. Gerasimidis K, Zafeiropoulou K, Mackinder M, Ijaz UZ, Duncan H, Buchanan E, et al. Comparison of clinical methods with the faecal gluten immunogenic peptide to assess gluten intake in coeliac disease. J Pediatr Gastroenterol Nutr. 2018;67(3):356–60.

63. Moreno ML, Sánchez-Muñoz D, Sanders D, Rodríguez-Herrera A, Sousa C. Verifying diagnosis of refractory celiac disease with urine gluten immunogenic peptides as biomarker. Front Med (Lausanne). 2020;7:601854.
64. Roca M, Donat E, Masip E, Crespo-Escobar P, Cañada-Martínez AJ, Polo B, et al. Analysis of gluten immunogenic peptides in feces to assess adherence to the gluten-free diet in pediatric celiac patients. Eur J Nutr. 2020.
65. Stefanolo JP, Tálamo M, Dodds S, de la Paz TM, Costa AF, Moreno ML, et al. Real-world gluten exposure in patients with celiac disease on gluten-free diets, determined from gliadin immunogenic peptides in urine and fecal samples. Clin Gastroenterol Hepatol. 2021;19(3):484-91.e1.
66. van den Broeck HC, van Herpen TW, Schuit C, Salentijn EM, Dekking L, Bosch D, et al. Removing celiac disease-related gluten proteins from bread wheat while retaining technological properties: a study with Chinese Spring deletion lines. BMC Plant Biol. 2009;9:41.
67. Carroccio A, Di Prima L, Noto D, Fayer F, Ambrosiano G, Villanacci V, et al. Searching for wheat plants with low toxicity in celiac disease: between direct toxicity and immunologic activation. Dig Liver Dis. 2011;43(1):34–9.
68. Appels R, Eversole K, Feuillet C, Keller B, Rogers J, Stein N, et al. Shifting the limits in wheat research and breeding using a fully annotated reference genome. Science. 2018;361(6403).
69. Greco L, Gobbetti M, Auricchio R, Di Mase R, Landolfo F, Paparo F, et al. Safety for patients with celiac disease of baked goods made of wheat flour hydrolyzed during food processing. Clin Gastroenterol Hepatol. 2011;9(1):24–9.
70. Rees D, Holtrop G, Chope G, Moar KM, Cruickshank M, Hoggard N. A randomised, double-blind, cross-over trial to evaluate bread, in which gluten has been pre-digested by prolyl endoprotease treatment, in subjects self-reporting benefits of adopting a gluten-free or low-gluten diet. Br J Nutr. 2018;119(5):496–506.
71. Liu YY, Lee CC, Hsu JH, Leu WM, Meng M. Efficient hydrolysis of gluten-derived celiac disease-triggering immunogenic peptides by a bacterial serine protease from burkholderia gladioli. Biomolecules. 2021;11(3).
72. Lamacchia C, Landriscina L, D'Agnello P. Changes in wheat kernel proteins induced by microwave treatment. Food Chem. 2016;197(Pt A):634–40.
73. Gianfrani C, Mamone G, la Gatta B, Camarca A, Di Stasio L, Maurano F, et al. Microwave-based treatments of wheat kernels do not abolish gluten epitopes implicated in celiac disease. Food Chem Toxicol. 2017;101:105–13.
74. Marino M, Casale R, Borghini R, Di Nardi S, Donato G, Angeloni A, et al. The effects of modified versus unmodified wheat gluten administration in patients with celiac disease. Int Immunopharmacol. 2017;47:1–8.

Non Celiac Wheat Sensitivity

Carlo Catassi, Giulia Guelzoni, and Giulia N. Catassi

1 Introduction

Non-celiac gluten/wheat sensitivity (NCGS/NCWS) is characterized by irritable bowel syndrome (IBS)-like symptoms and extra-intestinal manifestations, occurring in a few hours or days after ingestion of gluten/wheat-containing food, improving rapidly with gluten/wheat withdrawal and relapsing soon after gluten/wheat challenge. Pre-requisite for suspecting NCGS/NCWS is the exclusion of both celiac disease (CD) and wheat allergy (WA) when the patient is still on a gluten-containing diet [1].

The terminology of this disorder is still a matter of debate. Although the first cases of NCGS were reported in the 1970s [2, 3], this entity has been characterized only recently (year 2010) by Sapone et al. [4] who described the clinical and pathophysiological features of NCGS. Since then, the number of papers reporting on NCGS has grown exponentially, as well as the number of non-celiac individuals treated with the gluten-free diet (GFD) because of a wide array of symptoms or conditions. However, in recent years it has become clear that wheat components other than gluten, particularly so-called Fermentable Oligosaccharides, Disaccharides, Monosaccharides and Polyols (FODMAPs) [5] and Amylase-Trypsin Inhibitors (ATIs) [6], may elicit symptoms of NCGS. Since it is often impossible to establish which wheat component/s is/are the disease trigger/s, the disorder here described is best defined as NCWS. The major limitation of NCWS terminology is the exclusion of other gluten-containing grains, such as rye and

C. Catassi (✉) · G. Guelzoni · G. N. Catassi
Division of Pediatrics, DISCO Department, Polytechnic University of Marche, Ancona, Italy
e-mail: c.catassi@univpm.it

C. Catassi
Division of Pediatrics, "G. Salesi" Children Hospital, Via F. Corridoni 11, 60123 Ancona, Italy

© The Author(s), under exclusive license to Springer Nature Switzerland AG 2022   225
J. Amil-Dias and I. Polanco (eds.), *Advances in Celiac Disease*,
https://doi.org/10.1007/978-3-030-82401-3_16

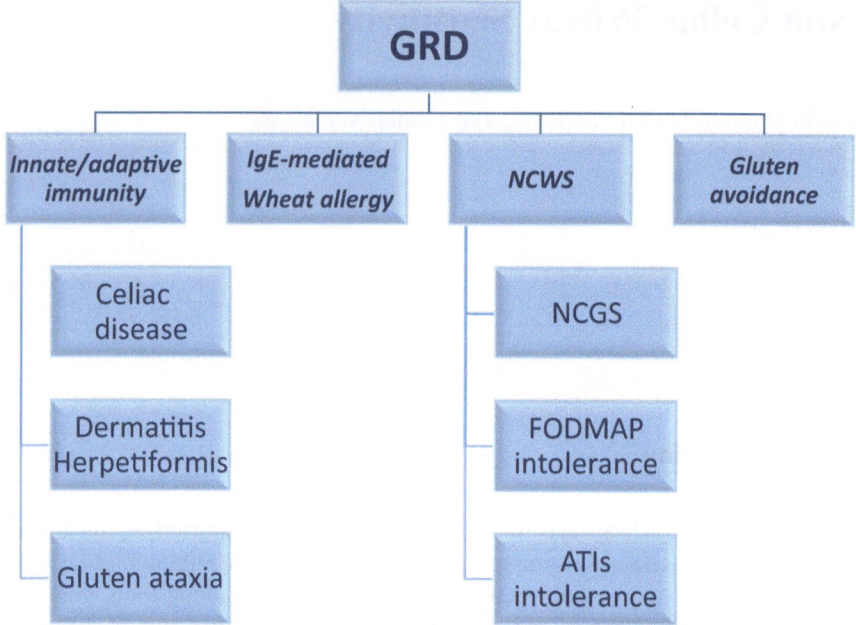

Fig. 1 Current classification of Gluten-Related Disorders (GRD). NCWS = non celiac wheat sensitivity; NCGS = non celiac gluten sensitivity; FODMAP = Fermentable Oligosaccharides, Disaccharides, monosaccharides and polyols; ATIs = Amylase-Trypsin Inhibitors

barley, that might trigger the disorder. Figure 1 shows an updated classification of NCWS and other gluten-related disorders.

2 Epidemiology of NCWS

Due to lack of a disease biomarker, the frequency of NCWS in the general population is still unclear. An early estimate from the Center For Celiac Research at the University of Maryland in Baltimore (US) suggested a NCWS prevalence of about 6% [7]. The limitation to this observation was that this is a tertiary centre seeing patients within a fee-paying system. Due to a selection bias, this may not accurately reflect international prevalence figures for NCWS. More recent data from the US National Health and Nutrition Examination Surveys (NHANES) indicate that the prevalence of NCWS increased significantly from 0.5% in 2009–10 to 1.0% in 2011–12 to 1.7% in 2013–14 in the general population [8].

NCWS has mostly been described in adults, particularly in females aged 30–50 years [9]; however, paediatric case series have also been reported [10, 11]. In a recent study conducted in Italy, Francavilla and coworkers investigated the prevalence of NCWS on a sample of 1114 children affected with functional gastrointestinal disorders (FGID) and found that 3.3% of them had "suspected" NCWS, but only 1.1% had the diagnosis confirmed by a gluten challenge [12]. Since FGID affect about 20% of children, the estimated prevalence of NCWS in the overall paediatric population is about 0.2–0.3%.

An emerging epidemiological issue is represented by self-reported gluten intolerance, i.e., people excluding gluten-containing food without a medical diagnosis of a specific gluten-related disorder. Many individuals perceive the GFD as healthy lifestyle practice, others erroneously believe that the GFD may help in losing weight or improving physical fitness. These "gluten avoiders" or lifestylers have nothing to do with true NCWS or any other gluten-related disorder, but are widely diffused in many countries, with a prevalence of 6.2–13% [13, 14].

In conclusion, current estimates indicate that the prevalence of NCWS is around 2% in the general population and 0.2–0.3% in children.

3   Clinical Picture

As noted previously, NCWS is characterized by symptoms that usually occur soon after wheat ingestion, disappear with wheat/gluten withdrawal, and relapse following wheat/gluten challenge within hours or days. Therefore, the latency between wheat ingestion/withdraw and the appearance/disappearance of symptoms is typically much shorter in NCWS (few hours/days) than in CD (weeks/months).

The 'classical' presentation of NCWS is a combination of irritable bowel syndrome (IBS)-like symptoms, including abdominal pain, bloating, bowel habit abnormalities (either diarrhoea or constipation), and minimal neurological manifestations such as 'foggy mind', headache and chronic fatigue. Other complaints may include joint and muscle pain, leg or arm numbness, dermatitis (eczema or skin rash), depression, gynaecologic problems (recurrent vaginitis and cystitis) and anaemia, and major neurological manifestations. When seen at a specialty clinic, many NCWS patients already report the causal relationship between the ingestion of gluten-containing food and worsening of symptoms. In children, NCWS usually manifests with IBS-like symptoms, such as abdominal pain and chronic diarrhoea, while the extraintestinal manifestations are less frequent [4, 10, 15, 16].

The prevalence of IBS worldwide is 10–20% [17, 18]. Approximately 50% of patients with gastrointestinal complaints seen in primary care have IBS-type symptoms [19]. Patients with IBS report a reduced quality of life and there is an associated economic and societal cost [20, 21]. Patients have always reported that food plays an important role in their IBS-type symptoms with estimates of up to 80% of patients having postprandial symptomology, and up to 40% reporting specific "food intolerances" [22–24]. Over the last 15 years there has been renewed interest in the concept of dietary interventions for FGID [25, 26]. IBS dietary research has focused on the role of two common components of the western diet, specifically FODMAPs and gluten in relation to the induction of IBS symptoms. To note, both these two components are found in large amounts in wheat that contains both gluten and fructans FODMAPs. Several randomized control trials have demonstrated the efficacy of the low FODMAP diet. There is overlap between NCWS and IBS-type symptoms [13, 15]. The fundamental difference between NCWS and IBS is that patients with NCWS self-report symptoms when consuming wheat and have identified or perceive wheat as the culprit. Conversely IBS patients do not report wheat as a specific stimulus for their symptoms. However, previously published literature has demonstrated that wheat is a commonly reported "food intolerance" when IBS patients are specifically questioned [22–24].

In recent years, several studies explored the relationship between the ingestion of gluten-containing food and the appearance of neurological and psychiatric disorders/symptoms, an issue that is analyzed in detail in the next two paragraphs.

4 Gluten Sensitivity and Autism

Research on the effect of diet on autistic spectrum disorder (ASD) has been increasing in recent decades. One of the most popular interventions for ASD is the gluten-free casein-free (GFCF) diet. Although an association between CD and autism has been anecdotally reported, the possible effect of the GFCF in children with autism is not due to underlying CD, but to an entirely different pathophysiological mechanism [27, 28]. It has been hypothesized that symptoms may be caused by opioid peptides formed from the incomplete breakdown of foods containing gluten and casein. Increased intestinal permeability, also referred to as the 'leaky gut syndrome', has been suspected in ASD to be part of the chain of events that allows these peptides to cross the intestinal membrane, enter the bloodstream, and cross the blood–brain barrier, affecting the endogenous opiate system and neurotransmission within the nervous system. The resulting excess of opioids is thought to lead to behaviours noted in ASD, and the removal of these substances from the diet could determine an improvement in autistic behaviours [29]. The leaky gut/autism connection has fuelled a strong debate within the scientific community, which is far from being settled.

One study reported a high percentage of increased intestinal permeability, [as established by the lactulose/mannitol (L/M) ratio], among patients with autism and their relatives compared with normal subjects. After starting the GFD, patients with autism had significantly lower intestinal permeability test values compared with those who were on an unrestricted diet and controls [30]. However, Robertson et al. [31] did not detect any changes in intestinal permeability in a small cohort of ASD children. In another small study, neither the L/M ratio nor behavioural scores were different between groups exposed to gluten/dairy or placebo [32]. The finding of IgG-class antibodies directed against food antigens is considered indirect evidence of increased intestinal permeability. Children with autism have significantly higher levels of IgG AGA (but not IgA) compared with healthy controls, particularly those with gastrointestinal symptoms [33]. Another study confirmed these findings and also reported an increase in antibodies directed to several other food allergens, including casein and whole milk [34].

A recent study reported that levels of serological markers of an impaired gut barrier (zonulin and intestinal fatty acid binding protein I-FABP) in children with ASD were similar to those found in healthy control. AGA IgG were found in 27.3% of ASD children while celiac specific antibodies (anti-TG2 IgA and anti-DPG IgG) were negative, but increased levels of antibodies against neural TG6 were found in 6.49% of children. According to this study, the mechanism of immune activation and gluten antibody response in ASD patients is probably not strictly connected with increased gut permeability [35].

Despite its popularity, the efficacy of the GFCF diet in improving autistic behaviour remains to be proven. A 2008 Cochrane review reported that only two small randomized controlled trials investigated the effect of the GFCF diet in children with ASD (n = 35). There were only three significant treatment effects in favour of the diet intervention: overall autistic traits, mean difference (MD) = −5.60; social isolation, MD = −3.20, and overall ability to communicate and interact, MD = 1.70. In addition, three outcomes were not different between the treatment and control group, while differences for ten outcomes could not be analyzed because data were skewed. The review concluded that the evidence for efficacy of these diets is poor, and large-scale good-quality randomized controlled trials are needed [36]. Similar conclusions were reached by a recently published systematic review on treatment of autistic children with the GFCF diet [37].

In a two-stage randomized controlled study of the GFCF diet in children with ASD, Whiteley et al. reported significant group improvements in core autistic and related behaviours after 8 and 12 months of diet. The results showed a less dramatic change between children having been on diet for 8 months and those on diet for 24 months, possibly reflective of a plateau effect [38]. Analyses indicated several factors to be potentially pertinent to a positive response to the dietary intervention in terms of symptom presentation. Age was found to be the strongest predictor of response, where those participants aged between 7 and 9 years seemed to derive most benefit from the dietary intervention [39]. The above data suggest that removing gluten from the diet may positively affect the clinical outcome in some children diagnosed with ASD, indicating that autism may be part of the spectrum of NCGS, at least in some cases.

On the other hand, a randomized controlled double-blind trial was performed on 74 children with ASD with severe maladaptive behavior and increased urinary intestinal fatty acids binding protein (I-FABP, i.e. a marker of enterocyte damage) to test the potential 'toxicity' of gluten/casein. Subjects on a regular diet were randomized to receive a gluten/casein or placebo supplement for 7 days. Administrating gluten/casein to children with ASD for 1 week did not increase maladaptive behavior, gastrointestinal symptom severity or urinary I-FABP excretion [40].

Another double-blind placebo-controlled study was conducted placing 14 children with autism on gluten-free/casein-free diet for 4–6 weeks. They did not find significant effects of the diet on autism symptoms or ASD-related behaviors [41].

Ghalichi et al. investigated the effect of the GFD on gastrointestinal symptoms and behavioral indices in children with ASD. The first group (40 children) was assigned to the GFD and reported a significant decrease both in gastrointestinal symptoms and behavioral disorders; the second group (40 children) on regular diet did not obtain significant results. The study concluded that some children with ASD can improve their stereotyped behaviors, communication and social interaction following a GFD [42].

In another study, a group of patients aged 3–8 years diagnosed with ASD on GFCF diet was compared with a patients on ketogenic diet and a control group (on regular diet). Both diet groups reported improvement in autistic manifestations [43].

In contrast, a randomized, controlled, single-blinded trial demonstrated that there were no differences between children with ASD after an 8-week GFD (n = 33) and those on a gluten-containing diet (n = 33) in autistic symptoms, maladaptive behaviors and intellectual abilities [44].

Gonzalez-Domenech et al. conducted a clinical trial to determine the influence of a GFCF diet not only on behavior disorders in patients with ASD, but also on urinary beta-casomorphin concentrations, which is a peptide with opioid activity resulting from an incomplete degradation of casein in the intestine. Each patient of the study consumed a GFCF diet for six months and a normal diet for another six months. GFCF diet did not induce a significant change in behavioral symptoms of autism and urinary beta-casomorphin concentrations [45].

A very recent systematic review and meta-analysis identified six RCTs investigating a GFCF diet compared to a regular diet in children aged 3 to 17 years with ASD. They reported no effect of a GFCF diet on clinician-reported autism core symptoms, parent reported functional level and behavioral difficulties. In addition, the study showed that a GFCF diet may trigger gastrointestinal adverse effects (RR 2.33). So, this study suggests caution in starting the GFCF diet in children with ASD [46].

A review published in 2021 [47] discussed the pathological mechanisms and the evidence on the use of a GFD in people with ASD. The result was that it is still unclear if the interaction between ASD and gluten is due to gluten specifically (opioid activity from improperly digested gluten, inflammation caused by oxidative stress or reactivity with anti-gluten antibodies) or it is a consequence of a generally raised autoimmune profile in ASD. In addition, authors concluded that there is not a proven benefit of the GFD in people with ASD (who do not have a clinical diagnosis of CD) according to the current literature.

In conclusion, further studies are needed to clarify the possible link between gluten ingestion and autism. Investigations are particularly required to identify possible phenotypes based on the best response and non-response to dietary modifications and to assess biological correlates before considering a dietary intervention.

5 NCGS and Other Psychiatric Disorders

An association between schizophrenia and CD was noted in reports spanning back to the 1960s [48]. In 1986, a double-blind gluten-free vs gluten-load controlled trial of 24 patients conducted by Vlissides et al. [49] showed changes in the symptom profile of schizophrenic patients in response to exclusion of gluten from the diet. On the other hand, a small blind study conducted by Potkin et al. [50] showed no differences in the clinical status of 8 schizophrenic patients on a 5-week gluten challenge in an inpatient setting, as measured by the Brief Psychiatric Rating Scale. A subsequent study by Storms et al. [51] tested 26 schizophrenic patients on a locked ward assigned to either a gluten-free or gluten-rich diet. No differences were found between the groups on their performance in a battery of psychological tests. A recent study using blood samples from the Clinical Antipsychotic Trials of Intervention Effectiveness (CATIE) found that 5.5% of the subjects with schizophrenia had a high level of anti-tTG antibodies (compared to 1.1% in the healthy control sample) and 23.1% had AGA IgG positivity compared with 3.1% of controls. Interestingly enough, a large proportion of transglutaminase (tTG)-positive subjects were endomysial antibody (EMA) negative, questioning the possibility that their tTG positivity was related to CD. Indeed, only 2% of schizophrenic patients fulfilled the CD diagnostic criteria (both anti-tTG and EMA positive), questioning the role of CD in schizophrenia [52]. Additional studies revealed that most of the tTG-positive subjects were tTG-6 positive, suggesting that these antibodies are more a biomarker of neuroinflammation than CD [53]. This study indicated the existence of a specific immune response to gluten in some of these patients, probably related to NCGS. Other studies confirmed the high prevalence of AGA among people with schizophrenia [54]; however, the exact mechanism underlying the observed improvement of symptoms with the GFD in some patients has remained elusive.

In 2019 Kelly et al. [55] performed the first double-blind clinical trial of gluten-free versus gluten-containing diets in patients with schizophrenia who were positive for AGA IgG and negative for CD. They noted improvement on the Clinical Global Impressions scale, in negative symptoms and in gastrointestinal symptoms in participants on the gluten-free diet. This study suggests that a subgroup of patients with schizophrenia could benefit from a GFD.

Another recent study compared concentrations of markers in patients with a first episode or chronic schizophrenia. The prevalence of increased AGA antibody titres was significantly higher in patients than in controls. In particular, chronic patients had significantly higher concentrations of AGA IgA antibodies. In this study, elevated AGA antibodies titres increased the risk of developing schizophrenia about four to seven times [56].

Another psychiatric disease that has been hypothesized to be associated with NCGS is depression. In an Australian study, a group of 22 patients with irritable bowel syndrome who had CD excluded underwent a double-blind crossover study with a placebo versus oral gluten supplementation after a GFD. The results showed that gluten induced depression scale worsening when compared to placebo and without producing effects on other symptoms, for example anxiety [57]. Similar results were reached by an Italian study: among the extraintestinal symptoms, depression was more severe in subjects with NCGS when they received gluten instead of placebo [58]. Porcelli et al. evaluated the presence of antibodies associated with gluten related disorders in patients with mood disorders. They considered IgA/IgG anti-gliadin antibodies, IgA/IgG anti-deamidated gliadin peptide antibodies and IgA/IgG anti-transglutaminase antibodies in patients with bipolar disorders or depressive disorders and controls. A significant difference was found only for anti-tTG IgG antibodies, in particular each unit increase in the anti-tTG IgG antibodies corresponded to about 5% increased risk of having a mood disorder [59].

Finally, some case-reports suggest a possible relationship between gluten ingestion and visual and auditory hallucinations [60, 61].

In conclusion, according to the literature, some subgroups of patients with psychiatric disorders could benefit from a GFD. So, further studies are needed on how to identify those patients with the altered response to gluten, in order to start the diet treatment in this disease subgroup.

6   Small Intestinal Biopsy Findings in NCWS

Unlike patients with active CD who show an increased number of intraepithelial lymphocytes (IELs) associated to a variable degree of villous atrophy and crypt hypertrophy (so called Marsh lesion grade 3a-3c) at the small intestinal biopsy, NCWS subjects show a normal to mildly inflamed mucosa (Marsh 0–1), sometimes with an isolated increase of duodenal IELs. In the Sapone's pioneer paper, CD patients had increased numbers of CD3 + IELs (>50/100 enterocytes) compared to controls, while NCWS patients had a number of CD3 + IELs intermediate between CD patients and controls in the context of relatively conserved villus architecture. The numbers of TCR-$\gamma\delta$ IELs were only elevated in CD subjects (>3.4/100 enterocytes), while in NCWS patients the numbers of $\gamma\delta$ IELs were similar to those in controls [4].

In a recent review, Sergi and coworkers indicate the following hystological features as suggestive, but not specific, of NCWS: (a) a "nearly" standard number of T lymphocytes (<25 for 100 epithelial cells); (b) a peculiar disposition of T lymphocytes in a small "cluster" of 4 or 5 cells in the superficial epithelium; (c) the linear distribution of T lymphocytes in the deeper part of the lamina propria of the mucosa over the *muscularis mucosae*, and (d) an increased number of eosinophils in the lamina propria (>5 cells per high-power field, HPF) [62]. Interestingly, a relevant eosinophilic infiltration has been found in the rectal mucosa of patients with NCWS, which was more intense in the rectum than in the duodenum, suggesting that NCWS might involve inflammation of the entire intestinal tract [63].

Unpublished results indicate a higher mast cell density in NCWS in comparison to healthy controls and CD patients. This increased mast cell number in NCWS seems to be closely related with the presence of IBS-like symptoms such as bloating, abdominal pain, and impaired bowel function. Moreover, the close vicinity of mast cells and nerve fibres observed in these patients may have a role in the generation of symptoms, e.g., abdominal pain, via a neuroimmune mechanism [64].

7 Searching for Biomarker/s of NCWS

Given the non-specific findings at the small intestinal biopsy in subjects with NCWS, the search for a non-invasive biomarker of this condition is currently very active.

By investigating a large number of Intestinal cell damage and systemic immune activation markers (anti-tTG IgA, anti-deamidated gliadin IgG and IgA, AGA IgG, IgA and IgM, lipopolysaccharide-binding protein (LBP), soluble CD14 (sCD14), endotoxin-core antibodies IgG, IgA and IgM, anti-flagellin IgG, IgA and IgM, and fatty acid-binding protein 2 (FABP2) in individuals reporting NCWS, Uhde et al.- found that Individuals with wheat sensitivity had significantly increased serum levels of soluble CD14 and LBP, as well as antibody reactivity to bacterial LPS and flagellin. Circulating levels of FABP2, a marker of intestinal epithelial cell damage, were significantly elevated in the affected individuals and correlated with the immune responses to microbial products. The principal component analysis showed that the "clouds" of NCWS, CD and control subjects were nicely separated on the graph, however none of these biomarkers alone showed enough sensitivity and specificity to be useful in clinical practice [65].

The most specific CD serological markers, such as IgA-class anti-tTG and EMA, are negative in NCWS patients by definition. However, IgG-class AGA directed against native gliadin (the first-generation AGA test) are found more frequently in these cases (about 50%) than in the general population. Therefore, the finding of isolated IgG-AGA positivity may be a clue to the diagnosis of NCGS. When initially positive, IgG AGA normalize more quickly in NCGS than in CD patients after starting treatment with the GFD [66]. Recently Uhde et al.showed that the AGA IgG antibody in NCWS is significantly different from CD in subclass distribution and in its relationship to intestinal cell damage. The observed increase in the gluten-reactive IgG2 and IgG4 subclasses and the correlation between the IgG4 subclass and FABP2 in NCGS may point to a protective response aimed at dampening the inflammatory effect of other antibodies and immune cells [67]. Again, due to poor sensitivity/specificity, IgG AGA determination may yield a clue but does not have a primary diagnostic role in clinical practice.

Zonulin is the protein prehaptoglobin 2, the only known human protein that can reversibly open the intestinal tight junctions. In CD, higher zonulin release correlates with increased epithelial permeability. Recently Barbaro et al. showed that serum zonulin levels were increased in patients with confirmed as well as self-reported NCWS over asymptomatic controls and patients with IBS-D. Zonulin was reduced with the elimination of wheat from the diets of participants with a genetic predisposition to CD. They developed a diagnostic algorithm based on zonulin serum levels, gender and abdominal symptoms and this provided a high performing diagnostic tool with an accuracy of 89.0% [68].

Finally, no association has so far been identified between NCWS and specific genetic markers. In NCWS subjects the prevalence of HLA-DQ2 and -DQ8, the genes that are strongly associated with CD, is comparable to that found in the general population (around 40%) and therefore has no significant positive or negative predictive value.

8 The "Salerno Diagnostic Criteria"

As anticipated in the introduction, the diagnosis of NCWS should be considered in patients with persistent intestinal and/or extraintestinal complaints showing a normal result of the CD and WA serological markers on a gluten-containing diet, usually reporting worsening of symptoms after eating gluten-rich food. Table 1 shows the major features of NCGS, CD, and WA that may help to differentiate these different gluten-related disorders. NCGS should not be an exclusion diagnosis only. Unfortunately, no biomarker is enough sensitive and specific of NCWS. Therefore, the diagnosis of NCWS is based on establishing a clear-cut cause-effect relationship between the ingestion of wheat/gluten and the appearance/ disappearance of symptoms.

Table 1 Major features of the different gluten-related disorders

	Celiac disease	NCWS	Wheat allergy
Time interval between gluten exposure and onset of symptoms	Months-years	Hours-days	Minutes-hours
Pathophysiology	Innate and adaptive immunity	Different triggers (gluten, FODMAPs, ATIs) Innate immunity	Allergic immune response
HLA	DQ2- and DQ8-restricted	No association with DQ2- and DQ8-	No association with DQ2- and DQ8-
Autoantibodies	Present	Absent	Absent
Enteropathy	Present	Absent/minimal changes	Absent
Symptoms	Both intestinal and extraintestinal	Both intestinal and extraintestinal	Both intestinal and extraintestinal
Co-morbidities and complications	Co-morbidities, long-term complications	No co-morbidities, no complications but long-term follow-up studies are needed	No co-morbidities, short-term complications (including anaphylaxis)

In the year 2014 a group of world experts on gluten-related disorders met in Salerno, Italy, to set up diagnostic criteria for NCGS. These criteria are known as the "Salerno diagnostic criteria" (SDC) [27]. It should preliminary be noted that these criteria can identify cases of NCGS, but do not necessarily fit for intolerances to other wheat components (e.g. FODMAP and ATIs).

Patients suspected of suffering from a gluten-related disorder should preliminarily undergo a full clinical and laboratory evaluation to exclude CD and WA while still on a gluten-containing diet, according to a previously outlined diagnostic protocol [29]. At baseline the patient has to be on a normal gluten containing diet for at least six weeks. A self-administered instrument incorporating a modified version of the Gastrointestinal Symptom Rating Scale (GSRS) is filled in. The patient identifies one to three main symptoms that will be quantitatively assessed using a Numerical Rating Scale (NRS) with a score ranging from 1 (mild) to 10 (severe). At time 0 the patient is switched to the GFD for 6 weeks (Step 1). Responders are defined as patients who fulfil the response criteria (>30% reduction of one to three main symptoms or at least 1 symptom with no worsening of others) for at least 50% of the observation time.

The diagnosis of NCGS is excluded in subjects failing to show symptomatic improvement after six weeks of GFD. GFD-unresponsive patients should be investigated for other possible causes of IBS-like symptoms, e.g., intolerance to FODMAPs or small bowel bacterial overgrowth.

In view of the high rate of perceived gluten sensitivity and the possible placebo/nocebo effect of any dietary intervention, a double-blind, placebo-controlled (DBPC) gluten challenge is a crucial step in the diagnostic algorithm of NCGS (Step 2). The gluten challenge includes a 1-week challenge followed by a 1-week washout of strict GFD and by the crossover to the second 1-week challenge. The duration of the challenge period may occasionally be longer than 1 week in patients showing fluctuating symptoms, such as headache or neurobehavioral problems. During the challenge, the patient will identify and report one to three main symptoms. A variation of at least 30% between the gluten and the placebo challenge should be detected to discriminate a positive from a negative result. The suggested dose of gluten for the challenge is 8 g. Gluten and placebo preparations must be undistinguishable in look, texture, and taste as well as balanced in nutritional components.

9 NCWS/NCGS Remains a Difficult Diagnosis

As previously anticipated, many individuals start a GFD based on a self-diagnosis and/or without an expert medical advice. How many of these self-reported gluten intolerants are indeed affected by true NCWS/NCGS. During these last years several studies addressed this issue by performing so-called gluten re-challenge in subjects with a provisional diagnosis of NCGS/NCWS. By meta-analyzing these results, Lionetti et al. [69] found that there was a considerable heterogeneity related to different sample size, type, and amount of gluten administered, duration of challenge and different type of placebo. The overall pooled percentage of patients with a diagnosis of NCGS relapsing after a gluten challenge was only 30%, ranging between 7 and 77%. Surprisingly, the meta-analysis showed a not significant relative risk (RR) of relapse after gluten challenge as compared to placebo (RR = 0.4; 95% CI = −0.15–0.9; p = 0.16). On the other hand, the overall pooled percentage of patients with a diagnosis of NCGS relapsing after a gluten challenge performed according to the Salerno criteria was significantly higher as compared to the percentage of patients relapsing after placebo (40 vs. 24%; p = 0.003), with a significant RR of relapse after gluten challenge as compared to placebo (RR = 2.8; 95% CI = 1.5–5.5; p = 0.002). Authors attributed the low percentage of diagnosis confirmation to several factors, such as a strong "nocebo" effect of the challenge, clinical overlapping with IBS, resolution of NCWS over time, and methodological issues. A causal relationship between gluten and relapsing symptoms was observed in 40% of patients when the Salerno criteria were adopted. Therefore the "Salerno" algorithm is recommended for confirmation of NCWS diagnosis, until a valid biomarker will be available. The poor performance of re-challenge studies based on purified gluten is a novel argument in favour of a possible role of NCGS triggers different from gluten itself.

10   NCWS Pathophysiology: One, Two or Many Triggers?

The pathogenesis of NCWS is likely to be multifactorial, with the innate immune response playing a key role. The first studies on NCWS assumed that gluten was the only wheat component responsible of triggering this disorder, since symptoms disappeared with the GFD. Studies have shown that gliadin can cause an immediate and transient increase in gut permeability. This permeating effect is secondary to the binding of specific undigestible gliadin fragments to the CXCR3 chemokine receptor with subsequent release of zonulin, a modulator of intercellular tight junctions [70]. Several studies have identified an altered expression of innate immune components in response to gluten consumption in NCWS individuals, including mucosal Toll-like receptor 2 (TLR2), PBMC-derived interleukin-10 (IL-10), granulocyte-colony stimulating factor (GCSF), transforming growth factor-α (TGF-α), and the chemokine CXCL-10 [64].

In 2011, Biesiekierski et al. [71] reported that gluten caused gastrointestinal symptoms in non-CD IBS subjects investigated by a randomized, double-blind placebo-controlled (DBPC) trial. However, in a subsequent study, the same research group reached different conclusions based on the results of a different DBPC crossover trial on 37 patients with IBS/self-reported NCGS [16]. Patients were randomly assigned to a period of reduced low-fermentable, poorly absorbed, short-chain carbohydrates (FODMAPs) diet and then placed on either a gluten or whey protein challenge. In all participants, gastrointestinal complaints consistently improved during reduced FODMAP intake but significantly worsened to a similar degree when their diets included gluten or whey proteins. The FODMAP list includes fructans, galactans, fructose, and polyols that are contained in several foodstuffs, including wheat, vegetables, and milk derivatives. In the small and large intestine, the FODMAP molecules exert an osmotic effect and are also rapidly fermented by colonic microflora producing gas. The increase in fluid and gas distends the bowel, with consequent sensation of bloating and abdominal pain or discomfort, and diarrhoea. Data of this study raised the possibility that the positive effect of the GFD in patients with IBS is an unspecific consequence of reducing FODMAP intake, given that wheat is one of the possible sources of FODMAPs [72]. However, it should be noted that FODMAPs cannot be entirely and exclusively responsible for the symptoms reported by NCGS subjects, since these patients experience a resolution of symptoms while on the GFD despite continuing to ingest FODMAPs from other sources, like legumes.

ATIs are a family of at least 11 structurally similar, small and compact mono-, di- or tetrameric wheat proteins, which serve as protective proteins in wheat and other cereals by inhibiting enzymes (amylase and trypsin-like activities) of wheat and some parasites. In the developing grain, ATIs are deposited together with gluten proteins in the endosperm and become associated with the starch granules. Encoded mainly by the B and D genomes, ATIs are high in most modern hexaploid bread wheats, and low in spelt (old hexaploid), tetraploid (durum wheat, emmer) and diploid (einkorn) wheat species. They are also present in other gluten containing cereals such as barley and rye. Long known as major allergens in baker's asthma, ATIs were identified as triggers of innate immune activation in intestinal myeloid cells via stimulation of Toll-like receptor 4 (TLR4). Notably, nutritional ATIs enhance intestinal inflammation in models of inflammatory bowel disease in mice, and immune activation is higher in the mesenteric lymph nodes than in the intestinal mucosa [73]. In intestinal tissues from control mice, ATIs induced an innate immune response by activation of Toll-like receptor 4 signalling to MD2 and CD14, causing barrier dysfunction in the absence of mucosal damage. Administration of ATIs to gluten-sensitized mice expressing HLA-DQ8 increased intestinal inflammation in response to gluten in the diet. Interestingly, ATIs are degraded by *Lactobacillus*, which reduced the inflammatory effects of ATIs [74]. However clinical data on the ability of ATIs to trigger symptoms of NCWS are still missing.

Finally, Carroccio et al. recently hypothesized that food antigens, including wheat proteins, initiate a Th2 response driving intestinal eosinophilia and suggested that NCWS is a form of non-IgE mediated food allergy [64]. They also reported that production of TNF-α by CD45+, CD3+, CD4+, and CD8+ cells and of IL-17 by CD4+ cells is higher in the rectal tissue of NCWS patients than in controls. Overall, these results suggest a significant role for an immune adaptive response in patients with NCWS [75].

In summary, these new studies seem to indicate that the pathophysiology of NCWS is much more complex than previously thought and may include different, non-mutually exclusive factors related to wheat consumption (gluten, FODMAPs and ATIs). The complex interplay between these dietary factors, the genetic

Pathophysiology of Gluten Related Disorders

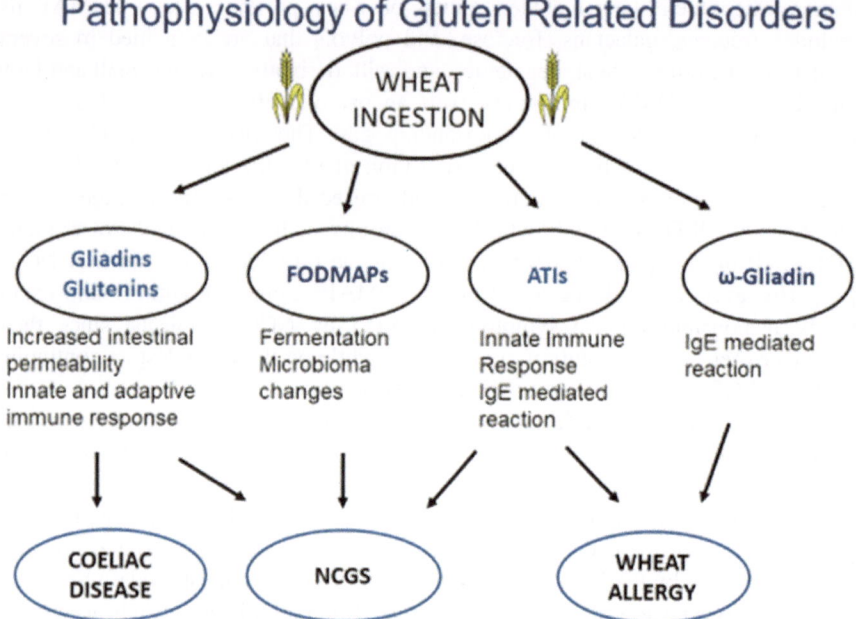

Fig. 2 The complex interplay between gluten and other wheat components leading to the different gluten/wheat-related disorders

background, the immune response and the intestinal microbiome drives the loss of tolerance to gluten/wheat and the development of different gluten/wheat-related disorders (Fig. 2).

11   Treatment

A strict gluten/wheat-free diet remains the only available treatment for NCWS. Since NCWS may be a transient condition, expert recommendation is to maintain the GFD for a given period, e.g. 12–24 months, and then test gluten tolerance again. Based on severity of symptoms, some gluten sensitive patients without CD may choose to follow a gluten-free diet indefinitely [70]. However, it is still unclear whether some basic principles of the "celiac" GFD, particularly the need of the complete exclusion of all wheat derivatives, holds for NCWS as well. For example it is theoretically possible that some NCWS patient may tolerate a small amount of gluten in their diet, particularly if clinical manifestations depend primarily on FODMAPs intolerance. Another interesting possibility is that a subgroup of NCWS might tolerate ancient wheat grains, e.g. einkorn, still containing gluten but much less of ATIs. Further studies are required to clarify these important treatment issues.

12 Conclusions

NCWS is a recently 'rediscovered' disorder that seems to be even more common than CD, at least in adults. In recent years, the number of both patients and publications on NCWS have increased greatly. The clinical picture of NCGS is variable and usually includes IBS-like gastrointestinal manifestations and minimal neurological complaints, such as foggy mind and headache. Treatment with the GFD may dramatically improve the quality of life of these patients, for which we have very little certainty and many knowledge 'black holes'. In view of the currently high rate of perceived gluten sensitivity and the possible placebo effect of any dietary intervention, the demonstration of a clear-cut relationship between gluten ingestion and symptoms appearance by means of the DBPC gluten (or wheat) challenge is a crucial step in the diagnostic algorithm of NCWS. Although validated biomarker(s) for the diagnosis of NCWS are not available yet, the diagnostic criteria summarized in this paper can help to recognize this disorder, optimize clinical care, avoid self-diagnosis, and advance the science of NCWS. The identification and validation of biomarker(s) will be instrumental to gain insights in NCGS pathogenesis, to investigate the epidemiology of this "new" disorder, and ultimately to improve the health and the quality of life of a large number of individuals.

Conflicts of Interests C. Catassi is a scientific consultant of Dr. Schär Food and Takeda. Other authors have no conflicts to declare.

References

1. Al-Toma A, Volta U, Auricchio R, Castillejo G, Sanders DS, Cellier C, et al. European Society for the Study of Coeliac Disease (ESsCD) guideline for coeliac disease and other gluten-related disorders. United European Gastroenterol J. 2019;7(5):583–613. https://doi.org/10.1177/2050640619844125.
2. Ellis A, Linaker BD. Non-coeliac gluten sensitivity? Lancet. 1978;1(8078):1358–9. https://doi.org/10.1016/s0140-6736(78)92427-3.
3. Cooper BT, Holmes GK, Ferguson R, Thompson RA, Allan RN, Cooke WT. Gluten-sensitive diarrhea without evidence of celiac disease. Gastroenterology. 1981;81(1):192–4.
4. Sapone A, Lammers KM, Mazzarella G, Mikhailenko I, Carteni M, Casolaro V, et al. Differential mucosal IL-17 expression in two gliadin-induced disorders: gluten sensitivity and the autoimmune enteropathy celiac disease. Int Arch Allergy Immunol. 2010;152(1):75–80. https://doi.org/10.1159/000260087.
5. Biesiekierski JR, Peters SL, Newnham ED, Rosella O, Muir JG, Gibson PR. No effects of gluten in patients with self-reported non-celiac gluten sensitivity after dietary reduction of fermentable, poorly absorbed, short-chain carbohydrates. Gastroenterology. 2013;145 (2):320–8 e1–3. https://doi.org/10.1053/j.gastro.2013.04.051.
6. Schuppan D, Zevallos V. Wheat amylase trypsin inhibitors as nutritional activators of innate immunity. Dig Dis. 2015;33(2):260–3. https://doi.org/10.1159/000371476.

7. Sapone A, Bai JC, Ciacci C, Dolinsek J, Green PH, Hadjivassiliou M, et al. Spectrum of gluten-related disorders: consensus on new nomenclature and classification. BMC Med. 2012;10:13. https://doi.org/10.1186/1741-7015-10-13.

8. Choung RS, Unalp-Arida A, Ruhl CE, Brantner TL, Everhart JE, Murray JA. Less hidden celiac disease but increased gluten avoidance without a diagnosis in the united states: findings from the national health and nutrition examination surveys from 2009 to 2014. Mayo Clin Proc. 2016.https://doi.org/10.1016/j.mayocp.2016.10.012

9. Catassi C. Gluten sensitivity. Ann Nutr Metab. 2015;67 Suppl 2:16–26.https://doi.org/10.1159/000440990

10. Francavilla R, Cristofori F, Castellaneta S, Polloni C, Albano V, Dellatte S, et al. Clinical, serologic, and histologic features of gluten sensitivity in children. J Pediatr. 2014;164(3):463–7 e1. https://doi.org/10.1016/j.jpeds.2013.10.007.

11. Feldman MF, Bird JA. Clinical, serologic, and histologic features of gluten sensitivity in children. Pediatrics. 2014;134(Suppl 3):S157–8. https://doi.org/10.1542/peds.2014-1817QQ.

12. Francavilla R, Cristofori F, Verzillo L, Gentile A, Castellaneta S, Polloni C, et al. Randomized double-blind placebo-controlled crossover trial for the diagnosis of non-celiac gluten sensitivity in children. Am J Gastroenterol. 2018;113(3):421–30. https://doi.org/10.1038/ajg.2017.483.

13. Aziz I, Lewis NR, Hadjivassiliou M, Winfield SN, Rugg N, Kelsall A, et al. A UK study assessing the population prevalence of self-reported gluten sensitivity and referral characteristics to secondary care. Eur J Gastroenterol Hepatol. 2014;26(1):33–9. https://doi.org/10.1097/01.meg.0000435546.87251.f7.

14. van Gils T, Nijeboer P, CE IJ, Sanders DS, Mulder CJ, Bouma G. Prevalence and characterization of self-reported gluten sensitivity in The Netherlands. Nutrients. 2016;8(11). https://doi.org/10.3390/nu8110714.

15. Volta U, Bardella MT, Calabro A, Troncone R, Corazza GR, Study Group for Non-Celiac Gluten S. An Italian prospective multicenter survey on patients suspected of having non-celiac gluten sensitivity. BMC Med. 2014;12:85. https://doi.org/10.1186/1741-7015-12-85.

16. Vazquez-Roque MI, Camilleri M, Smyrk T, Murray JA, Marietta E, O'Neill J, et al. A controlled trial of gluten-free diet in patients with irritable bowel syndrome-diarrhea: effects on bowel frequency and intestinal function. Gastroenterology. 2013;144(5):903–11 e3. https://doi.org/10.1053/j.gastro.2013.01.049.

17. Dalrymple J, Bullock I. Diagnosis and management of irritable bowel syndrome in adults in primary care: summary of NICE guidance. BMJ. 2008;336(7643):556–8. https://doi.org/10.1136/bmj.39484.712616.AD.

18. Lovell RM, Ford AC. Global prevalence of and risk factors for irritable bowel syndrome: a meta-analysis. Clin Gastroenterol Hepatol. 2012;10(7):712–21 e4. https://doi.org/10.1016/j.cgh.2012.02.029.

19. Canavan C, West J, Card T. The epidemiology of irritable bowel syndrome. Clin Epidemiol. 2014;6:71–80. https://doi.org/10.2147/CLEP.S40245.

20. Whitehead WE, Palsson O, Jones KR. Systematic review of the comorbidity of irritable bowel syndrome with other disorders: what are the causes and implications? Gastroenterology. 2002;122(4):1140–56. https://doi.org/10.1053/gast.2002.32392.

21. Akehurst RL, Brazier JE, Mathers N, O'Keefe C, Kaltenthaler E, Morgan A, et al. Health-related quality of life and cost impact of irritable bowel syndrome in a UK primary care setting. Pharmacoeconomics. 2002;20(7):455–62. https://doi.org/10.2165/00019053-200220070-00003.

22. Morcos A, Dinan T, Quigley EM. Irritable bowel syndrome: role of food in pathogenesis and management. J Dig Dis. 2009;10(4):237–46. https://doi.org/10.1111/j.1751-2980.2009.00392.x.

23. Bohn L, Storsrud S, Tornblom H, Bengtsson U, Simren M. Self-reported food-related gastrointestinal symptoms in IBS are common and associated with more severe symptoms and reduced quality of life. Am J Gastroenterol. 2013;108(5):634–41. https://doi.org/10.1038/ajg.2013.105.

24. McKenzie YA, Thompson J, Gulia P, Lomer MC. British Dietetic Association systematic review of systematic reviews and evidence-based practice guidelines for the use of probiotics in the management of irritable bowel syndrome in adults (2016 update). J Hum Nutr Diet. 2016;29(5):576–92. https://doi.org/10.1111/jhn.12386.

25. Jones VA, McLaughlan P, Shorthouse M, Workman E, Hunter JO. Food intolerance: a major factor in the pathogenesis of irritable bowel syndrome. Lancet. 1982;2(8308):1115–7. https://doi.org/10.1016/s0140-6736(82)92782-9.

26. King TS, Elia M, Hunter JO. Abnormal colonic fermentation in irritable bowel syndrome. Lancet. 1998;352(9135):1187–9. https://doi.org/10.1016/s0140-6736(98)02146-1.

27. Catassi C, Elli L, Bonaz B, Bouma G, Carroccio A, Castillejo G, et al. Diagnosis of Non-Celiac Gluten Sensitivity (NCGS): The Salerno Experts' Criteria. Nutrients. 2015;7 (6):4966–77. https://doi.org/10.3390/nu7064966.

28. Batista IC, Gandolfi L, Nobrega YK, Almeida RC, Almeida LM, Campos Junior D, et al. Autism spectrum disorder and celiac disease: no evidence for a link. Arq Neuropsiquiatr. 2012;70(1):28–33. https://doi.org/10.1590/s0004-282x2012000100007.

29. Marcason W. What is the current status of research concerning use of a gluten-free, casein-free diet for children diagnosed with autism? J Am Diet Assoc. 2009;109(3):572. https://doi.org/10.1016/j.jada.2009.01.013.

30. de Magistris L, Familiari V, Pascotto A, Sapone A, Frolli A, Iardino P, et al. Alterations of the intestinal barrier in patients with autism spectrum disorders and in their first-degree relatives. J Pediatr Gastroenterol Nutr. 2010;51(4):418–24. https://doi.org/10.1097/MPG. 0b013e3181dcc4a5.

31. Robertson MA, Sigalet DL, Holst JJ, Meddings JB, Wood J, Sharkey KA. Intestinal permeability and glucagon-like peptide-2 in children with autism: a controlled pilot study. J Autism Dev Disord. 2008;38(6):1066–71. https://doi.org/10.1007/s10803-007-0482-1.

32. Navarro F, Pearson DA, Fatheree N, Mansour R, Hashmi SS, Rhoads JM. Are "leaky gut" and behavior associated with gluten and dairy containing diet in children with autism spectrum disorders? Nutr Neurosci. 2015;18(4):177–85. https://doi.org/10.1179/ 1476830514Y.0000000110.

33. Lau NM, Green PH, Taylor AK, Hellberg D, Ajamian M, Tan CZ, et al. Markers of celiac disease and gluten sensitivity in children with autism. PLoS One. 2013;8(6):e66155.https:// doi.org/10.1371/journal.pone.0066155

34. de Magistris L, Picardi A, Siniscalco D, Riccio MP, Sapone A, Cariello R, et al. Antibodies against food antigens in patients with autistic spectrum disorders. Biomed Res Int. 2013;2013:729349.https://doi.org/10.1155/2013/729349

35. Jozefczuk J, Konopka E, Bierla JB, Trojanowska I, Sowinska A, Czarnecki R, et al. The occurrence of antibodies against gluten in children with autism spectrum disorders does not correlate with serological markers of impaired intestinal permeability. J Med Food. 2018;21 (2):181–7. https://doi.org/10.1089/jmf.2017.0069.

36. Millward C, Ferriter M, Calver S, Connell-Jones G. Gluten- and casein-free diets for autistic spectrum disorder. Cochrane Database Syst Rev. 2008(2):CD003498. https://doi.org/10.1002/ 14651858.CD003498.pub3.

37. Mari-Bauset S, Zazpe I, Mari-Sanchis A, Llopis-Gonzalez A, Morales-Suarez-Varela M. Evidence of the gluten-free and casein-free diet in autism spectrum disorders: a systematic review. J Child Neurol. 2014;29(12):1718–27. https://doi.org/10.1177/0883073814531330.

38. Whiteley P, Haracopos D, Knivsberg AM, Reichelt KL, Parlar S, Jacobsen J, et al. The ScanBrit randomised, controlled, single-blind study of a gluten- and casein-free dietary intervention for children with autism spectrum disorders. Nutr Neurosci. 2010;13(2):87–100. https://doi.org/10.1179/147683010X12611460763922.

39. Pedersen L, Parlar S, Kvist K, Whiteley P, Shattock P. Data mining the ScanBrit study of a gluten- and casein-free dietary intervention for children with autism spectrum disorders: behavioural and psychometric measures of dietary response. Nutr Neurosci. 2014;17(5):207– 13. https://doi.org/10.1179/1476830513Y.0000000082.

40. Pusponegoro HD, Ismael S, Firmansyah A, Sastroasmoro S, Vandenplas Y. Gluten and casein supplementation does not increase symptoms in children with autism spectrum disorder. Acta Paediatr. 2015;104(11):e500–5. https://doi.org/10.1111/apa.13108.
41. Hyman SL, Stewart PA, Foley J, Cain U, Peck R, Morris DD, et al. The gluten-free/casein-free diet: a double-blind challenge trial in children with autism. J Autism Dev Disord. 2016;46(1):205–20. https://doi.org/10.1007/s10803-015-2564-9.
42. Ghalichi F, Ghaemmaghami J, Malek A, Ostadrahimi A. Effect of gluten free diet on gastrointestinal and behavioral indices for children with autism spectrum disorders: a randomized clinical trial. World J Pediatr. 2016;12(4):436–42. https://doi.org/10.1007/s12519-016-0040-z.
43. El-Rashidy O, El-Baz F, El-Gendy Y, Khalaf R, Reda D, Saad K. Ketogenic diet versus gluten free casein free diet in autistic children: a case-control study. Metab Brain Dis. 2017;32 (6):1935–41. https://doi.org/10.1007/s11011-017-0088-z.
44. Piwowarczyk A, Horvath A, Pisula E, Kawa R, Szajewska H. Gluten-free diet in children with autism spectrum disorders: a randomized, controlled, single-blinded trial. J Autism Dev Disord. 2020;50(2):482–90. https://doi.org/10.1007/s10803-019-04266-9.
45. Gonzalez-Domenech PJ, Diaz Atienza F, Garcia Pablos C, Fernandez Soto ML, Martinez-Ortega JM, Gutierrez-Rojas L. Influence of a combined gluten-free and casein-free diet on behavior disorders in children and adolescents diagnosed with autism spectrum disorder: a 12-month follow-up clinical trial. J Autism Dev Disord. 2020;50(3):935–48. https://doi.org/10.1007/s10803-019-04333-1.
46. Keller A, Rimestad ML, Friis Rohde J, Holm Petersen B, Bruun Korfitsen C, Tarp S, et al. The effect of a combined gluten- and casein-free diet on children and adolescents with autism spectrum disorders: a systematic review and meta-analysis. Nutrients. 2021;13(2). https://doi.org/10.3390/nu13020470.
47. Croall ID, Hoggard N, Hadjivassiliou M. Gluten and autism spectrum disorder. Nutrients. 2021;13(2). https://doi.org/10.3390/nu13020572.
48. Dohan FC. Cereals and schizophrenia data and hypothesis. Acta Psychiatr Scand. 1966;42 (2):125–52. https://doi.org/10.1111/j.1600-0447.1966.tb01920.x.
49. Vlissides DN, Venulet A, Jenner FA. A double-blind gluten-free/gluten-load controlled trial in a secure ward population. Br J Psychiatry. 1986;148:447–52. https://doi.org/10.1192/bjp.148.4.447.
50. Potkin SG, Weinberger D, Kleinman J, Nasrallah H, Luchins D, Bigelow L, et al. Wheat gluten challenge in schizophrenic patients. Am J Psychiatry. 1981;138(9):1208–11. https://doi.org/10.1176/ajp.138.9.1208.
51. Storms LH, Clopton JM, Wright C. Effects of gluten on schizophrenics. Arch Gen Psychiatry. 1982;39(3):323–7. https://doi.org/10.1001/archpsyc.1982.04290030055010.
52. Cascella NG, Kryszak D, Bhatti B, Gregory P, Kelly DL, Mc Evoy JP, et al. Prevalence of celiac disease and gluten sensitivity in the United States clinical antipsychotic trials of intervention effectiveness study population. Schizophr Bull. 2011;37(1):94–100. https://doi.org/10.1093/schbul/sbp055.
53. Cascella NG, Santora D, Gregory P, Kelly DL, Fasano A, Eaton WW. Increased prevalence of transglutaminase 6 antibodies in sera from schizophrenia patients. Schizophr Bull. 2013;39 (4):867–71. https://doi.org/10.1093/schbul/sbs064.
54. Dickerson F, Stallings C, Origoni A, Vaughan C, Khushalani S, Leister F, et al. Markers of gluten sensitivity and celiac disease in recent-onset psychosis and multi-episode schizophrenia. Biol Psychiatry. 2010;68(1):100–4. https://doi.org/10.1016/j.biopsych.2010.03.021.
55. Kelly DL, Demyanovich HK, Rodriguez KM, Cihakova D, Talor MV, McMahon RP, et al. Randomized controlled trial of a gluten-free diet in patients with schizophrenia positive for antigliadin antibodies (AGA IgG): a pilot feasibility study. J Psychiatry Neurosci. 2019;44 (4):269–76.

56. Dzikowski M, Juchnowicz D, Dzikowska I, Rog J, Prochnicki M, Koziol M, et al. The differences between gluten sensitivity, intestinal biomarkers and immune biomarkers in patients with first-episode and chronic schizophrenia. J Clin Med. 2020;9(11). https://doi.org/ 10.3390/jcm9113707.

57. Peters SL, Biesiekierski JR, Yelland GW, Muir JG, Gibson PR. Randomised clinical trial: gluten may cause depression in subjects with non-coeliac gluten sensitivity—an exploratory clinical study. Aliment Pharmacol Ther. 2014;39(10):1104–12. https://doi.org/10.1111/apt. 12730.

58. Di Sabatino A, Volta U, Salvatore C, Biancheri P, Caio G, De Giorgio R, et al. Small amounts of gluten in subjects with suspected nonceliac gluten sensitivity: a randomized, double-blind, placebo-controlled, cross-over trial. Clin Gastroenterol Hepatol. 2015;13(9):1604–12 e3. https://doi.org/10.1016/j.cgh.2015.01.029.

59. Porcelli B, Verdino V, Ferretti F, Bizzaro N, Terzuoli L, Cinci F, et al. A study on the association of mood disorders and gluten-related diseases. Psychiatry Res. 2018;260:366–70. https://doi.org/10.1016/j.psychres.2017.12.008.

60. Genuis SJ, Lobo RA. Gluten sensitivity presenting as a neuropsychiatric disorder. Gastroenterol Res Pract. 2014;2014:293206.https://doi.org/10.1155/2014/293206

61. Lionetti E, Leonardi S, Franzonello C, Mancardi M, Ruggieri M, Catassi C. Gluten psychosis: confirmation of a new clinical entity. Nutrients. 2015;7(7):5532–9. https://doi.org/10.3390/ nu7075235.

62. Sergi C, Villanacci V, Carroccio A. Non-celiac wheat sensitivity: rationality and irrationality of a gluten-free diet in individuals affected with non-celiac disease: a review. BMC Gastroenterol. 2021;21(1):5. https://doi.org/10.1186/s12876-020-01568-6.

63. Carroccio A, Giannone G, Mansueto P, Soresi M, La Blasca F, Fayer F, et al. Duodenal and rectal mucosa inflammation in patients with non-celiac wheat sensitivity. Clin Gastroenterol Hepatol. 2019;17(4):682–90 e3. https://doi.org/10.1016/j.cgh.2018.08.043.

64. Volta U, De Giorgio R, Caio G, Uhde M, Manfredini R, Alaedini A. Nonceliac wheat sensitivity: an immune-mediated condition with systemic manifestations. Gastroenterol Clin North Am. 2019;48(1):165–82. https://doi.org/10.1016/j.gtc.2018.09.012.

65. Uhde M, Ajamian M, Caio G, De Giorgio R, Indart A, Green PH, et al. Intestinal cell damage and systemic immune activation in individuals reporting sensitivity to wheat in the absence of coeliac disease. Gut. 2016;65(12):1930–7. https://doi.org/10.1136/gutjnl-2016-311964.

66. Caio G, Volta U, Tovoli F, De Giorgio R. Effect of gluten free diet on immune response to gliadin in patients with non-celiac gluten sensitivity. BMC Gastroenterol. 2014;14:26. https:// doi.org/10.1186/1471-230X-14-26.

67. Uhde M, Caio G, De Giorgio R, Green PH, Volta U, Alaedini A. Subclass profile of IgG antibody response to gluten differentiates nonceliac gluten sensitivity from celiac disease. Gastroenterology. 2020;159(5):1965–7 e2. https://doi.org/10.1053/j.gastro.2020.07.032.

68. Barbaro MR, Cremon C, Morselli-Labate AM, Di Sabatino A, Giuffrida P, Corazza GR, et al. Serum zonulin and its diagnostic performance in non-coeliac gluten sensitivity. Gut. 2020;69 (11):1966–74. https://doi.org/10.1136/gutjnl-2019-319281.

69. Lionetti E, Pulvirenti A, Vallorani M, Catassi G, Verma AK, Gatti S, et al. Re-challenge studies in non-celiac gluten sensitivity: a systematic review and meta-analysis. Front Physiol. 2017;8:621. https://doi.org/10.3389/fphys.2017.00621.

70. Leonard MM, Sapone A, Catassi C, Fasano A. Celiac disease and nonceliac gluten sensitivity: a review. JAMA. 2017;318(7):647–56. https://doi.org/10.1001/jama.2017.9730.

71. Biesiekierski JR, Newnham ED, Irving PM, Barrett JS, Haines M, Doecke JD, et al. Gluten causes gastrointestinal symptoms in subjects without celiac disease: a double-blind randomized placebo-controlled trial. Am J Gastroenterol. 2011;106(3):508–14; quiz 15. https://doi.org/10.1038/ajg.2010.487.

72. Catassi G, Lionetti E, Gatti S, Catassi C. The low FODMAP diet: many question marks for a catchy acronym. Nutrients. 2017;9(3). https://doi.org/10.3390/nu9030292.

73. Catassi C, Alaedini A, Bojarski C, Bonaz B, Bouma G, Carroccio A, et al. The overlapping area of Non-Celiac Gluten Sensitivity (NCGS) and wheat-sensitive Irritable Bowel Syndrome (IBS): an update. Nutrients. 2017;9(11). https://doi.org/10.3390/nu9111268.
74. Caminero A, McCarville JL, Zevallos VF, Pigrau M, Yu XB, Jury J, et al. Lactobacilli degrade wheat amylase trypsin inhibitors to reduce intestinal dysfunction induced by immunogenic wheat proteins. Gastroenterology. 2019;156(8):2266–80. https://doi.org/10.1053/j.gastro.2019.02.028.
75. Mansueto P, Di Liberto D, Fayer F, Soresi M, Geraci G, Giannone AG, et al. TNF-alpha, IL-17, and IL-22 production in the rectal mucosa of nonceliac wheat sensitivity patients: role of adaptive immunity. Am J Physiol Gastrointest Liver Physiol. 2020;319(3):G281–8. https://doi.org/10.1152/ajpgi.00104.2020.

Index